D1281387

THE STEAM LAUNCH

THE STEAM LAUNCH

RICHARD M. MITCHELL

International Marine Publishing Company
Camden, Maine

To Lucile

Contents

Preface

In all of our lives there comes a time to look back, reflect on what we have done, and dream of the enjoyable things we wish we could do again. Having reached that time, I would like to share with the readers of this book my pleasant experiences of 40 years with steam launches. I'd like to pass along the things I have learned and the best of the steam launch drawings and photographs that I've collected.

When I was a child I did not know what a steam launch was. Some still existed on New England lakes, but most of the old hulls were rotten and useless and much of the machinery that had escaped the junkman was hidden away in barns and boathouses. There were few steam hobby publications in America then and no regularly issued magazine that enthusiasts could use to communicate with one another. It was not until after World War II that steam launch hobbyists became numerous enough to begin exchanging information with each other nationwide.

Steam engines of the types that powered locomotives, factories, and sawmills abounded in my early years, and I loved them with a passion. While I was in high school in the 1930s my brothers and I strayed far enough to ride large steamboats on Lake Champlain and along the coasts of Massachusetts and Connecticut.

I was well acquainted with early gasoline boats, which were plentiful on the Connecticut River at Brattleboro, Vermont — my childhood town — and on nearby Spofford Lake in New Hampshire. In addition to the noisy and temperamental outboards, there were all kinds of inboard boats, from those with little 1-cylinder, 2-cycle engines to the powerful speedboats of the period. My two brothers and I became involved in any and every old boat that could be bought for $15. Like other kids who grew up in the depression years, we had no money but a lot of fun, and we were covered with grease and dirt most of the time.

In 1941, I heard that a small steam launch, which had operated on Spofford Lake years before, still existed in nearby Keene, New Hampshire. The old hull was beyond repair, but all the mechanical components were sheltered intact in a barn. I was able to buy this unit, and through a railroad man who lived on my street, I soon learned what it was all about, as we fired up the little marine plant in my folks' garage.

In 1943, I was given a 22-foot Fay & Bowen hull, and I began installing the Spofford Lake steam plant. In June 1944, the launch *Lourick* became a reality. The maiden cruise on the Connecticut River began 25 of the most pleasant years of my life.

While I was assembling *Lourick* I actually thought that, aside from the man who had originally owned her machinery, I was the only person in history to own and operate a small private steamer! I knew nothing of the little steamers that had been built and used in every corner of the world.

My boat ran nicely and I wanted to tell others about her, so in 1945 I wrote my first story, which appeared in *Motor Boating*. This was the first of more than 50 steam launch stories I have written, and the publication of these led to the exchange of thousands of letters with hundreds of correspondents.

After seven years of cruising in *Lourick,* I dismantled her and sold her parts, in order to focus my resources on building a new home by the Connecticut River. When the house was finished, I again turned to steam launching. With the help of a good friend, Lowell Patch, I spent five years of spare time building the 23-foot launch *River Queen* in my backyard. The new steamer, my "boat of a lifetime," was launched on July 4, 1957. For the next 16 years, I churned the waters of the Connecticut River with family and friends, while the whistle echoed from the hills of Vermont and New Hampshire. Ill health finally forced the sale of this good boat, but it can still be seen on Lake Winnipesaukee as the *Barbara C.,* now owned by Vincent Callahan.

I no longer have a boat of my own, but I keep active in the steam launch hobby helping others in one way or another, especially in gathering and distributing information. My interest in steam-powered craft has taken my wife Lucile and me across the Atlantic to see some of the finest launches in the world at Lake Windermere in England and to ride beautiful old paddle steamers in Scotland. We have traveled to Puget Sound and been aboard many of the steam launches active in that area. During these travels I have taken photographs constantly, and through correspondence have collected pictures of hundreds of varied launches — rarely two alike — both past and present.

Unlike the steam locomotive, which in its day affected the lives of more people than any machine known before, or the automobile, which later came to the door of every house in America, small steamboats were never enjoyed by large numbers of people. The Golden Age of steam launching lasted scarcely 30 years, and it occurred too early (roughly 1870 to 1900) to find a potential market among large masses of prosperous and mobile people. Nevertheless, steam launches were once present in the thousands on the rivers, lakes, and harbors of America and England. Thousands more were shipped around the world to serve remote areas where water transport generally preceded roads or railways.

The original little steamers all but disappeared in the first decades of this century, abandoning lakes and waterways to gasoline launches and runabouts. In their absence, recollections of the little boats faded, too.

This book tells about the early history of steam launches, their development and many uses a century ago, and their revival as a hobby since 1940. I cannot hope to make the book all-inclusive or, with regard to modern steamers, completely up-to-date. Modern hobby steamers keep switching owners or engines, or changing beyond recognition, so no description of them can be final. Much steam launch history and lore is lost forever, and much remains undiscovered to delight my successors.

Richard M. Mitchell
May 1982 Brattleboro Road, Box 153
Hinsdale, New Hampshire 03451

Acknowledgments

One person can never know all about a subject. Through much of my life I have held in my mind the idea of doing this book, but making it a reality depended on the help and generosity of many friends. It is safe to say that nearly every modern steam launch owner has in some measure contributed to this effort. A great deal of research was necessary, much of it in publications from long ago. I wrote hundreds of letters, not only to steam buffs in this country, but also to friends and museums in Canada and Europe. Some problems seemed insurmountable at times, but through the encouragement and devotion of others, these problems all faded away.

I am forever grateful to the following, who have helped in so many ways:

Everett A. Arnes; William A. Baker; Frederick G. Beach; Clifford Blackstaffe; Professor Evers Burtner; Hugh Cawdron; Wilbur J. Chapman; Artú Chiggiato; Edward O. Clark; John S. Clement; David Crockett; Bernard F. Denny; Louis D. De Young; Bill Durham; Jonathan Eaton and the staff of International Marine Publishing Co.; Julius B. Emmert; Weston Farmer; Robert B. Fearing; Castle Freeman; Milton Gallup; Jerry Heermans; The Henry Ford Museum; Sidney Herreshoff; Harcourt and Ellen Hervey; Brian E. Hillsdon; H. Hobart Holly; Arthur E. Hughes; Art Knudsen; Steven Lang; Henry Luther; Robert W. Merriam; Edward Middleton, Jr.; C.B. Mitchell; Gary L. Mitchell; Lucile Mitchell; Peter E. Moale; Richard Morrison; Editors of *Motor Boating & Sailing;* Mystic Seaport; The National Maritime Museum, Greenwich, England; Temple Nieter; Morgan North; Frank A. Orr; Steven Pope; Howard C. Rice, Jr.; Editors of *Rudder;* The Science Museum, London; Frederick H. Semple; William U. Shaw; Sheridan House; The Smithsonian Institution; Staten Island Historical

Society; Steam Boat Association of Great Britain; The Steamship Historical Society of America; Bror Tamm; Thomas G. Thompson; U.S. Naval Institute and Naval Photographic Center; Robert W. Valpey; Muriel Vaughn; William W. Willock, Jr.; Editors of *Yachting*.

Part One

The Steam Launch Past and Present

1
The Rise of Steam Navigation

Steam-powered vessels first stemmed the currents of rivers in France and America in the late 1700s, but 100 years were to pass before small steamers reached their peak of economic usefulness and popular interest. The experimental boats that presaged the ascendance of steam navigation were frequently no bigger than the mature steam launches of three or four generations later, but they were ungainly, fragile, and inefficient. Much of the long evolution of steam navigation had to be effected through ships, because the crude new technology required economies of scale. The innovations that spurred steamship development from the first tentative designs to the refined ships of 1885 also made possible the steam launch configurations of 1860 to 1900. Those configurations — which are still used today — were in most respects a miniature of the first fully evolved propeller steamships with modern engines, which appeared between 1855 and 1870, and of screw-tug design of the 1870s and 1880s.

Today, many steam launch operators have a smattering of knowledge of "the rise of steam navigation" and enjoy anecdotes that illustrate the strangeness of the early boats, but they do not identify their own steamers with this antique strangeness. The steam launch "Golden Age" flowered late, in the same years that brought the earliest development of the next technological bloom, the Otto four-stroke-cycle gasoline engine.

EARLY DEVELOPMENTS — GREAT BRITAIN

No single individual invented steamboats. The idea was proposed in detail a lifetime before the fact, and after the 1770s dozens of men were building or planning steam-propelled vessels.

The "fire engine" had been an object of speculation since antiquity, but its first large-scale practical applications occurred in the second half of the 17th century. From the time Charles II ascended to the

English throne in 1660, he encouraged his subjects to pursue the sciences and mechanical arts. One of the principal problems for British miners of the day was removing the vast quantities of water that were halting progress in ever-deepening mine shafts, and early efforts were directed toward this end. When Louis XIV required steam pumps for his new palace at Versailles, they were available from England.

Thomas Savery, a military engineer, called his 1698 pump "the Miner's Friend." Steam was admitted to a chamber, then condensed by pouring cold water on the chamber's outer surface; the resultant vacuum was used to draw water up a long pipe. Edward Somerset, the second Marquis of Worcester, had spent the better part of 40 years, beginning in 1628, and a considerable fortune developing similar engines, but he never won public acceptance for his invention. Savery was more persistent — his pumps found uses in estates and cities but were not widely employed in mines. Extremely high first costs, the expense of frequent repairs, and low power output were principal drawbacks. Savery patented an arrangement for propelling vessels in calm weather that involved paddle wheels driven by a muscle-powered capstan. Steam-propelled boats were still in the distant future.

The piston-and-cylinder principle that was employed in all steamers before 1900, when steam turbines began to come into use, was patented by Denis Papin, a trained medical doctor, in 1690. The cylinder in Papin's engine was in effect a boiler, half-filled with water. In operation a fire was alternately placed under the cylinder, generating steam, and removed. A piston would be thereby driven up and down; condensation of the steam, causing the piston to fall, produced the working stroke. Papin calculated that a cylinder of two-foot diameter and four-foot stroke, working at one stroke per minute, would generate one horsepower, but he acknowledged the difficulty of building such a cylinder with the metals and techniques then available. He soon reverted to the Savery-type, pistonless engine.

Thomas Newcomen, a blacksmith, combined the ideas of his predecessors in his single-acting "atmospheric engine" of 1705, in which the condensation of steam in a cylinder enabled atmospheric pressure to push down a piston, and with it one end of a counterweighted walking beam. Newcomen's later, improved engines made 10 or 12 strokes per minute. This fuel-wasteful device was widely used for pumping water (mainly out of deep mines) for two generations.

By 1775 there were scores of steam engines totaling thousands of horsepower, with some engines of as much as 75 h.p. (six-foot bore, nine-foot stroke). This quantity of power was clearly sufficient to propel a ship of several hundred tons, once the details of transforming a pumping engine into a marine propulsion engine were worked out. A 20-horsepower engine might cost $2,000 (equivalent to $500,000 today); operating expenses were equally stiff, as this passage from Robert H. Thurston's *A History of the Growth of the Steam Engine* (1878) makes clear:

> Smeaton found 57 engines at work near Newcastle in 1767, ranging in size from 28 to 75 inches in diameter of cylinder, and of, collectively, about 1,200 horse-power. Fifteen of these engines gave an average of 98 square inches of piston to the horse-power, and the average duty was 5,590,000 pounds raised 1 foot high by 1 bushel (84 pounds) of coal. The highest duty noted was 7.44 millions [or about 20 pounds of coal per horsepower-hour]. The most efficient engine had a steam-cylinder 42 inches in diameter; the load was equivalent to 9¼ pounds per square inch of piston-area [derived mostly from the imperfect vacuum, not from "steam pressure"], and the horse-power developed was calculated to be 16.7.
>
> Price, writing in 1778, says, in the Appendix to his "Mineralogia Cornubiensis": "Mr. Newcomen's invention of the fire-engine enabled us to sink our mines to twice the depth we could formerly do by any other machinery. Since this invention was completed, most other attempts at its improvement have been very unsuccessful; but the vast consumption of fuel in these engines is an immense drawback on the profit of our mines, for every fire-engine of magnitude consumes £3,000 worth of coals per annum. This heavy tax amounts almost to a prohibition."

After 1767 James Watt introduced a series of improvements that greatly increased the economy and potential usefulness of steam engines. Chief among these were a condenser, which was distinct from the cylinder and thus circumvented the energy-wasteful practice of alternately heating and cooling the cylinder surfaces; a mechanism for enabling the steam engine to drive a revolving shaft; the use of double action (introducing steam and vacuum to each side of the piston alternately); and the use of steam's ability to do work expansively by cutting steam admission to the cylinder partway through a stroke and allowing expansion to bring the stroke to completion.

STEAMSHIPS BEFORE 1840

By 1770 any "mechanical philosopher" could figure out that the new power that was contributing so much to industry would soon be used to propel ships. It remained to be seen who would be first, and what device would transmit the engine's power to the water. There were formidable obstacles — the difficulty of finding wealthy sponsors for an expensive and extremely risky undertaking, and machine production techniques that could completely cripple a good design concept — but they were practical obstacles, made to be overcome. There was a half-mad, prophetic gleam in the eyes of most of the pioneers of marine steam power. They knew with certainty that their chosen revolution was as inevitable and far-reaching as the other revolutions of the time, and their main preoccupation often was keeping other good men down, or beating them to the mark. A scramble for patent protection of mechanical design and other monopoly advantages was part of the cause for the quarter-century lag between the first mechanically successful steamer (1783) and the first commercially profitable boat (1807). Even the crankshaft principle was patent-protected and not available to experimenters!

The early evolution of practical steamboats, approximately 1785 to 1825, was nurtured by the needs of the times. Inland water transport was more important than in later generations, after railroads and highways were developed; most early steamboat hulls were derived from the barges, lighters, and canalboats they were intended to replace. A few were novel designs, usually multihull, invented to go with the novel power source. By 1820, increasing numbers of European steamers adopted sailing-ship hull forms, for service in open water. By that time also, after a bewildering early assortment of proposed solutions, the favored propulsion design was a rotating crankshaft and side wheels.

The early inventors, experimenters, and would-be entrepreneurs were scattered in England, France, and America; Americans frequently looked toward Europe for the technological expertise to bring their ideas to fruition.

A steam towboat for helping vessels in and out of harbors was planned by Jonathan Hulls in Evesham, England, as early as 1737. Hulls proposed to use a Newcomen engine for the purpose, and may actually have built an unsuccessful prototype. He became an object of derisive doggerel, and no further steamboat developments of note took place in Britain for 50 years, despite that country's technological and economic advantages.

Several French noblemen, led by the Comte d'Auxiron and the Chevalier Charles Monnin de Follenai, took the initiative in 1772, forming a company that advanced funds for a steamer. D'Auxiron secured a 15-year monopoly from the prime minister for the use of steam in river navigation, with the provision that the experimental boat should perform successfully. Unfortunately the vessel sank just before it was completed, and the company dissolved after absorbing a loss of 15,000 francs.

In 1775 Jacques C. Perier, a competent power engineer, built a little steamer that worked, but without enough force to be of any practical use. Neither of these failures discouraged the Marquis de Jouffroy, who had been party to the planning of the latter venture but had failed to convince Perier that he should use more power in his boat. D'Auxiron's monopoly was transferred by the King to Jouffroy, and in 1778 he completed a 40-foot by 6-foot steamer with the innovations of double (single-acting) cylinders and hinged-flap propulsion. This design was ineffective, so the Marquis began work on a larger boat, with a single-cylinder, double-acting engine, and ratchet-driven side wheels. *Pyroscaphe* moved steadily upstream against the current of the Saone River on July 15, 1783, before an audience of thousands, but on a trumped-up technicality the government withdrew the promised monopoly. Unable to raise capital without the monopoly, and with his boat's rickety hull and engine ruined by the 15-minute trial, Jouffroy gave up his schemes. France's early lead in steamboat development soon evaporated in the social disorder that accompanied the French Revolution. The builder's scale models and drawings of *Pyroscaphe* are preserved in Paris.

After 1785 dozens of Americans and Britons were thinking and talking about steamboats, and seeking patronage from people with enough wealth and power to support ambitious plans. In many cases the history of early developments is confused by conflicting accounts, assertions, and accusations. Contested monopolies and alleged patent infringements were frequent occurrences, and collaborations and rivalries constantly shifted. These events are described in more detail than space allows here, in Thurston's book and in *Steamboats Come True,* by James Flexner.

America, far behind the mother country in steam engineering, took the lead in steamboats. There was an obvious need for mechanical propulsion on America's great rivers, and many ingenious and ambitious mechanics to fill that need. In England, the large and growing steam power industry was fully committed to heavy industrial uses — and James Watt wasn't interested in steam-powered boats.

John Fitch, a Revolutionary War gunsmith, one-time surveyor and silversmith, ingenious inventor, and social misfit, had a 34' x 8' steamer running at Philadelphia in 1786. Six paddles dipped and stroked, canoe-style, on either side of the boat, for a speed of 3 m.p.h. In 1790 Fitch and his associate, Henry Voight, a German clockmaker, launched a second steamer, *Thornton,* on the Delaware River. The 60' x 9' x 4' boat was propelled by three dipping paddles at the stern, driven by an 18-inch engine. *Thornton* was the first steamer to run on a regular schedule. In June 1790, Fitch advertised that the boat would depart Philadelphia on Mondays, Wednesdays, and Fridays, and run to Burlington, Bristol, Bordentown, and Trenton. Return trips were on Tuesdays, Thursdays, and Saturdays.

Failing to profit on the Delaware, Fitch took his steamboat ideas and experience to France in 1793, in the face of a revolution that made commercial advance impossible. By 1796 he was drifting from his overweening hopes of a decade earlier, downward toward oblivion, experimenting with screw propulsion in a ship's boat on a small pond in Manhattan.

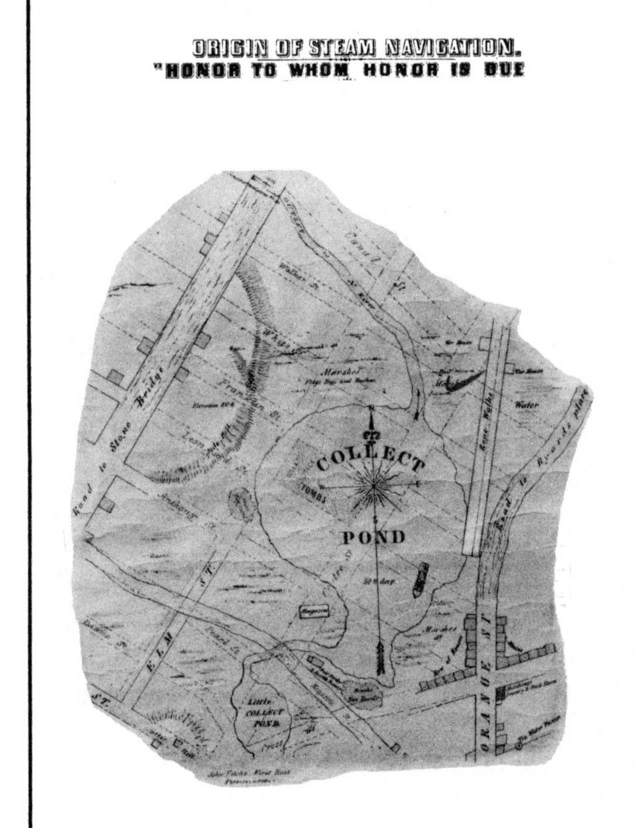

There used to be a quarter-mile-wide pond near the southern end of Manhattan, centered where the Tombs prison was later built. America's chief "neglected genius," John Fitch, ran one of his steam launches here, on Collect Pond, in the summer of 1795, '96, or '97. A neighborhood boy, John Hutchings, enjoyed the excitement and later helped pull the boat apart for firewood after Fitch abandoned it. Almost 50 years later, Hutchings published an appreciation of Fitch, having witnessed how history had heaped acclaim on Fulton but forgotten his wild-eyed forerunner. Fitch built little privately conceived steamers while almost unaware that steam technology was advancing on a broad industrial front in England. His Collect Pond boat's cylinders were made of wooden staves; the boiler was an iron pot with a plank cover clamped over it. (Crown Copyright. Science Museum, London)

Steamboat hobbyists still like to quote Fitch's summation of his career in steam: "I know of nothing so perplexing and vexatious to a man of feelings as a turbulent wife and steamboat building."

James Rumsey, of Virginia, had more sophisticated design ideas, and more influential friends, than his rival Fitch, but he was no more successful in the end. His first steamer, in 1787, employed a water-tube boiler and jet propulsion and carried two tons of cargo against the currents of the Potomac at 3 or 4 m.p.h. With the encouragement of friends such as General Gates, General Washington, and Benjamin Franklin, Rumsey went to England in 1788. He obtained comprehensive English patents, negotiated with Boulton, Watt, and Company of Birmingham to use their good engines for boat propulsion in America, and had a 100-ton steamer built for service on the Thames. The cost of the *Columbia Maid* was beyond his resources, and Rumsey had to struggle to keep his enterprise afloat from 1790 to 1792. He died of apoplexy shortly before his steamer succeeded in plying the Thames at 4 knots — not a competitive speed.

In 1798 Nicholas Roosevelt, who held an American patent for 90-degree crank separation in a 2-cylinder engine, ran a 3-m.p.h. steamer on the Passaic River. Roosevelt's backers were Chancellor Robert Livingston and Colonel John Stevens. (Stevens would later become a steam engineer and steamboat builder of considerable standing.)

In 1790 Samuel Morey, of Orford, New Hampshire, is said to have operated a steamboat on Fairlee Pond, Vermont. By 1793 he had built a 4-m.p.h. steamer, and this is reputed to have attained 5 m.p.h. when exhibited in New York. In Philadelphia in 1797 he demonstrated a greatly improved boat, which was driven by two paddle wheels, one on either side, and a connecting crankshaft. He had many influential backers and was offered $100,000 in capitalization by Chancellor Livingston if he could produce an 8-m.p.h. boat, but this he could not do. By 1803 Morey had the American patent for crank motion in steamboating.

Elijah Ormsbee and David Wilkinson, of Cranston, Rhode Island, said that they "assembled" a little steamer, using a borrowed longboat, in the 1790s — no more is known. There were undoubtedly other unremembered steamboat builders and dreamers off in the backwoods of France, America, and Great Britain, hopefully fumbling their way toward the Great Invention that would make them rich and famous. Some of the American inventors had never seen a steam engine, relying on the *Encyclopaedia Britannica* and newspaper accounts for their technical knowledge.

In 1788 the Scottish engineer and inventor William Symington built a Newcomen steam engine for a 25-foot, double-hulled paddle boat. This was financed by Patrick Miller, an Edinburgh banker, who had experimented with multihull designs employing hand-turned paddle wheels between the hulls. The boat made 5 m.p.h., and Miller gave Symington more money for a larger version. In 1789 this boat traveled at 5 m.p.h. or better. However, Miller lost patience with Symington's workmanship and sought instead to enter into a partnership with Boulton and Watt for the building of steamboats. The engine builders politely turned him down.

In 1801 Symington built the engine for the steam tug *Charlotte Dundas,* often called the first practical steamboat. The boat performed successfully as designed, towing two 70-ton barges nearly 20 miles at better than 3 m.p.h. into a stiff wind, but she was not put into regular service.

"Poor John Fitch" died impoverished, a suicide on the Kentucky frontier in 1798. Ever-successful Robert Fulton found Fitch's steamboat plans in the U.S. Consul's files in France, and some say that Fulton's epochally successful North River steamboat of 1807 was really designed by Fitch in 1791.

Robert Fulton was a trained and accomplished artist who traveled to England in 1786 to continue his studies. There he met Francis Edgerton, Duke of Bridgewater, and Charles, Earl Stanhope, both of whom were keenly interested in steam power and steamboats. He also met James Watt and studied the literature on steamboats available in England. The artist became a self-taught mechanical engineer, specializing in canal and bridge design and, later, submarine and torpedo warfare. Given Fulton's diligent researches, and his many influential and wealthy friends, it is not surprising that he established the first commercially successful steamboat line.

In 1803, with the financial backing of Chancellor Livingston, who was then the American minister to

William Symington's 1788 marine engine (left) was installed in a "pleasure boat" on Dalswinton Lake, near Dumfries, Scotland. The 4" x 18" Newcomen cylinders spun a drum back and forth. From this, chains drove the paddle wheels through pawls and ratchets. The engine provided close to 1 h.p., for 5 m.p.h. By neighborly happenstance, one of the passengers on the boat was Robert Burns. The boat, much smaller than the commercially intentioned French and American steamers of the period, could reasonably be called "the first steam launch." (Crown Copyright. Science Museum, London)

France, Fulton built a 66.4' x 10.4' steamer in Paris. The 17.7" x 31.5", 8-h.p. engine was built for Fulton by J.C. Perier, who had built and run his own steamer on the Seine 28 years earlier.

Fulton ordered a new, larger engine from the Boulton and Watt factory and returned to America in 1806 with a clear idea of how to build a serviceable steamboat and operate a profit-making steamboat business. He possessed the great advantages of ample financing by Chancellor Livingston and a 30-year monopoly of steam navigation on the Hudson.

In New York, Fulton ordered a 133' x 13' hull (he had also become an expert on low-resistance hull design) and a 20' x 7' x 8' externally fired, rectangular, copper boiler. Power transmission was by bell-crank and spur-geared flywheel. *The Steamboat* (later renamed *Clermont)* attained 4.7 m.p.h. on her first trip to Albany in 1807 and ran profitably for another seven years.

By 1817 more than 20 Fulton-designed steamers were in service, including a warship and ferries. The *Chancellor Livingston,* built in 1813, was a fast (10 m.p.h.) and luxurious packet, and the first American coalburner.

Steamboats were firmly established in Britain

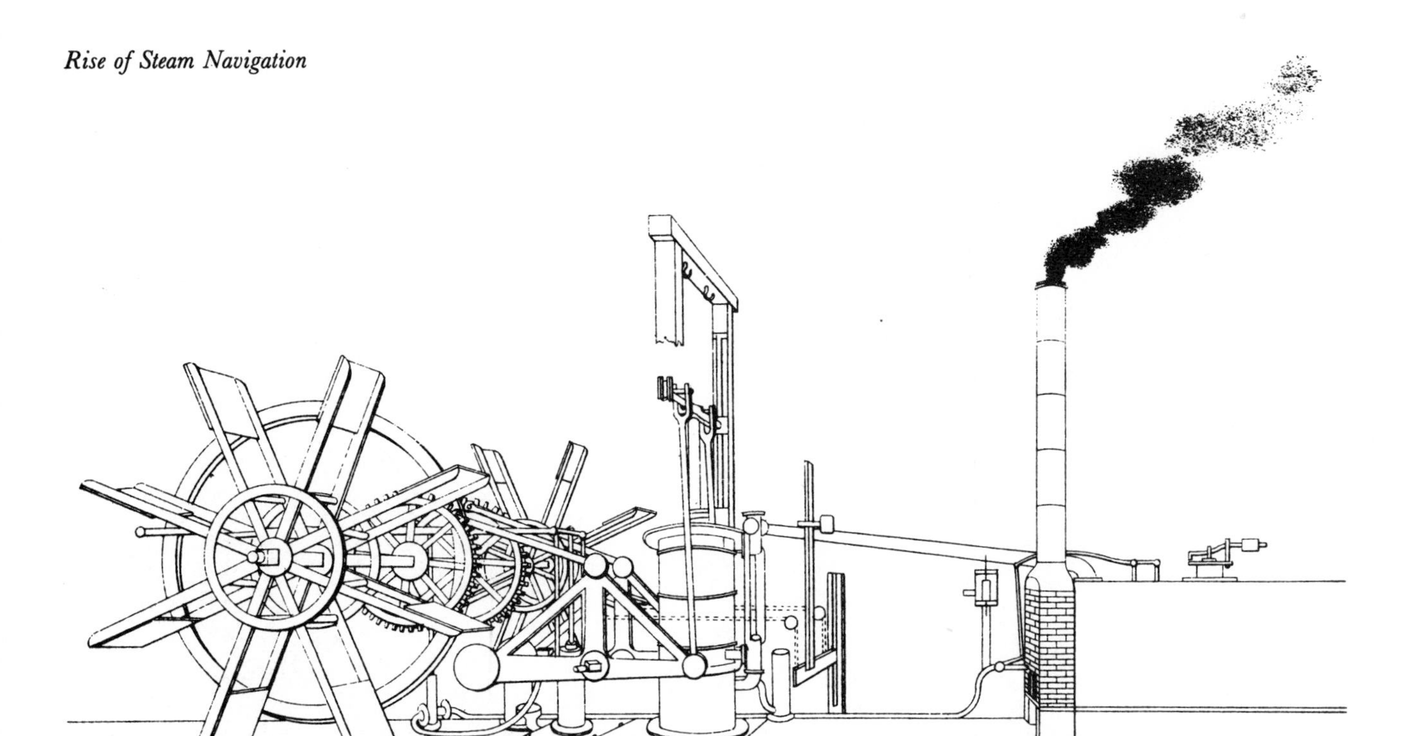

Robert Thurston believed that this drawing, "made by Fulton's own hands," was a representation of The Steamboat*'s machinery. A 19-h.p. (24" x 48") engine drove the 133' x 13' passenger boat of perhaps 60 tons displacement. Robert Fulton achieved success through the study of many previous steamboat designs and the combining of numerous elements — politics, business, finance, patents, legislation, hull design, engine, boiler, wheels, fuel, crew, rates, scheduling, passengers, publicity, etc. — in a balanced system capable of earning a profit. Many earlier steamboat builders had workable ideas, but their engines were often poorly executed or too small for the job. Fulton's bell-crank and cogwheel drive from the Boulton and Watt engine to the side wheels was copied in a number of early steamers, both American and British, but soon gave way to more refined mechanical linkages. (Crown Copyright. Science Museum, London)*

after 1812, the year that Henry Bell's *Comet,* a small side-wheeler, initiated excursion trips on the Clyde between Glasgow and Helensburgh. *Comet* immediately spawned several imitators, and the successful and safe operation of these Clyde steamers helped clear the way for Britain's commercial steamships.

The first steamers on the Thames were the Glasgow-built, 70-ton *Margery,* and the *Richmond,* of 50 tons and 10 h.p. Neither of these boats was very successful, the principal problems being mechanical. The *Thames,* built in 1813, had steam-heated cabins, a stewardess, and a library. By 1822, 25 steamers served the river between Richmond and Ramsgate, and the centuries-old system of passenger transport by barge, wherry, and hoy was in disarray.

The 3- and 4-m.p.h. steamboats built during the 18th century could not compete with rowed, sailing, or horse-drawn boats, which cost only a fraction as much. The 5- to 10-m.p.h. steamers of 1807 to 1820, even though they were very costly and fuel-wasteful, quickly made significant inroads into the passenger services for coastal and inland waters. For the first time, passengers could relax in a (sometimes) comfortable conveyance while advancing as much as 100 miles in a day along rivers, canals, and other waterways that paralleled the terrible stage roads. For this privilege, in the years before railroads began to compete seriously, one had to pay dearly. In America usual steamer fares were two or three times higher than fares for slow canalboats or miserable stages. A 5-cents-per-mile passenger fare in the 1820s would be equivalent to $2 a mile for a modern wage earner. The London-to-Margate fare on the *Thames* was 15 shillings — a week's wages.

At this juncture in America's development a vigorous, westering population was looking over the Ohio River threshold, eager to make money in the territories beyond. Nicholas Roosevelt, in association with Livingston and Fulton (John Stevens was

Henry Bell paid John Robertson, of Dempster St., Glasgow, £165 to build the engine (right) that Bell designed for his Comet, *of 1812. The use of half-side-levers, return connecting rods, and a crankshaft make this the prototype for "grasshopper" engines, some of which remained in service in the 20th century. The 12.5" x 16" engine developed nearly 4 h.p. with steam a little above atmospheric pressure (perhaps 8 or 10 p.s.i. effective pressure). The boat worked regularly on various Scottish routes until she was driven ashore in a storm in 1820. By then there were dozens of British and European steamers, principally fast passenger packets. The photograph of a replica of* Comet *was taken in 1962. (Crown Copyright. Science Museum, London)*

no longer a partner, as he was by this time fighting Fulton's monopoly on the Hudson), built the first western boat at Pittsburgh in 1811 and took her down the Ohio and Mississippi rivers to New Orleans. This early packet lacked the power for steaming upriver on a regular schedule, so she was put into service between the Gulf and Natchez.

Henry Shreve built the first real "Western Rivers" boat in 1816. She was flat-bottomed, high-pressure, and had an upper deck, and — prophetically — had a boiler explosion that killed 13 in her first year. Shreve fought Fulton and Livingston's claimed monopoly on the Western rivers to a standstill, and river transport — and America — entered into 40 years of unprecedented growth.

Steamboating spread like wildfire through America's sheltered bays, sounds, and estuaries, and especially throughout the vast Ohio-Mississippi-Missouri river system. Before 1855, a river boat, *Eclipse,* was 365 feet long, with 42-foot wheels and 3' x 11' cylinders. There was no need to develop steam launches as auxiliaries for American steamboats, as they were mainly shallow draft and highly maneuverable. However, knowledge of simple steamboat technology (later, steam locomotive also) became widespread in the West, so when demand for steam launches arose after 1880, the western states were as well equipped to meet the need as the Atlantic coast.

Steamboat building spurted so far ahead in America between 1807 and 1825 that the most advanced boat in Britain in 1826, a 200-h.p. 160-footer built in Scotland by David Napier, failed to surpass the qualities embodied in Fulton's *Chancellor Livingston* 13 years earlier. After Napoleon's surrender at Waterloo, however, an unprecedented quantity of new capital and talent became available in Britain for manufacturing civilian goods. Many new steamboat builders set up shop, especially on the Thames, and British steamboat builders began to catch up with the Americans. The positions were finally reversed after 1840 as British iron "propellers" (propeller-driven ships) began crossing oceans, and the Americans chose to continue with wooden sailing vessels.

STEAMSHIP TECHNOLOGY, 1800 TO 1840s

By 1800, many people knew how to design a serviceable steamboat, but they had to wait decades for manufacturing techniques and materials to catch up with their imaginations. Low steam pressure necessitated large working surfaces and massive mechanisms in order to generate moderate amounts of power. In his *History of Marine Engineering,* John Guthrie concluded that the precise metalworking skills long known to instrument makers, model-makers, silversmiths, and gunsmiths were unavailable to engine builders, both because of the vastly larger scale of the machines and because engine building was carried out by illiterate mill-wrights and blacksmiths.

A pumping-engine cylinder bored at Philadelphia by waterwheel power in 1800 required 4½ months to turn, attended by two machinists and a tool grinder day and night. The bore turned out to be ⅜" wasp-waisted and had to be redone. The boiler was made of 38" x 32" wrought-iron plates imported from England. The largest American plates were 36" x 18", and irregular.

The machinery of most early steamboats would be an affront to a modern mechanic's or designer's eye. Design and finish were equally crude. Cast iron was employed wherever possible. Where cast iron would be too brittle, parts were forged from wrought iron, a material of uneven texture and uncertain strength. Wooden cylinders and working parts were not unknown. Wooden driving rods (pitmans) survived in stern-wheelers for more than a century.

The early engines were handmade. They had no fair surfaces and were held together by wedges, keys, lag screws, and dogs. Manufactured machine screws did not exist. The engines were built by millwrights familiar with wooden windmill and watermill construction and blacksmiths who knew how to shape iron. Piston rods and shaft journals were hand filed to their finished shape. Later refinements, such as forked connecting rods, were impossible to manufacture. Cast-iron cylinders were sanded to approximate bore, and even after horse-driven cylinder-boring machines were developed, no cylinder was truly cylindrical.

Early boilers and early iron hulls suffered under the same handicaps. There were no plates of standard size and quality. Surface finish was rough and variable. Hulls and boilers both leaked steadily through caulked seams and soft packing — or through flaws in the plate itself. Advances in the metallurgical industries made reliable hulls and

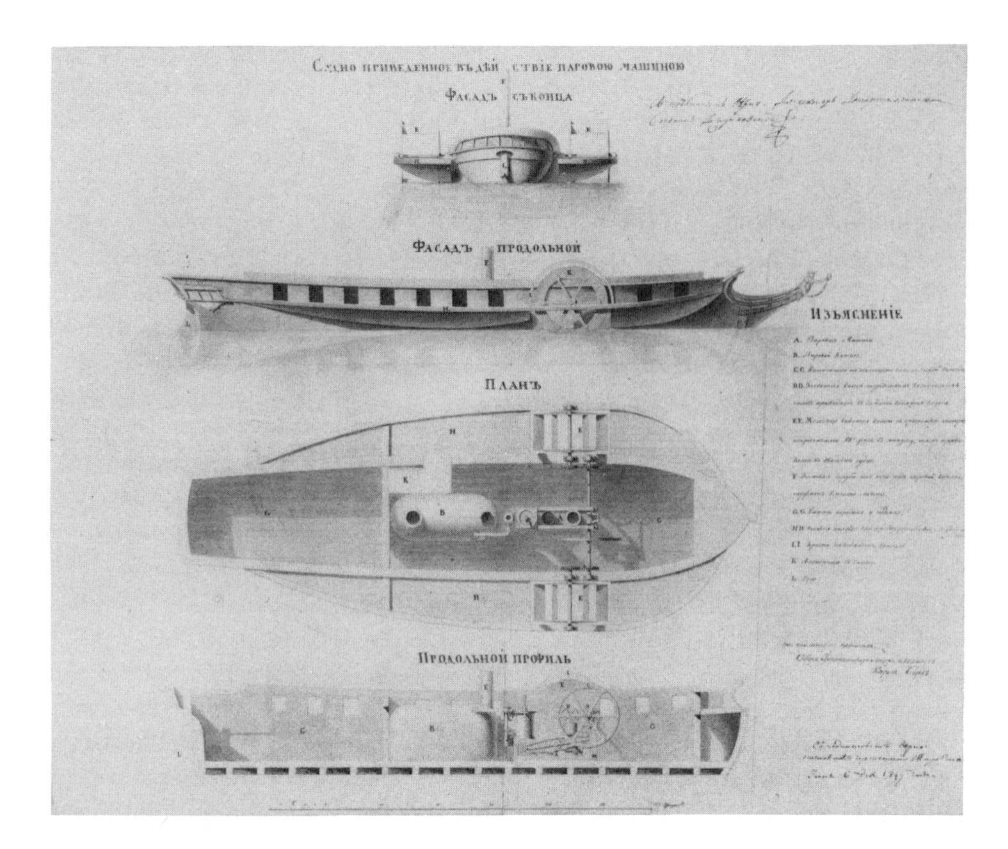

This Russian river steamer of 1817 was graced with a figurehead, quarter ornaments, and guards sponsoned out to an extreme degree. At first, each new steamer was a novelty on the face of the earth, and the designers had to fill in the blank spaces on their drawing boards with hubris and imagination. (Crown Copyright. Science Museum, London)

boilers possible by the 1850s, more emphatically so 15 to 20 years later with the introduction of inexpensive steel. Mild steel is nominally only one-third stronger than wrought iron, but the uniform quality available after 1870 gave it at least twice the structural value of the iron used in early steamers.

Early steam engines had derived most of their power from the vacuum created by condensing the steam in the cylinder. The steam needed only enough pressure to cause it to move into the cylinder and displace the air, and 1½ to 3 p.s.i. was enough for this. After Watt's improvements, the ability of steam to perform work while expanding was incorporated in engine design, yet early, low-pressure steamers operated with steam pressures only a few pounds above atmospheric pressure (which is about 15 p.s.i., absolute). Watt himself favored 2½ to 3 pounds' steam pressure, so exhaust-condensing and vacuum-enhancing arrangements were essential. A French scientist, Sadi Carnot, published a correct theory of heat engines in 1824, but practical engineers continued to delude themselves for many more years.

Some contemporaries of James Watt recommended steam pressures 50 to 80 times higher than those he advised. In America, John Stevens liked 50- to 100-p.s.i. steam and employed it in his 1804 water-tube-boiler, geared-propeller launch, *Little Juliana.* Oliver Evans, innovative Philadelphia mechanic and steam engineer, thought 200-pound steam about right. These pressures were dangerously excessive for 1800, but they were prophetic. On the Western rivers high steam pressure was recognized almost from the beginning as the only practical means to float sufficient horsepower to go upstream in boats shallow enough for the service. Locomotive working pressures (50 to 90 p.s.i.) were employed in boilers and engines that were many times larger than locomotive machinery, but much lighter than the condensing power plants of low-pressure steamers.

For 40 years safety standards on the Western rivers were dictated by the owners' consciences, and the loss of life to fire and explosion was astounding. This undoubtedly resulted from frontier attitudes, not from insurmountable technical obstacles to safe design. On the Atlantic coast 25-pound boilers were mounted out on the paddle guards to reassure lily-livered passengers, and mechanical failures caused public debate, litigation, and loss of patronage. The

The year 1804 was far too early for screw propulsion, a water-tube boiler, and high steam pressure (noncondensing) to be successful commercially. John Stevens didn't know this, and he gave his twin-screw steam launch, Little Juliana, *all of these features. The meshed cogwheels drove opposite-turning screws. The propellers were correctly designed for the service, with high pitch and large blade area to make good use of the low r.p.m. available. James Renwick (who was 12 when he saw the* Little Juliana, *operated by Stevens' sons, leave the Battery and head across the Hudson toward Hoboken) later described the boat as "about the size of a Whitehall rowboat." A replica (shown here) of the steam launch, built by Stevens' sons in 1844, was 25 to 32 feet long and 5½ to 6 feet in beam. The 1804 power plant was installed in the 1844 hull, and the boat made about 8 m.p.h. in a speed trial. (Below, courtesy of Babcock & Wilcox Company; photo at left, Samuel C. Williams Library, Stevens Institute of Technology)*

majestic beam or side-lever paddle engines, with their immense 15-p.s.i. boilers, remained popular on salt water until "modern" vertical, propeller engines and tubular boilers began to appear in the 1860s and '70s. On the Western rivers (where every man carried a pocket pistol, and risks were to savor), grand dining saloons were built atop 90-pound boilers, and gruesome disasters became grist for jokes, brags, and Western-style stories.

Throughout the first half of the 19th century the locomotive working pressures employed in the Western river packets were a radical departure from convention. Steam pressures in customary marine practice were much lower, though rising slowly: 3 p.s.i. in 1815; 5 p.s.i., 1830; and 15 p.s.i. by 1840. Steamers of 100 tons and upward could run profitably with heavy, crude engines operating inefficiently on 2 to 20 pounds' steam pressure, but scale effects made such inefficiency a critical obstacle for steam launches. They were not practical for commercial uses until steam pressures approached 50 p.s.i., permitting lighter machinery and more efficient use of fuel.

Another prerequisite for the success of steam launches was screw propulsion. The advance to screw propulsion depended on attaining higher r.p.m., supplying power to a fore-and-aft shaft in the bottom of the hull, and designing a serviceable thrust bearing, stern bearing, and stern gland. The perfection of the "common screw" propeller during the 1840s, after hundreds of years of tinkering with the idea, permitted neat, compact, relatively fast-turning machinery and a propelling device well suited to rough water use.

In the Science Museum in London is a beautiful tiny engine from the river launch *Firefly* of 1840. One wonders why such a charming mechanism did not come into immediate demand among people with

Only 28 years separated the blacksmithed crudities of the pioneer Comet *engine from the sparkling delicacy of the* Firefly *engine of 1840.* Firefly, *22' x 4.75' x 1.67', ran from Blackfriars (on the Thames) to Putney and return at an average speed of 7¾ knots, equipped with a 22", six-bladed Ericsson propeller. The two 3" x 6" cylinders oscillated, receiving steam through elbow pipes with packed glands. Sixty-p.s.i. steam was supplied by a small locomotive-type boiler. The "busyness" of the engine's mechanical motion is mitigated by its 200-r.p.m. service speed. The engine was equal-speed spur-geared to the propeller shaft, since it was too wide to set low in the hull. Several amateur machinists have contemplated building replicas of this pretty boat engine, 140 years old in 1980, but none is in service yet. (Crown Copyright. Science Museum, London)*

unlimited money for playthings. The technical capacity to build beautiful, practical, even exciting small steam launches existed in the 1840s and 1850s, but the concept of using mechanical power on a small, personal scale, or for private pleasure, did not yet exist — and what could a steam launch engine do that could not be done by oarsmen earning two shillings a day?

TECHNOLOGICAL ADVANCES, 1840 TO 1885

After 1840, steamer evolution began to accelerate, especially because of the development of machine tools and a rapid increase in the numbers of trained mechanical engineers. Between 1855 and 1885, the power plants in new main-line ships were becoming obsolete within 10 years, so re-engining and re-boilering constituted a major part of shipbuilding. By 1885 ship design was fully evolved, with efficient engines (1½ pounds of coal per horsepower-hour), steel hulls and machinery, refrigerated stores and cargo, incandescent electric lights, and as much power as the owners cared to pay for. (A cross-Channel ferry of 1888, the *Princesse Henriette,* had 6,500 h.p. on 1,400 tons — twice as high a power density as a modern businessman would choose to put in a ship for this service.)

The curious engines that drove sailing ships out of the passenger business between 1820 and 1860 were side-lever, steeple, return-piston-rod, Siamese, annular, oscillating, inclined, geared, disc, disc-and-vibrating lever, return-connecting-rod, trunk, horizontal direct-acting, beam, etc. There were precious few steam launches in the world before 1860, so steam launches escaped having to accommodate the myriad varieties of engines that propelled ships on every sea before the stabilization of marine engine design in the 1860s and 1870s.

The necessary and logical direct-acting marine steam engine configuration was first employed in Britain's HMS *Gorgon* in 1837, with the crankshaft above the cylinders. Some *Gorgon*-type engines began to be "inverted" 10 years later, with cylinders mounted on massive iron castings. The 2-cylinder simple-expansion engines of many ships of the 1850s and 1860s would look very familiar today, as they were no more than large versions of the double-simple steam launch engines that have survived ever since. With working steam pressures of only 20 or 30 p.s.i. during the 1850s, these hulking, broad-shouldered brutes required cylinder bores of four or five feet to develop 1,000 h.p.

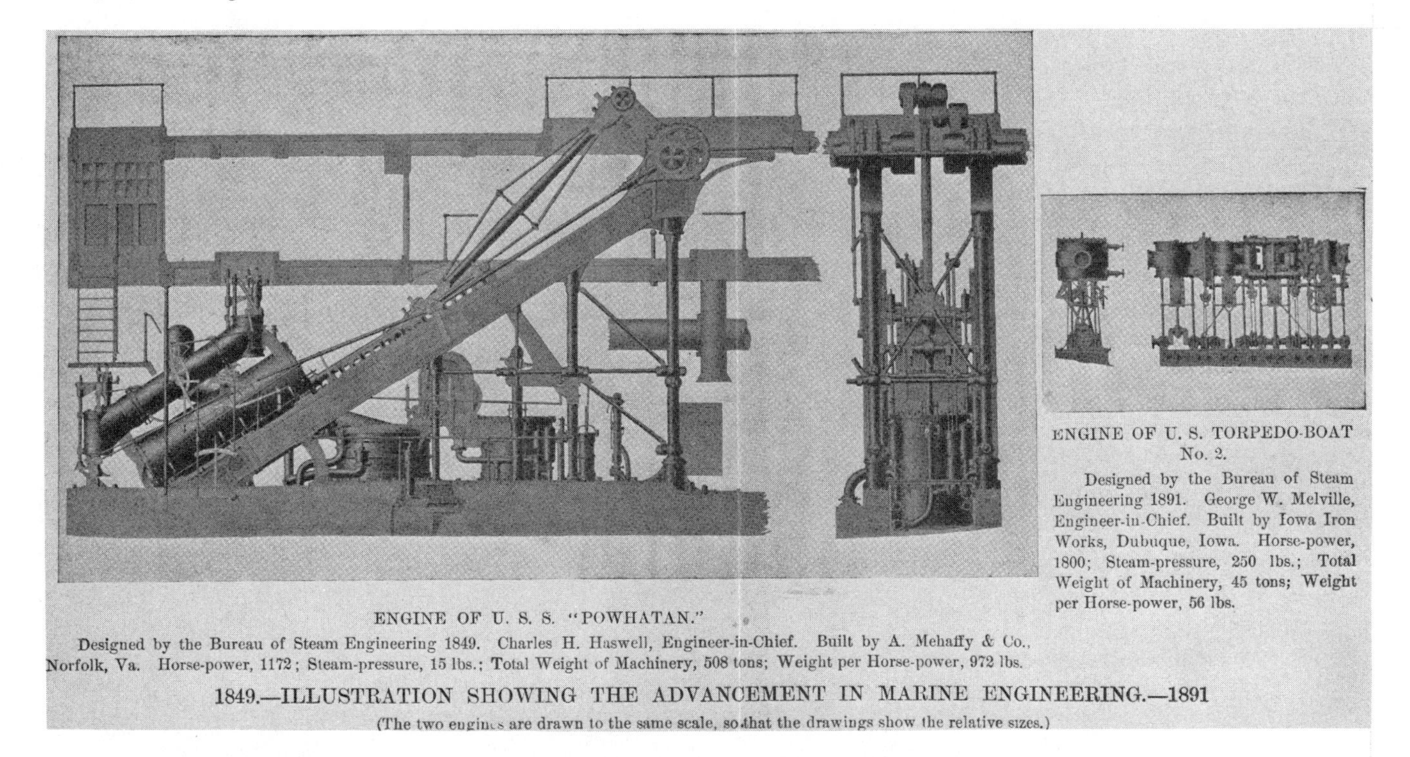

The U.S. Navy's Bureau of Steam Engineering prepared this drawing "... to show at a glance the progress of marine engineering during the past 40 years [1849 to 1891]. The engines of the Powhatan, *the last paddle-wheel ship in the navy, and the engines of the* Ericsson, *the newest torpedo-boat, form a contrast so striking as to be almost ridiculous . . . ." Details of the* Powhatan*'s inclined engine recall the engines of the Western rivers packets. The modern engine was built in Iowa.*

Through the 1850s and subsequent decades, the increase in steam pressures in customary marine practice continued: 25 p.s.i. in 1855; 35 p.s.i., 1865; 60 p.s.i., 1870; 170 p.s.i., 1885. Cylindrical-tubular marine boilers to accommodate locomotive working pressures were developed during the 1850s, and engineers and manufacturers started down the road that would culminate in multiple-expansion engines designed for working pressures up to 220 p.s.i. Seawater boiler feed had to be abandoned, but before this was possible, surface condensers had to be developed and tallow lubrication replaced. Boiler water chemical treatment had to be invented, and so on. All the required theory was available by 1860; many practical steamship builders and operators were ready to accept these advances in the state of the art by 1870.

All other developments were crowned by the triumph of the marine compound steam engine in the 1870s. World commerce was revolutionized to the same degree that jet planes revolutionized passenger travel 90 years later. Steamers could now profitably carry low-value cargo to any port on earth. The ships engaged in hauling grain from San Francisco to Europe, 1881 to 1885, included 11 British and one French steamship along with 418 American and 198 British wood sailing ships and 761 British iron sailing ships. (At 30,000 round-trip miles, this was then the longest bulk-cargo route.)

Sailing ships continued to haul bulk commodities for a few more years, relying on cheap labor and subsidies from governments that could not compete with British steamships. British iron and steel propellers soon dominated every sea, to a degree not seen before or since.

Many of the momentous advances between 1840 and 1885 may be sorted into 15-year intervals: 1840 to 1855 saw perfected screw propellers and iron hulls; 1855 to 1870 brought vertical-inverted, direct-connected engines, Scotch boilers, adequate stern bearing and thrust bearing design, freshwater boiler feed, and boiler water treatment; 1870 to 1885 saw compound engines, refrigeration and electric lights, steel hulls and boilers, and the rise of the triple-expansion engine.

The development of several processes for making cheap steel of uniform excellence was the principal stimulant of the rapid advances of the 1870s and '80s. In 1860 steel had cost more than five times as much as wrought iron and was rarely used in large construction. After 1875 steel construction became cheaper than iron. Boilers suited to the triple engines could not have been made with the wrought iron of a few years earlier.

Between 1815 and 1885, steamship thermal efficiencies increased by as much as 1,000 percent, and customary steam pressures increased by more than 5,000 percent. The steam launches of the Golden Age were a small offshoot from the most advanced branch of the "old" industrial revolution of coal, iron, and mechanical power.

2

A History of Steam Launches

Small steamers, scattered and often experimental, existed from the early years of steam navigation, so although steam launches never amounted to much before the 1870s, their lineage is certainly older than that. Some American boats built during the first decade of commercial steamboating were of scarcely more than steam launch size. *Stoudinger,* 1816, 47' x 12' x 4.8', was used in experimental work by James P. Allaire, then owner of the largest steam engine factory in America. Renamed, the boat became Cornelius Vanderbilt's *Mouse in the Mountain. Emeline,* 1818, was 47' x 15' x 5'. It is unfortunate that there are not more descriptions or illustrations of early small steamboats and their machinery.

Steam navigation grew steadily and machinery design was refined through the 1820s and '30s. Paddle tugs became an important adjunct to sailing ships, and small working steamers began to appear on the rivers of the far-flung British Empire. Rivers, lakes, and sounds were the cheapest and most reliable highways over most of America. Not enough is known of the small local steamers that served most American communities on water until rails (1840s to '80s) put them out of business.

After the 1830s locomotive boiler pressures were available to small boats, but steam launches remained very scarce through the next 20 to 30 years, when they "might have been." The river launch *Firefly* of Britain, 1840 (see Chapter One), did not signal the advent of the steam launch Golden Age; there was no recreational market for mechanically powered small boats, and sails and oars were still cheaper for commercial purposes.

Before steam launches could flourish, two conditions had to be met; one was technical, the other, economic. The steam plant had to become light, fast-turning, and fuel-thrifty enough to suit small craft. This occurred at approximately 50 p.s.i. working pressure, 500 pounds' weight of machinery per horsepower, 200 r.p.m., and coal consumption well under 10 pounds per horsepower-hour.

For many commercial uses of small launches to

become competitive, capitalization of a steam launch horsepower had to cost less than equivalent powering by oarsmen (perhaps four oarsmen per 10-hour day, 12 for continuous service). Hiring oarsmen for a horsepower-day would cost nearly $1,000 now, vs. $3.00 for 24 gasoline horsepower-hours. Before 1860 the cost difference was in favor of muscle power.

STEAM YACHTS

Inexpensive, compact power plants were not essential to large yachts and their wealthy owners, and larger boats were not victim to the scale effects that delayed steam launch development. There were steam yachts in England as early as 1830. Descended from oceangoing steamers, fast sailing ships, and opulent sailing yachts, these had a different family tree than the humble launches. They are deserving of mention in a book about steam launches, for they were first to establish the concept of harnessing steam in a vessel for private consumption.

According to Reginald Crabtree in *The Luxury Yacht from Steam to Diesel* (1974), the first accredited steam yachtsman in England was Asheton Smith of Andover, who built a 120-foot paddle-wheeler driven by double side-lever engines in about 1830. A few years later, the first Royal steam yacht, *Victoria and Albert,* was christened by Queen Victoria and launched at Pembroke, Wales.

When steam launches were expected only to be more reliable than sailboats, not more powerful, they could be as lithe and rakish as any Baltimore clipper. The iron screw steamer Fairy *was 146' x 21'; few today would call her a steam launch, but she was merely a tender to the side-wheel Royal yacht* Victoria and Albert. Fairy, *of 1845, was no doubt observed with great interest by young men who became the designers of great propeller yachts a generation or two later. (Crown Copyright. Science Museum, London)*

It was Queen Victoria, more than any other individual, who set the tone of the "Great Age of Steam," that half-century when steam served so many needs so well that it would have been difficult to convince anyone that electricity or explosive engines might someday be equally useful. On a sailing-yacht trip to Scotland in 1842, Victoria noticed that common steam coasters outdistanced the Royal yacht. She immediately relegated sailing yachts to frivolous racing uses, and for the following 59 years lent her authority and interest to the encouragement of steam power.

Asheton Smith ordered eight additional steam yachts before 1850, and he gave them names such as *Glow Worm, Fire King, Water Care,* and *Sea Serpent. Fire Queen III,* a steeple-engined propeller yacht of 1846, attained 16 knots and was the world record holder for a time. None of the long, rakish blockade runners sold to the Confederacy more than 15 years later could top this speed.

In 1839 Smith offered to race *Fire King* against any steamer afloat for 5,000 guineas — a sort of melding of the idle profligacy of the Regency with the commitment to steel and steam that characterized Victoria's reign.

The first steam yacht in America was Commodore Vanderbilt's *North Star,* a 270-foot, plumb-stemmed, wooden side-wheeler. Vanderbilt was a steamship magnate; so, not surprisingly, *North Star* was reminiscent of a commercial steamer. In 1853 she cruised to several landfalls in Europe and back within a four-month interval. However, she was soon converted to a cargo vessel and served between the United States and Nicaragua until she was condemned and broken up in 1870.

The New York Yacht Club registered its first steam yachts, *Clarita* and *Bijou,* in 1864. The first U.S. presidential steam yacht was *River Queen,* which Abraham Lincoln chartered to provide mobility around the Virginia war front, and to house conferences away from the turmoil in Washington.

In *Captain Nat Herreshoff,* L. Francis Herreshoff refers to the growing popularity of steam yachts as evidenced by their numbers in England: 30 in 1863; 140, 1873; and 466, 1883. In America there were fewer, but by the 1870s these floating monuments to success were not an unusual sight between Philadelphia and Boston.

FROM THE BACKWATERS TOWARD THE MAINSTREAM

Steam launches first became practical where their high costs were justified by their ability to perform services beyond the capabilities of rowing and sailing boats. By the time of the Crimean War and the War Between the States, navies required a few steam launches to provide power and endurance. The British Admiralty tendered the construction of a 27-foot steam pinnace to John Samuel White in 1861. She had a four-bladed propeller, a maximum draft of 2.5 feet, and a speed of close to 7 knots, and she proved herself extremely seaworthy during a surveying expedition along the African coast.

The year 1860 provides a convenient peg in what was really a continuum; one can say that after that year, advances in marine engineering made small steamers practical. They grew lighter, trimmer, thriftier, and more numerous, and as their marketability increased, competent builders began to produce a few, at the same time making further refinements.

British recreational launches continued the tradition of gentlemen's water vehicles, with steam power substituted for the oarsmen of earlier centuries. These river launches were the last flowering of an ancient line of ceremonial and pleasure boats. Commercial steam launches evolved out of rowed boats and small working steamers and were shaped not by custom but by determined efforts to make money with engines in boats. The classic American fantail launch (a design that enables a vessel with a fine run to swing a large, slow-turning propeller and is only incidentally beautiful) appeared a little later, soon after the lines of screw tugs evolved, about 1870.

The 1860s marked the first efforts of two young British marine engineers, John I. Thornycroft and Alfred Yarrow, the record of whose early work is inseparable from steam launch developments of the time. Thornycroft's first launch, built in 1859, was 36 feet long, and her two-cylinder steam engine and locomotive-type boiler gave her 9.5 knots. The 40-foot *Ariel* of 1863 steamed at 12.2 knots. In 1864 Thornycroft completed his studies in mechanical engineering under Professor William Rankine, and with a sound knowledge of steam engine technology

Of the three great names in steam launch building, John Thornycroft was first in the water with Nautilus, *launched in 1862. He had laid her keel in 1859, when he was an experienced model-engine builder of 16. Thus he had a lead of several years on his contemporary, Alfred Yarrow, and almost a decade on N.G. Herreshoff, who was five years his junior. (Courtesy Vosper Thornycroft, Ltd., Paulsgrove, Portsmouth)*

he set out to improve his line of launches in a works at Church Wharf, Chiswick, London. His tenth launch, *Miranda* (1871), measured 49.75 feet by 6.5 feet and with her 64-h.p. engine was capable of 18.4 m.p.h., a shocking speed for her time and her size. Her performance forced experts to reformulate their theories concerning the dependence of maximum speed on waterline length.

Thornycroft's chief English rival was the young genius Yarrow, possessor of patents for a steam plow and steam carriage while still in his early twenties. It is indicative of the growing popularity of steam launches in England that when Yarrow's struggling engineering firm turned in desperation to manufacturing the little craft in 1868, financial problems were ended. The following paragraphs from Eleanor Barnes' *Alfred Yarrow, His Life and Work* describe this beginning:

This class of work appealed strongly to Yarrow, and on deciding to follow it up he began advertising as follows:

"STEAM-LAUNCHES. — Anyone wanting a steam launch would be well served if they came to Yarrow and Hedley, Isle of Dogs."

This was the last resource. Without capital, and with the business considerably in debt, but with increased experience, Yarrow awaited developments. Within three days of the appearance of this advertisement, an old gentleman, Colonel Halpin, came to the yard. He placed an order for a steam-launch, 24 feet in length, to be fitted with a small cabin to hold four people. It was to have a single-cylinder engine, and a vertical boiler of 3 horse-power. The price was to be £145. This little boat was built in three months, and cost £200 , so that the financial outlook had not yet improved. Yarrow, however, was delighted with the boat, and himself steered her up to Colonel Halpin's house at Islesworth one Saturday afternoon. Somewhat later than was expected, owing to having lost the tide through many finishing touches being required at the last moment, he arrived in a begrimed and oily condition; but the Colonel, who had constantly visited the yard during the construction of his boat, and whom Yarrow always referred to as a "charming old gentleman," received him very warmly and insisted on his going with him into the house, black as he was, to have some luncheon. This steam-launch was quite a show boat that summer, and it may be of interest to record that, although Yarrow made a loss in building her, he bought her back after the summer season for £100, and sold her the same day for £200. After the following summer she was again purchased by Yarrow for £100, and sold for £300 to a Russian nobleman, who took her to Petrograd.

Yarrow pushed this new business for all it was worth. He started one day from Oxford with a number of framed photographs of Colonel Halpin's launch, landed at every inn on the Thames, and asked permission to hang one up in some conspicuous place. This led to many launches being ordered.

Steam launch speeds and refinements in their power plants progressed steadily through the 1860s, through the efforts of men like Thornycroft and Yarrow, even before receiving the hearty boosts in speed, power, and efficiency given by cheap steel and multiple expansions of steam. The fast launches of this period sometimes carried 100 or 125 p.s.i. — locomotive pressures — while new oceangoing ships were advancing from 30 p.s.i. to the 65 p.s.i. supplied to compound engines. Working pressures in launches had to be higher if power plants were to be small enough to suit small hulls. Even at that, the weight of machinery in Thornycroft's *Miranda* was greater than half her total displacement, and there was precious little room left over for creature comforts.

More sedate launches were far more common, though they attracted less public attention. A photograph of Yarrow's yard shows low-power launches of easy lines that would look familiar to any enthusiast today. Yarrow built 350 such launches between 1868 and 1875, at first for gentlemen and their ladies along the Thames, later to be used as ship's boats on oceangoing vessels.

There came in the 1870s a worldwide burgeoning of steam launches. In America this was related to the impetus the Civil War gave to manufacturing and travel. Boating — rowing, mainly — was a popular activity after the Civil War, and the extension of this interest to powered boats developed naturally.

For the first two or three generations of the industrial revolution the milords and robber barons had taken their greedy cut of the new wealth generated by machinery, and not much had been left for ordinary people. From the 1870s on, however, increasing numbers of wage-earners and small businessmen had enough surplus income to look around (after a sweaty 60- or 70-hour week) and think about what they would like to buy. Small steam launches were the first mechanically powered

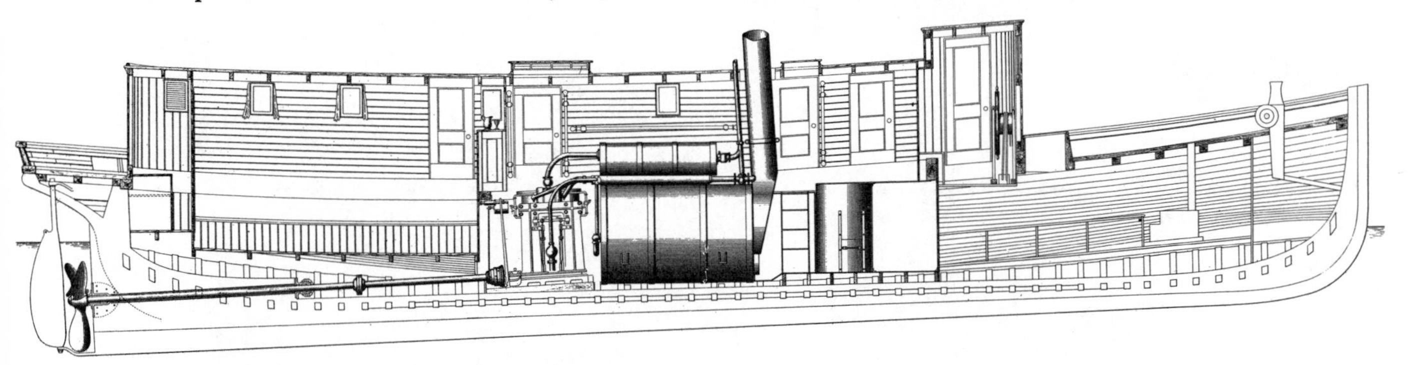

By the 1860s, steam launches had clearly become a new kind of vessel, having a distinctive character and providing useful services available from no other craft. At last it was possible to deliver supplies or carry mail or passengers on an assured schedule, with a small crew and without concern for which way the wind blew. The Mare Island Naval Shipyard (California) was established in 1853; their Steam Launch No. 4, shown here, was 50' x 10' and had a cruising speed of 8 knots. (From Transactions of the Institute of Naval Architects, *1872)*

LADY WOODSUM.

Advertisements.

DAVID J. LEWIN,

ENGINEER, STEAM YACHT AND LAUNCH BUILDER,

POOLE, DORSET

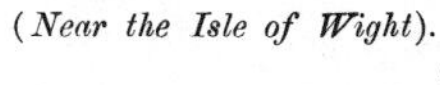

(*Near the Isle of Wight*).

SPECIALTIES :—

FAST STEAM YACHTS.
SEA GOING LAUNCHES. LIFE BOAT LAUNCHES.
STEAM LAUNCHES FOR HOISTING ENTIRE INTO YACHTS' DAVITS.
FAST RIVER LAUNCHES OF DIFFERENT CLASSES.
HIGH CLASS MACHINERY, ADAPTED FOR YACHTS AND LAUNCHES.

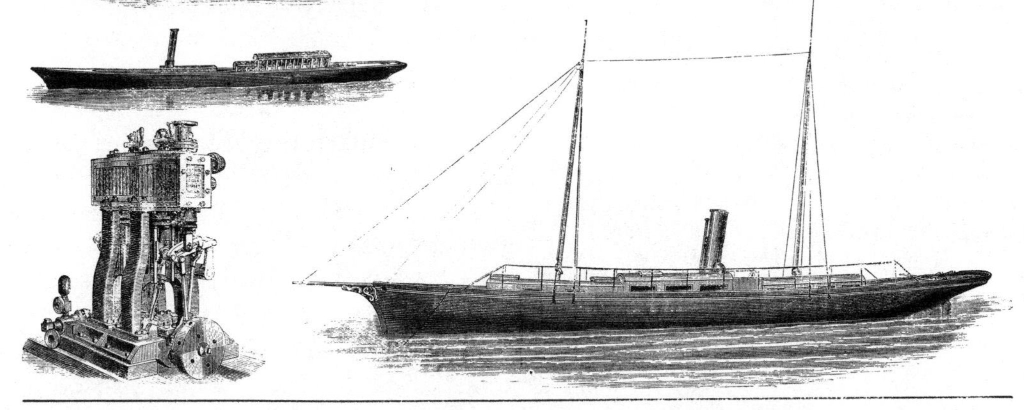

MAKERS TO THE ADMIRALTY OF BERTHON'S PATENT COLLAPSIBLE LIFE BOATS.

R. HEDLEY AND CO.,

ENGINEERING WORKS,

WHARF ROAD, ISLE OF DOGS, POPLAR, LONDON.

The Firm of YARROW and HEDLEY having been dissolved, the above Works have been established by Messrs R. HEDLEY &CO. especially to manufacture

STEAM YACHTS, LAUNCHES, TUGS, AND BARGES,

FOR CANAL, RIVER, AND COASTING TRAFFIC,

And, having secured several valuable Patents for the improvement of such Vessels, they are prepared to give estimates for any description of craft, varying from 20ft. long to 100ft.; and having had upwards of 40 years' experience as Engineers and Steam Launch Builders, they hope, therefore, to be able to secure a share of public patronage and support.

N.B.—ENGINES MADE FOR BOATS BUILT ABROAD.

Above: *From Dixon Kemp,* Yacht Architecture, *1876.*
Opposite top: *An early (probably around 1870) photograph of Yarrow's Isle of Dogs yard. (From Eleanor Barnes,* Alfred Yarrow, His Life and Works, *1924)*
Opposite below: *Much of America's steam launch history was centered on the sylvan lakes near rail lines where the summer resort industry was concentrated before automobiles opened up the whole landscape.* Lady Woodsum, *of 1876, met the Boston trains and distributed the summer folks and their trunks to all the grand hotels around the shore of Lake Sunapee, New Hampshire. The steamer lasted into the 1920s — not long enough for a grievous shadow on her early career to be forgotten. Her boiler exploded in 1877, killing one of the Woodsum brothers.* Lady Woodsum*'s engine is shown on page 269. (Courtesy* New Hampshire Profiles *— Charles Hill Collection)*

playthings offered to a large number of potential customers (private trains and steam yachts were not for everyman).

By 1878 Robert Thurston, in *A History of the Growth of the Steam Engine,* had this to say:

> The introduction of *Steam-Launches* and small pleasure boats driven by steam-power is of comparatively recent date, but their use is rapidly increasing. Those first built were heavy, slow, and complicated; but, profiting by experience, light and graceful boats are now built, of remarkable swiftness, and having such improved and simplified machinery that they require little fuel and can be easily managed. Such boats have strong, carefully-molded hulls, light and strong boilers, capable of making a large amount of dry steam with little fuel, and a light, quick-running engine, working without shake or jar, and using steam economically.

After the mid-1870s the name Herreshoff became intimately associated with steam launches. Nathanael Herreshoff, one member of this gifted family, studied engineering at Massachusetts Institute of Technology from 1866 to 1869, then set out to supplement education with experience by going to work as a draftsman for the Corliss Steam Engine Company of Providence, which at the time was the leading engine builder in the country, if not the world. Nine years and hundreds of engine designs, indicator diagrams, and valve adjustments later, Herreshoff was ready to apply his accumulated engineering skills, knowledge of hull design and building materials, and native abilities to the creation of steamers as a full-time occupation. Evenings while working with the Corliss firm he had been designing hulls and engines for his brother J.B.'s manufacturing concern in Bristol, Rhode Island. *Anemone,* the first Herreshoff steam launch, was built in 1870. In 1878 the brothers entered into a partnership as the Herreshoff Manufacturing Company, and until 1890 their work was confined almost entirely to steam launches and small steam yachts.

A third brother, James, in 1873 invented the coil boiler, which became standard in Herreshoff power plants until 1885. With a weight of about 70 pounds per horsepower, coil boilers stimulated interest and experimentation both in America and in Europe. Their production was finally discontinued by the Herreshoffs because scale deposits on the inner sur-

Early Herreshoff steam launch. (Courtesy Mrs. Muriel Vaughn)

face of the single long, curved tube proved to be an insurmountable obstacle to long-term service.

Light boilers and light hulls facilitated the hoisting of steam launches on the davits of steam yachts and steamships, and the Herreshoff Manufacturing Company built many small launches for that market. Others were built for several government agencies — the U.S. Navy, the Ordnance Department, the U.S. Coast Survey, and the U.S. Fish Commission. Different models for series production were tried, but the power plant, described in *Captain Nat Herreshoff,* was fairly standard.

> The power plant of the small launches between 1876 and 1880 showed a small coil boiler over a circular fire box and ash pit, the whole enclosed in a cylindrical casing not so dissimilar in shape from that of a donkey boiler but certainly much lighter, and a boiler capable of making steam in a few minutes. The engine was a single cylinder one, I think, three and one half inches bore and seven inches stroke, and developed about five horsepower . . . . The whole business was neat, simple, and light, and, of course, reliable and quite economical.

The smallest engine built had a bore of two and one-half inches and a five-inch stroke. Herreshoff engines are almost instantly recognizable for their light, uncluttered, and deceptively delicate appearance.

Every steam launch component was subject to ceaseless experimentation on both sides of the Atlantic, at least partly because radical innovations were not as potentially costly as they would have been if applied directly to large and expensive ships. Herreshoff was developing progressively lighter steam launch hulls despite frequent criticism. The critics were wrong — the light, supple hulls had longer working lives than their heavier counterparts, and Herreshoff would become an acknowledged authority on light construction.

Like other American engineers, he kept abreast of developments on the other side of the Atlantic, and he found much of interest in the *Goethe,* a Clyde-built Atlantic ferry that he took to England in 1874. With 3,000 horsepower on 3,600 tons, she was quite high powered. The 60'' & 104'' x 48'' compound engine turned a 19' x 27' propeller at 52 r.p.m. — 27 feet toward England at each turn of the screw! It is heroic images such as this that make lifelong fans of steam power. After Herreshoff returned home, he designed a series of steam launch propellers, all of 1.5 pitch-to-diameter ratio.

By the late 1870s the world's navies were clamoring for ever-faster small craft that, with their speed and maneuverability, could outflank enemy warships and get close enough to sink them by torpedo. Thornycroft built the first one, the 58-foot, 17.5-m.p.h. *Rasp,* for the Norwegian navy in 1873. By 1877 Yarrow was in the business; the Herreshoffs followed the next year. The performance demands of torpedo boats laid much of the groundwork for the high-speed yachts and launches of the 1890s. A builder might earn return business with a high-quality product; with a good turn of speed he earned an international reputation.

A Herreshoff compound engine. The Herreshoff Manufacturing Co. did build some engines for stationary use, and it is impossible to determine from the picture whether this is a stationary engine ready for shipment or a marine engine on a test stand. (Courtesy Mrs. Muriel Vaughn)

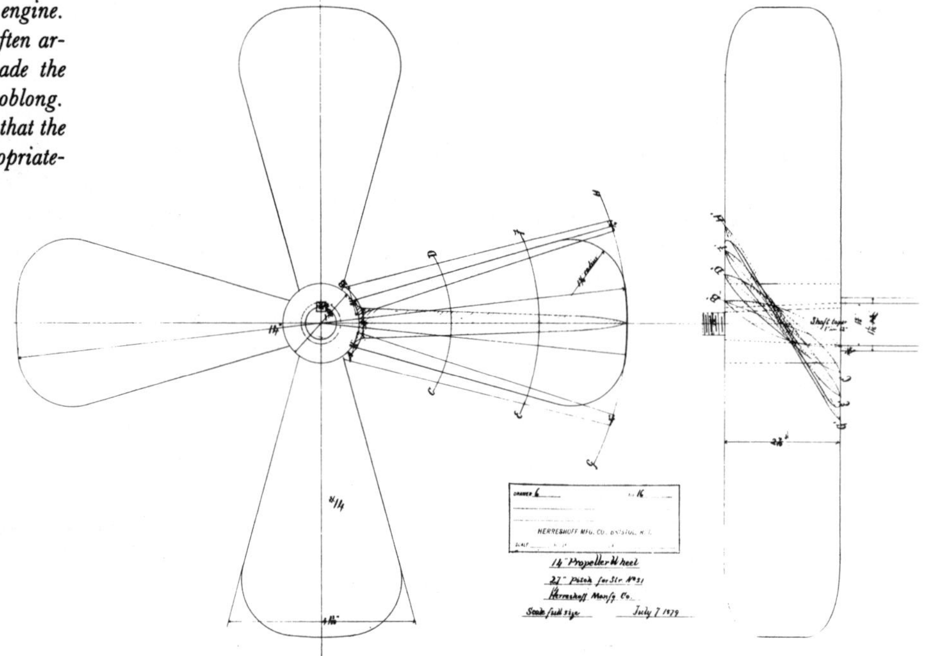

Nat Herreshoff designed this 14" x 27" propeller in 1879 for Herreshoff "Str. No. 51," a 17' x 4.4' launch for the yacht Lurline. *A 20" x 22" coil boiler supplied steam to a 2.5" x 5" engine. Supremely confident in all his spare and often arrogantly arbitrary designs, Herreshoff made the lateral profile of the propeller a plain oblong. Those who wish to can find joy in the fact that the blade sections in this specimen are, inappropriately, streamline instead of lifting-foil. (Hart Nautical Museum, M.I.T.)*

1880s — THE HALCYON YEARS

The 1880s were fine, expansive years. New steam launch designs were confident and settled — not always as interesting as the tentative or eccentric boats of the 1870s. Private pleasure launches were becoming common, though not yet outnumbering commercial boats, and numerous manufacturers sprang into being to supply the visible market.

In America the Herreshoff Manufacturing Company was a thriving exception to the country's standard practice. Most American launches were created by country boatbuilders, distilled from an amalgam of manual skills and a good eye for a boat. Plans were sketched simply, if at all, on the proverbial piece of brown wrapping paper. Engines were ordered from an engine builder, boilers from the local boiler works. Even if the country boatbuilder grew into a sizable manufacturer of steam launches, the design department remained minimal, and perhaps closer to sales than to production. (There are very few builders' plans of American steam launches in public collections.)

British boats were built by slower, more painstaking craftsmen, at lower wages, to the high standards that corporate or autocratic employers could enforce. In Britain, steam launches were designed and built by professionals, often in great shipbuilding yards as a minor sideline to building oceangoing vessels. Engines and boilers were frequently manufactured on the same premises as the hulls. The whole paraphernalia of professional design was usual — hull lines, plan and profile drawings, detailed machinery drawings, cost and performance estimates. Among the half-million ship plans in the National Maritime Museum, Greenwich, are many detailed plans of steam launches — naval, commercial, and private.

English-speaking people built nearly all of the great steam yachts and most of the steam launches, but this should not obscure the fact that many fine launches were designed and built by others. Japan had a varied lot of little naval and civilian steamers, though few for private recreation, and Chinese and Russian launches were not exclusively the products of British yards. The 48' x 6.5' x 1.6' *Mab* was built all of brass at St. Petersburg in 1874, and at 16.5 knots she was the fastest on the Neva River. Italy, Switzerland, and Germany built steam launches as good as any; most workaday Scandinavian launches were locally built. A great deal remains to be discovered about the history of the steam launch worldwide.

In England there were concentrations of beautiful boats on the upper Thames and on The Solent; the Thames was a sort of world center for "steam

The steam launch Maude Lorie, *launched in the summer of 1886. (Photo by Hodge, courtesy The New England Wireless and Steam Museum, East Greenwich, Rhode Island)*

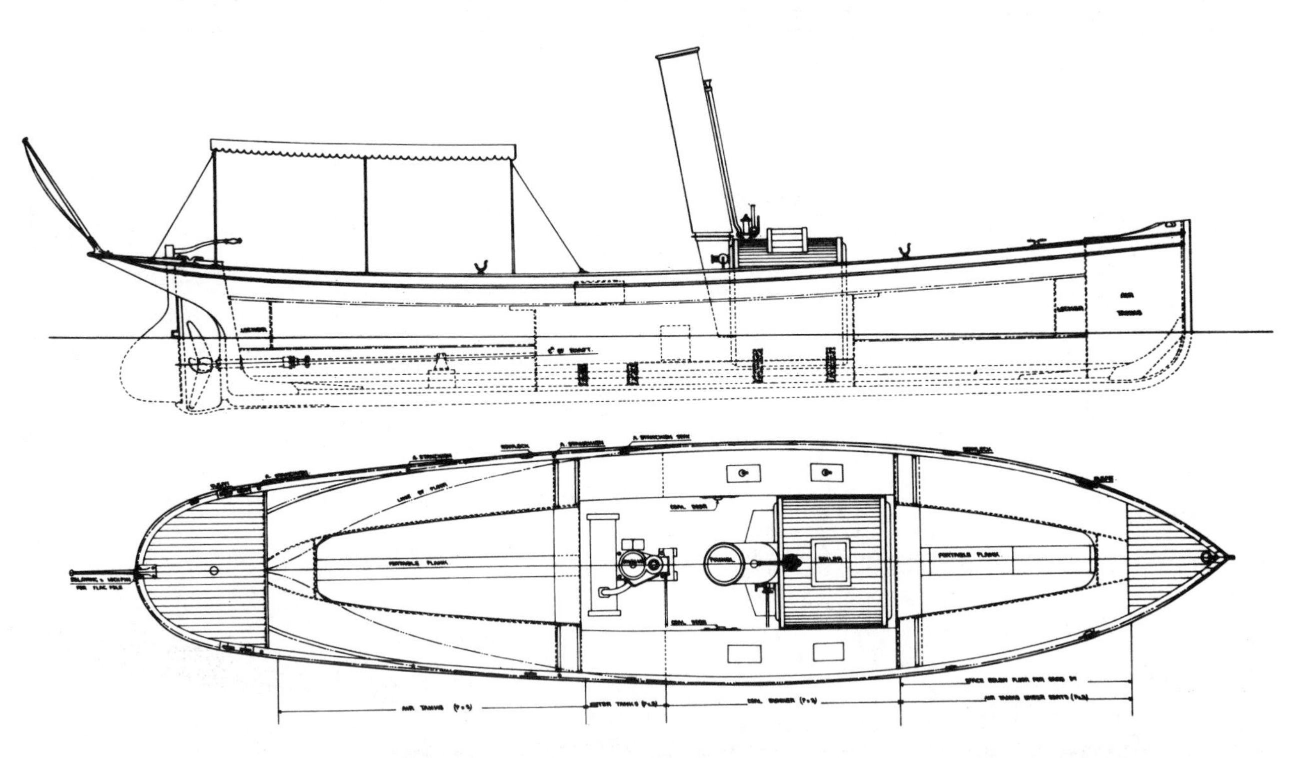

One of the few wooden vessels out of the Denny yard (Govan, Scotland), the 39' x 9' Santa Maria *was contracted for in 1894 for £640, to make 8.75 knots and to be delivered in two months. On trials, at 8.53 tons displacement, she made 8.38 knots with 35.15 indicated horsepower at 325 r.p.m. (The engine was 5.5" & 10" x 6.5".) Upon delivery in Spain, the boat made only 7.5 knots with an inexperienced crew. The builders remitted five percent of contract price for speed deficiency, but still they netted a £12 profit on the job. (National Maritime Museum, London)*

When hundreds of steam yachts congregated on the south coast of England, there also had to be hundreds of yachts' hoisting launches. Early yacht tenders were not unlike the plain steam dinghies and cutters designed for hoisting on board commercial or naval ships. Later, they became flashily opulent, with no visible surface that was not of precious wood, costly fabric, or brass buffed by the servants 'til it shone like a new sovereign. The example shown here (from Simpson, Strickland and Company, the principal builder of the type) was built around 1912. LOA, 25 feet; beam, 6 feet; engine, 23 indicated horsepower (triple expansion); oil-fired boiler; speed, 11 to 12 m.p.h.; displacement, about 1.6 tons. (From The Yachting Monthly and Marine Motor Magazine, *1913)*

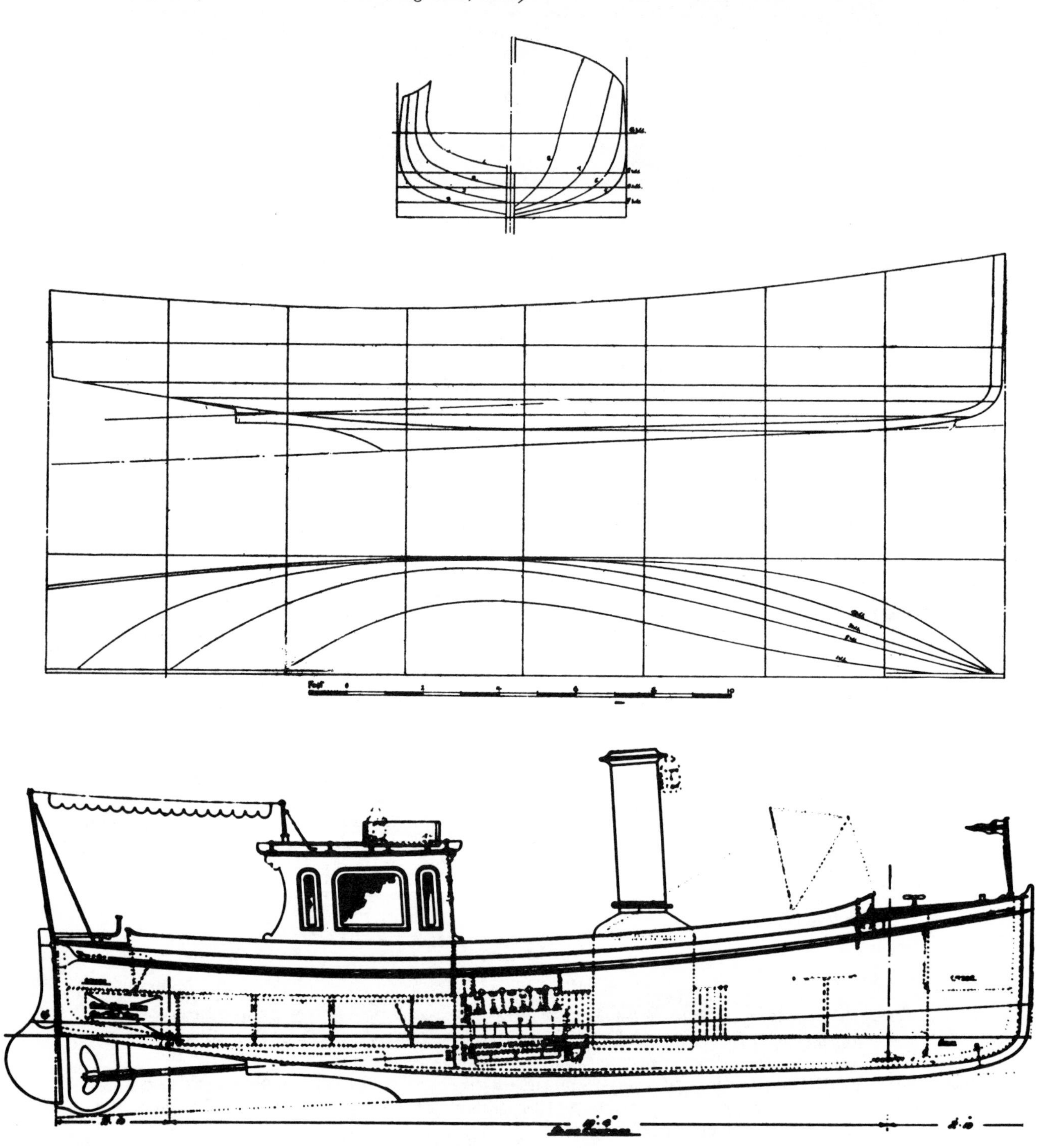

In the western world, good land transport by rail was developed while steam launches were still in the bud, so the use of small steamers for public transport occurred mostly in such faraway places as India, Paraguay, Oregon and China. Venice was the European exception. There the city buses were steam launches, 65 to 75 feet long. This 1896 photo is of an 1882-class steam water bus on the Grand Canal in Venice. The first two of these were ordered in Nantes, France, but all subsequent "vaporetti" were Italian built. (Courtesy Artú Chiggiato)

launching" as a pastime, as Cowes and The Solent were for steam yachts. The grimy crews of working launches in a hundred Empire ports, and on the Thames itself, were in a world apart from the starched crowd on board "The River's" slender and patrician boats.

In the United States and Canada a larger number of recreational steam launches were scattered on numerous resort lakes in New England and Ontario, on the St. Lawrence, and on every manner of waterway in the Midwest. There were tidewater steam launches, too; these were more likely to be powerful small yachts, yachts' launches, or workboats.

Most American launches of the 1870s had been built to haul passengers or to do other tasks for hire. Few had much brightwork or upholstery, or a fringe on the canopy. By 1890, late in the brief heyday, a large proportion were private pleasure boats, but they continued to reflect middle-class values in their fundamental design. They tended not to be as stylish as the launches of the British gentlemen, which were rakish with their alfresco engine rooms. The operators of the Thames River launches suffered the rain and the sun; the operator of an American launch shared the same shelter as the passengers — and perhaps owned the boat as well.

American manufacturers were supplying series-produced hulls and power plants of standard design to anyone in the world who could make the down payment and contract for shipment. Many stock American launches were kept within 33 feet overall length — the limiting size of flatcars then. Soon after the transcontinental rails reached Great Salt Lake, Brigham Young ordered a pretty 33-footer, *Lady of the Lake,* shipped from New York, 2,000 miles away. Steam launch catalogs quoted shipping rates to places as diverse as Spokane and Pernambuco (from Michigan, Wisconsin, Illinois, New York, and New England, mainly). A few were even sold in Europe, by undercutting British prices.

The merchandisers of American mail-order steam launches employed just about every marketing gim-

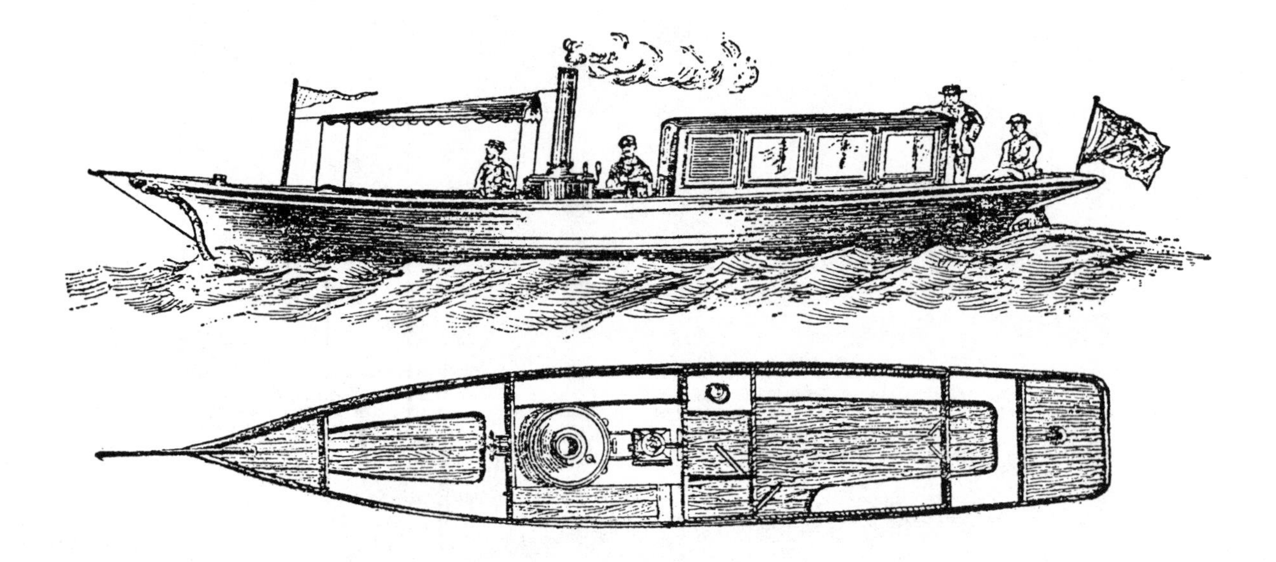

Steam Launch "Sphinx."—The general particulars of the "*Sphinx*" class of boat built as a gentleman's pleasure launch for river service are as follows :—Length, over all, 33 ft. 0 in. ; beam, outside, 6 ft. 0 in.; depth, inside of skin to gunwale, 3 ft. 0 in. ; carvel built of pitch pine ; copper fastened, with keel, stem, stern post, and timbers, of American elm ; swan stem, and counter stern; lined with fir down to the benches ; cuddy deck at each end of the boat, with doors to form lockers ; seats arranged fore and aft, with cross benches in stern sheets and forward of boiler ; coal bunkers amidships on each side of the boiler; rudder fitted outside of propeller, and with galvanized iron tiller. The engine, a single vertical inverted cylinder, direct acting, high-pressure—5¼ in. dia. by 6 in. stroke, and fitted with link motion, reversing gear, and feed pump. The boiler, of the vertical cross tube type—2 ft. 2 in. dia. by 3 ft. 0 in. high, with 6 cross tubes, tested to 150 lbs. per sq. in., and designed to work at 60 lbs. per sq. in. The boiler is felted, and lagged with mahogany, and brass bands; also fitted with all necessary furnace and steam mountings, spring balance safety valve, with escape pipe ; steam pressure gauge, with double dial facing fore and aft ; a double set of water gauges, blow-off cock, steam whistle, blower cock, and hinged chimney. The propeller is of steel, 30 in. dia. ; screw shaft of wrought iron, 1½ in. dia. ; stern tube, bushed with gun-metal, and fitted with collar thrust bearing ; steam and feed pipes of copper ; exhaust pipe of iron. All necessary connections, such as steam starting valve, back pressure valve, and suction cock ; a set of stoking tools, spanners, and oil can, engineer's tools, and coal hammer ; mop-bucket, scrub brush, and life buoy, with name painted on. The cabin, 10 ft. long, is fitted aft, framed of mahogany, with framed and glazed sashes, seats, lockers, and table ; a door opens forward, and a pair of doors opens into after cock pit. The fore part of cabin is divided off to form—w.c. and a passage to pass from cabin forward. The fore part of the launch forms an open cock-pit, with seats and lockers, and small deck in head to form chain cable locker ; cushions are fitted to the cabin, and seats in forward cock-pit. An awning is fitted to cover the engine room and forward cock-pit, with light stanchions and tricing lines. A No. 3 Korting's universal feed injector is fitted to the boiler, with the valves and connections ; also an ejector to expel bilge water ; a silent exhaust box is applied to the engine ; a galvanized Trotman's anchor, with 15 fms. of chain fitted into the chain cable locker in the bows ; a steering wheel is fitted forward and a spare tiller aft ; mast and lug sail, with all necessary tackle and running gear and galvanized iron shrouds, are fitted ; port and starboard side lights, mast head light, riding-light, cabin lamp, and engineer's lamp, with screens, are supplied ; a w.c. is fitted in the forward compartment of cabin with wash-stand, basin, &c.

Steam Launch of Steel, 55 ft. by 6 ft. 6 in. by 2 ft. 6 in., does 16 miles ; price £525. No. 97.

Steam Launch of Wood, cabin forward, 30 ft. by 6 ft., dft. 3 ft. ; price £90. No. 98.

Steam Launch, diagonally built of wood, rebuilt 1882, 37 ft. by 10 ft. 5 in., by 4 ft. dft., engines 2 cylinders of 6 in. by 8 in. stroke, boiler, steel return tube, almost new 1882, speed 9 miles, decked round 2 open cockpits ; price £315. No. 80.

Steam Launch of Steel, 51 ft. by 6 ft. 6 in. by 3 ft. 6 in., dft. 2 ft. 6 in., 2 cylinders 6 in. and 6 in. stroke, loco. boiler brass tubes, cabin aft. 12 ft., speed 18 miles ; price £550. No. 85.

Steam Launch of Steel, 45 ft. by 6 ft. 6 in., dft. 2 ft. 6 in., 2 cylinders of 5 in. by 6 in. stroke, loco. boiler, cabin aft, speed 20 miles ; offer. No. 86.

Steam Launch of Wood, coppered, and schooner rigged, 45 ft. by 10 ft. 6 in. by 4 ft. 6 in., cylinder 5 in. by 6 in. stroke, loco. boiler, cabin 6 ft. 6 in. aft. with w.c. and lavatory attached ; price £500. No. 84.

Steam Launch, 51 ft. by 12 ft. by 5 ft. 6 in., dft. 5 ft., 2 cylinders of 8½ in., 8 in. stroke, upright boiler, bunkers 3 tons, hull iron, coated with pitch pine planks 1883; good little tug or yacht, having forecastle, saloon and cabin with w.c. attached ; price £500. No. 106.

Steam Launch of Wood, 42 ft. by 7 ft. by 4 ft., dft. 3 ft., 2 cylinders, 7 in. stroke 6 in., return tube marine boiler, speed 12 miles, forecastle and cabin forward ; price £400. No. 108.

Steel River Launch, 35 ft. 6 in. by 6 ft. 9 in. by 3 ft. 4 in., dft. 2 ft. 2 in. ; 2 cylinders 4 in., stroke 5 in., loco. boiler, speed 10 miles, cabin ; price £255. No. 141.

Steel River Launch, 62 ft. by 7 ft. 8 in. by 4 ft. 10 in., dft. 2 ft. 10 in, 2 cylinders of 6 in., stroke 8 in., loco. boiler, cabin, speed 11 miles; price £600. No. 142.

Steel River Launch, new, open, 31 ft. 6 in. by 5 ft. 5 in. by 2 ft., dft. 1 ft. 3 in., 2 cylinders 4 in., stroke 4 in., loco. boiler, speed 11 miles ; price £350. No. 143.

Steel River Launch, new, 40 ft. by 6 ft 1 in. by 2 ft. 8 in., dft. 1 ft. 10 in., 2 cylinders 6¼ in., stroke 4 in., loco. boiler, cabin; £560. No. 144.

Steel River Steam Yacht, 53 ft. 6 in. by 6 ft. 8 in. by 3 ft. 1 in., dft. 2 ft. 1 in. ; 2 cylinders 6 in., stroke 6½ in., loco. boiler, cabin, speed 18 to 20 miles ; price £1,000. No. 145.

Steel River Launch, new, 52½ ft. by 6 ft. 8 in. by 3 ft. 2 in., dft. 2 ft. 1 in., 2 cylinders 6 in., stroke 6½ in., loco. boiler, cabin ; speed 18 to 20 miles ; price £1,000. No. 146.

Steel Steam Yacht, new, for sea, 52 ft. by 9 ft. 9 in. by 4 ft. 10 in., dft. 3 ft. 6 in., cylinder 11 in., stroke 9¼ in., multitubular boiler, 2 cabins and lavatory, speed 18 to 20 miles ; £1,450. No. 147.

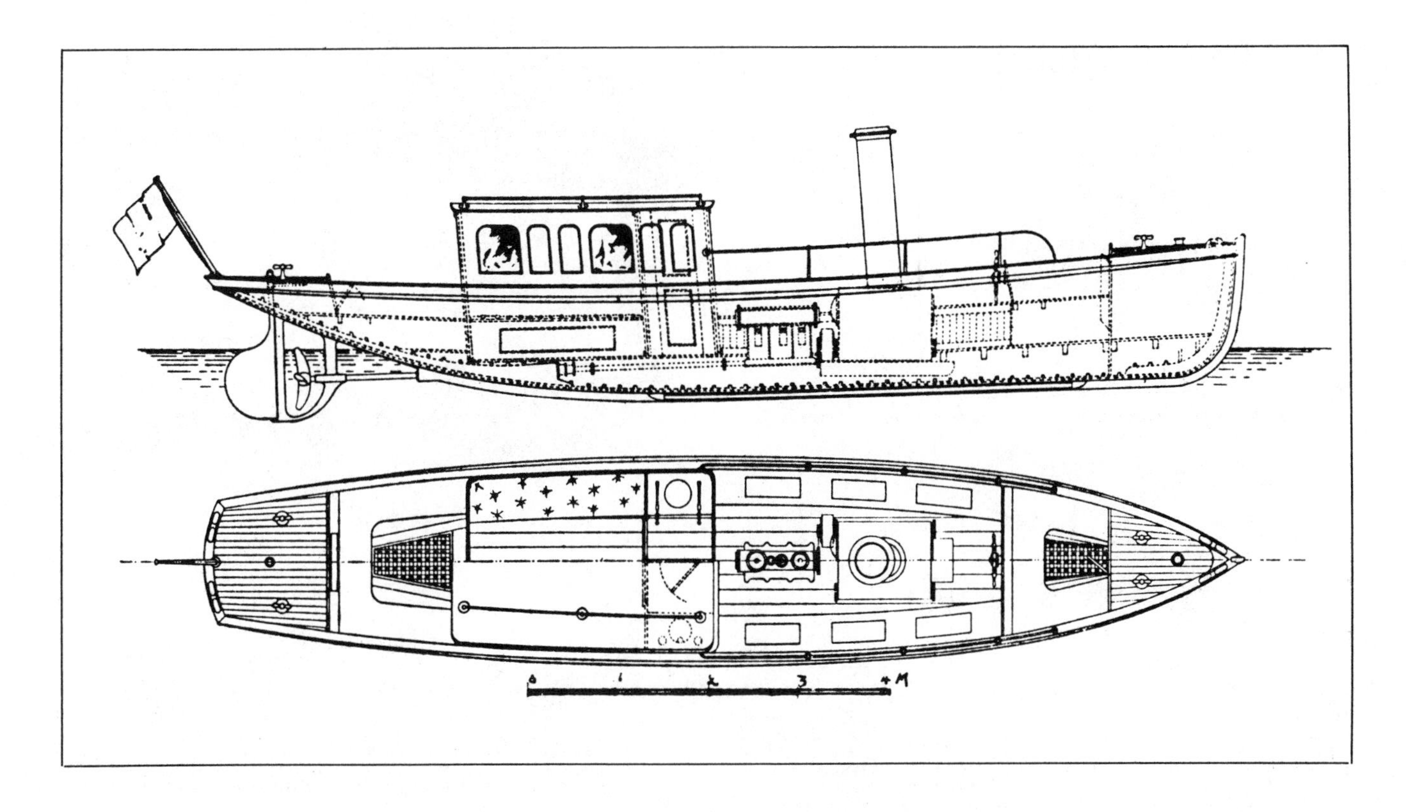

No American boats equaled the stylishness of English "cruising launches," which were designed to tickle the fancy of old-money boat men (or new-money men who hoped to be mistaken for old). The short boxlike cabin, deposited in what is otherwise a large open boat occupied by servants, honors old traditions. Antonio Canaletto's paintings of 18th-century London show numerous gentlemen's barges, which lack only a funnel to pass for a Simpson, Strickland cruising launch. The smallest cruising launches could be managed by "a hand and a boy," or even by a single paid hand. (From the facsimile edition of Simpson, Strickland & Co.'s Catalogue No. 5)

mick that we have since come to associate with automobile huckstering. There was a "model" to suit every purse and family, "grades" ranging from plainest pine and black iron to mahogany and birch, copper-fastened. "Options" included various levels of powering, awnings, condensers, and boating equipment of all kinds.

American steam launches were rather inexpensive, relative to the costs of labor and other products of the time (a good bicycle cost $100 when a schoolteacher's annual salary was $250). This resulted from simplicity of design and structure, cheap forest and foundry products, and the prevalence of fast, skilled mechanics in boatbuilding and engine shops. The rapid growth and healthy condition of the industry must have made it difficult indeed for anyone to foresee the imminent replacement of reciprocating steam engines by another technology. By 1885 the steam launch was better than halfway through its fleeting Golden Age.

Left: *Quite a bit of information about steam launches was published in the 1880s, while they were the only powered vehicles that were small-scale and often used for private recreation. They received less notice during the 1890s, as popular interest in technical things shifted to cars, electricity, motorboats, and chemistry. By reading contemporary advertisements and technical descriptions, we can savor the atmosphere and vocabulary of the best years of "steam launching."* Sphinx *was a gentleman's pleasure launch for the Thames. Its description and the classified advertisements are from James Donaldson's* Practical Guide to . . . Small Steamers, *London, 1885.*

In issuing this enlarged list of Steam Launches, and machinery for small steamboats, we desire to call the attention of those into whose hands it may come, who have not given particular attention to the subject, to the many uses to which these Launches can be put, and how exceedingly serviceable and convenient they are for many places that are at present without any means of transportation except over bad roads, which are at their worst at a season of the year when there is most produce to be transported, and when the rivers and streams are at their best, and most available for transporting goods by water.

It seems needless to enumerate the purposes for which small steamers of this kind can be used; but we might mention their use on lakes and rivers for pleasure purposes and for the transportation of mails in districts where roads are very bad, or actually impassable, at many seasons of the year; also for ferry boats or for carrying fishing and hunting parties on lakes and rivers to fishing or hunting grounds, and particularly for towing barges laden with lumber, salt, coal, stone, or other products of the country, on lakes, rivers or canals.

STEAM LAUNCHES.

(COMPLETE.)

SIZES AND PRICES.

NUMBER.	1	2	3	4
Length of Boat, over all	21	25	30	33
Breadth Outside	5	6½	7	7½
Depth Amidship	2.4	2.9	3.0	3.0
Draft, Aft	1.9	2.3	2.6	2.6
No. of Persons will carry with comfort and safety	9	15	25	30
Approximate Speed, miles per hour	8	9	10	11
Approximate Consumption of Soft Coal per 12 hours, lbs	150	250	350	350
Actual Horse Power, with 70 pounds steam	3	5	9	9
Engine, Diameter Cylinder	3	5	6	6
" Length of Stroke	4	7	8	8
No. of Revolutions per minute	300	275	250	260
Vertical Boiler, Diameter,	22	30	36	36
" " Height	36	48	50	50
Propeller Wheel, Diameter	20	24	30	30
Wheel Shaft, Diameter	1¼	1¾	1 15/16	1 15/16
Price, complete and ready to run	$490 00	$750 00	$875 00	$925 00

Both Chas. P. Willard & Co. and Thomas Kane & Co. offered several lines of steam launches in their catalogs of the mid-1880s. The former company, to attract customers, explained why more people should buy steam launches; the latter explained why people who did buy should purchase from Kane. Chas. P. Willard & Co.'s smallest engine was 3" x 4", weighed 150 pounds, stood 30 inches in height, and sold for $95. Their wrought-iron boilers were designed for 150 p.s.i. steam; steel could be substituted by request. "Kerosene oil" fuel and automatic power plants were standard for the steam launches built at the Racine, Wisconsin, works of Thomas Kane & Co. Steering wheels, freshwater tanks, mahogany joinery, awnings, carpets, and upholstery were extras. The company was prepared "to fit up [their] Yachts as handsome and elegant as [could] be desired."

ELLA D.

Ella D. is 27 ft. 6 in. long, 6 ft. beam. A dozen people find comfortable space as can be guessed from the cut and the small amount of room that the seven people aboard occupy. Speed, about nine miles.

Hull, Grade A..............$445 00 | Hull, Grade B..............$365 00 | Hull, Grade C.............$300 00
Fitted with canopy, steering wheel, 4 H. P. outfit—Grade B..........................$825 00

For 1890 wage-earners, Ella D*'s $825 price was equivalent to $15,000 to $20,000 in modern terms. Most purchasers who were not wealthy expected to hire their boats out occasionally, to help defray the cost. Merwin, Hulbert & Company (26 West 23rd Street, New York, New York) boats were built in Waukegan, Illinois, where they could be shipped to mid-continent or Atlantic, Gulf, or Pacific coast customers with equal facility. The firm's 1-h.p., 16½-foot steam launch,* Baby Mine, *sold for as little as $310, including engine, oars, and rowlocks. By 1890, several steam launch manufacturers were attempting to bridge the gap between the popular $40 rowboats and the $800, 25-foot launches, but powerboating on the scale of* Baby Mine *did not become widespread until gasoline engines swept the market, after 1900.*

HIGH-SPEED STEAMERS

Late in the history of steam launches, around 1890, large, powerful, and very fast commuting launches appeared around New York City. These provided daily transport for men who shook the earth on Wall Street during waking hours but preferred to sleep far from the city. There was no English equivalent for these, as 500-h.p., 25-knot launches would have been intolerable on the upper Thames.

One of the first of the breed was the cabin launch *Stiletto,* built in 1885 by the Herreshoff brothers as a manifestation of their growing prosperity. *Stiletto* was fast and showy — her 12" & 21" x 12" compound engine pushed her 94' x 11.5' hull easily at 20 m.p.h., and she was reported to have maintained 26.5 m.p.h. for up to eight hours. She was used to humiliate the *Mary Powell,* a packet steamer on the Hudson that had been, for a few years around 1880, perhaps the fastest vessel in the world.

The Herreshoffs must have guessed at the probable benefits of the excitement and free press coverage generated by such exploits in a world new to the possibilities of mechanical speed. In the next few years, numerous orders for high-speed launches and yachts were placed with the Herreshoff Manufacturing Company. The 48-foot *Henrietta* (1885), the 86-foot *Now Then* (1887), and the 138-foot *Say When* (1888) went to Norman L. Munroe, a New York publisher. *Now Then* was a radical departure from previous design, with a straight run aft and a wide, flat stern to prevent squatting at high speeds, and an engine that benefited from Herreshoff's experience in the extensive use of light forged steel. Her maiden run from Bristol to New York, a distance of 130 miles, was made in just over seven hours. No other American steam yacht dared answer Munroe's challenge to race on uncorrected time, but the precedent had been set.

During the winter of 1890-91, Edwin D. Morgan had Herreshoff create the 98-foot, 26-m.p.h. *Javelin.*

Now Then *was five years old and her thin mahogany planking already failing when George E. Whitney had a photographer come over to record his presence as a consultant on some refitting. The boat had made her everlasting mark in history, and Whitney had a deep feeling of personal involvement in the onward flow of history. (George Whitney collection)*

Vamoose, built alongside *Javelin* for William Randolph Hearst, could steam at just over 27 m.p.h.

While the Herreshoffs were doing very well, they were far from having a corner on the high-speed market. Two years earlier, Charles D. Mosher and George Manson had designed and constructed the high-speed launch *Buzz* in Amesbury, Massachusetts, incorporating Mosher's considerable experience with water-tube boilers. Fifty feet long, with a 6.5-foot beam and 30 inches of draft, *Buzz* cut a measured mile at 29.6 m.p.h. in 1892. Later renamed the *Yankee Doodle,* she was the fastest boat in the world for a time.

Buzz brought Mosher other orders, including a 63-footer commissioned by Norman Munroe. *Norwood*'s power plant exemplified the high powering of the fast launches. Her 36-inch, three-bladed propeller was driven by a 9'' & 14.5'' & 22'' x 9'' triple-expansion engine that weighed one ton and could develop 400 h.p. at 600 r.p.m. With 26 square feet of grate surface, 1,000 square feet of heating surface, and 200 pounds of steam in her 2.5-ton boiler, she was capable of 30.5 m.p.h. The completed boat cost $20,000. (Munroe had sold the Herreshoff-built *Say When* for $46,000 in 1888.)

Mosher's most famous enterprise, undertaken several years later, was a consequence of a history-making technological breakthrough on the other side

Charles D. Mosher's 50' x 6.5' Buzz *made 29.6 m.p.h. in 1892. That was as fast as any private vehicle could travel then, unless you owned a railroad. (Courtesy Professor Evers Burtner)*

of the Atlantic. In 1894 a company was formed in England by Charles A. Parsons "to provide the necessary capital for efficiently and thoroughly testing the application of Mr. Parsons' well-known steam turbine to the propulsion of vessels"; by the end of the year the 100-foot *Turbinia,* the first turbine-driven vessel in the world, had begun her speed trials. Early results were disappointing — several designs and combinations of screws were tried on the single propeller shaft, but the maximum attained speed fell short of 20 knots. In 1896 the single turbine was replaced by three turbines, in series, driving three propeller shafts (each carrying three propellers), and *Turbinia* eventually attained the previously unheard-of speed of 35 knots.

In a *Rudder* article of September 1958, L. Francis Herreshoff discussed the subsequent development in America that involved Mosher.

This naturally made an American want to acquire even greater speed, so Mr. Charles R. Flint, who was a very adventurous man in many ways, thought he would make a try at a world record. Thus he had the famous *Arrow* built with very great care and expense. Her hull and steam plant were designed by Charles D. Mosher who made a specialty of fast steam launches and light boilers. *Arrow* was very much the same shape or model as the torpedo boats designed for the French Navy by Normand, and ran with her forefoot in the water. Her hull was built by S. Ayers and Sons of Nyack, N.Y. and was a fine job. She was of light construction; her frames were steel below water and aluminum above. She was double-planked mahogany; her power plant consisted of two water tube boilers and two quadruple expansion engines of 2,000 h.p. each, so she compared with *Turbinia* as follows:

	LOA	Beam	HP	Weight	Claimed Speed	Weight Per HP	Year
Turbinia	100'	9'	2,000	44.5 tons	35 knots	49 lbs.	1897
Arrow	130'	12.5'	4,000	66 tons	39 knots	36 lbs.	1902

A new concept was born in 1897 when Charles Parsons drove the first turbine vessel, Turbinia, *to over 30 knots at the Spithead Jubilee Review.* Turbinia *can be seen today in Newcastle upon Tyne, England. (Courtesy Museum of Science and Engineering, Newcastle upon Tyne, England)*

The weight per horsepower of these vessels was much less than any before. For instance, the weight per horsepower of the best torpedo boats was in the neighborhood of 120 pounds, and the best racing motorboats did not get down to a weight of less than these figures until around 1905. Hence, in spite of their rather queer model they were undoubtedly fast, and I do not doubt that the *Arrow* went very fast for a few minutes one day. Right after her record run one of her boilers was removed and she never was seen to go very fast thereafter. However, *Arrow* had pretty good accommodations in her large stern and was used as a yacht for several years.

Such contests assume a lasting place in American mythology. *Turbinia,* as the name would imply, was an experimental vessel, built to prove a point. For large ships the steam turbine was a rising star. In small craft the reciprocating steam engine had another, far more serious adversary.

END OF THE STEAM LAUNCH ERA

By the turn of the century, the prospective buyer of a pleasure launch had many choices to make, including the choice of motive power. Would it be a graceful naphtha launch? An "explosive motor" launch? To most observers it was apparent that steam launches would soon lose completely their diminishing share of the pleasure boat market, but there were those who still felt otherwise. In a 1901 issue of *The Rudder,* an article predicted a rosy future for steam launches employing gasoline-fired, fully automatic power plants: "The field for a steam engine of this nature is unlimited and it is only a question of time when they will be found wherever power boats are used." A.J. Kenealy, in an article that appeared in *The Rudder* later that year, preferred to hedge his bets, and wasn't picking any winners: "The launch type of yacht [is] . . . propelled by

HELEN.

The above illustration, reproduced from a photograph, is an exact representation of our No. 2 yacht, and it will be seen at a glance that she has excellent lines and is as pretty a model as one could wish for. The machinery being placed aft makes it a very roomy and comfortable boat, without sacrificing speed or sea worthiness. This model was on exhibition at the World's Columbian Exposition, and received highest award of merit, which speaks for itself.

SPECIFICATIONS.

Length, 20 feet; beam, 5 feet; draught, 18 inches with load; seating capacity for 10 persons. Is fitted with a 1-horse power outfit, and is rated at 7 miles per hour.

Price, Fancy Grade......$825.00 Grade A......$600.00 Grade B......$510.00 Grade C......$405.00 Grade D......$380.00

Helen, *20' x 5', was "rated at" 7 m.p.h., with 1 h.p. on an 18-inch screw, and she carried up to 10 passengers. She was a prize-winner at the World's Columbian Exposition of 1893 (and no wonder!), but the Racine Boat Co. knew she might soon become a relic in the rapidly transforming marketplace of the 1890s. Their 1894 brochure mentioned that they built "several sizes of electro vapor [gasoline] engines." Within a year or two manufacturers were finding that the new device, which promised phenomenal convenience and fuel economy, was outselling steam power plants.*

This photograph of a 20-knot steam launch, builder unknown, was the frontispiece in an early book on gasoline motorboats. The editors knew with certainty that the future belonged to motorboats, but in 1900 no gas boat could approach the spectacular performance of some steam launches, so a steamer was used to illustrate the speed and excitement that motorboats would soon provide. (From C.D. Mower, How to Build a Motor Launch, *Rudder Publishing Company, 1901)*

steam, naphtha, gas power and electricity. These are numerous everywhere and increasing in popularity every year.''

After 1900, 1½-to-5-horsepower gas engines were built by hundreds of machine shops all over America. These could make every steam launch in a county obsolete in one season if the gas-engine builder was any kind of a salesman (and his product was any good). The bicycle craze of the 1890s eroded the dominance of boats in the recreation-equipment market, and the motorboat boom of 1905 to 1915 killed steam launches abruptly. They went into hibernation in America about the time when Ford cars and Evinrude outboards came out — although the navy, always traditionalist, had some in service in the same quarter-century that witnessed jet planes and satellites. (In England, pleasure steam launches survived competition with gas boats 10 or 15 years longer than in America, and they offered numerous 20th-century refinements in hull design and steam technology before they died away.)

The manifest advantages of gas engines extinguished small steamers so quickly that a kind of bewildered ambivalence was widespread among boatmen who had known steam. The little steamers were so warm and charming and companionable, so quiet and full of interest — but only a fool would carry a ton of iron and a cord of stovewood in his boat to accomplish what a hundred-dollar engine could do with 10 gallons of gas.

The boating magazines contributed quite a bit to the demise of steam. Hundreds of gas-boat and engine, spark plug, coil, magneto, and carburetor manufacturers bought advertising space during the pre-World War I motorboat boom. The magazines ignored the non-advertising steam launch sector of boating, the editors writing as if all steamers were already dead and buried. The numerous steam launches changing hands through the magazines' classified ad sections, up to 1915, gave the lie to the editorial position.

Production of new pleasure steam launches ceased before 1910 (except in England); most of the relicts were gone by 1920. Steam launches had been able to claim near-parity, and even a few advantages over the immature, pre-1900 gas boats. They were out of the running against boats powered by cheap and lightweight automobile-type engines after 1920. At

---

No. 6503—37 feet x 7½ feet x 3 feet draught; oak frames, copper fastened, tiller aft, wheel forward, steeple comp. engine 4 in. x 8 in. x 5½ in.; water tube boiler, 5 square feet G. S. and 200 square feet H. S.; 150 pounds steam; galvanized water tanks, brass keel condenser, ball-bearing thrust block, 26-inch bronze propeller; steam feed pump, metropolitan injector, steam bilge ejector, hand pumps, cushions, ice chest, water cooler, steam whistle, anchor, ship's light, life belts, dinkey, etc. Fast boat, steams very easily, bunkers hold about 600 pounds. Price, $400. Address, J.J. Butcher Thompsonvile, Conn.

Boating magazine ads of the early 1900s, feeling the first impact of the gasoline revolution, showed an almost panicky eagerness to sell steam launches while they still had some value. The $400 price for the 37-foot Anglo-American *was equivalent to the price of four bicycles in 1901. The steam machinery had only scrap-metal value. (From* The Rudder, *1901)*

the same time, thousands of people who had enjoyed steam launches before 1910 or in the navy during World War I, and millions of people who still worked with steam daily in power plants, factories, ships, or locomotives continued to regard steam launches as the most appealing of small boats.

STEAM LAUNCH REVIVAL

A few stubborn people decided not to buy what everyone else was buying in 1900 and 1910, and perhaps one steam launch in a hundred remained in service until she wore out or the old man died. During the long hiatus in steam launch use, there were always at least a few boat men among the wild-eyed believers in the second coming of steam. Many people found space to store a retired steam launch engine — "such a pretty thing" — after the hull received the inevitable gas engine and the boiler went to the scrap heap. Still, steam launches and their components were scarce and scattered by the time a revival of interest got underway after World War II.

Little steamers began to return to public notice during the 1950s, with the revival most clearly signaled by the founding of the Semple Engine Company and the plentiful nationwide publicity given to Captain McCready's interesting launch *Little Effie* and the author's *River Queen.* The revival had abundant human ties to the earlier Golden Age. Nearly all enthusiasts building new launches during the 1950s and '60s encountered old men who had operated steam launches long ago and could proffer much advice — some of it useful. Once a new launch steamed up, she became a magnet attracting a wide variety of people who had "always been interested" in some aspect of the steam launch mystique — history, engineering, noiseless boating, or perhaps coal smoke or whistles.

There were several regional organizations for steam launch fanciers by 1960, all of them easygoing and informal. "Steam launch meets" in New England in the 1950s were hosted by Fred Semple at

Quite a few marine engineers working on American freighters of 20,000, 30,000 or 40,000 horsepower remember when their instructor in steam engineering at the U.S. Merchant Marine Academy used to talk about his steam launch, Little Effie. *The 26-foot motor-whaler hull that Commander Lauren S. McCready chose for his 1953 launch was designed for a 25-h.p. diesel and was well suited to the complete and powerful steam plant he installed in it. (Courtesy Rear Admiral Lauren S. McCready, U.S.M.S.)*

Kezar Lake, Maine, and by the author at Hinsdale, New Hampshire.

Steam launch meets continue to be almost entirely noncommercial in tone, and their organizers try to limit attendance. Every year there are several colorful gatherings, for picnicking and racing and amiable chatter about propellers, furnace draft, or the proper rake of a smokestack, but these events are not very noticeable amid the din created by eight million gas boats.

David Thompson hosts a steam launch event each September in Moultonboro, New Hampshire, on Lake Winnipesaukee and the Pump House Steam Museum, in Kingston, Ontario, sponsors the Steamboat Regatta in midsummer. The annual Northern California Steamboat Meet is held at the B & W Resort Marina on the San Joaquin River delta, near the town of Isleton. There is a very active group on Puget Sound, but around the Great Lakes, where steam launches were once so common, steamboat operators are only poor relations of the large traction engine and steam railroad interest groups.

Many expected the steam launch groups to change into something like the steam car cult as the old-timers dropped away. It hasn't happened that way. Today there are more youngish owners of steam launches than people who remember working steam, and the emphasis on pure "marine steam" is as strong as ever. The expansion in America was partly stimulated by many magazine articles and by the distribution between 1960 and 1963 of *Steamboats and Modern Steam Launches,* a magazine sponsored by Morgan North in California and edited by Bill Durham of Seattle. Since then, steam launches have caught on strongly in Great Britain, and England is again the world center for "steam launching," with the most effective clubs and the most interesting new engines and boats.

Steam launches have never been a central part of the yachting scene. The rich used to possess great steam or sailing yachts, and used steam launches incidentally, as utility vehicles. Working people had rowboats. Steam launches were used by small-scale entrepreneurs to make money, or employed recreationally by practical people who admired machinery or who valued their time too highly to wait for the wind to waft them hither and yon.

Now that millions of people can afford boats of several hundred horsepower — which only a few millionaires owned a lifetime ago — steam launch people are still a separate bunch. They burn different fuels and go at a different pace. A modern steam launch is emblematic of the whole sweep of 19th-century technology in a way that no other small, privately owned "hobby" machine can be. The refinements in metals and the experiments with constantly rising working pressures and rotative speeds are all faithfully mirrored in steam launch evolution. In 1981 the design of the little vertical marine engine in most hobbyists' boats was 125 years old.

3

CONTEMPORARY STEAM MEN

A psychologist has suggested that much of the appeal of steam launches arises from the fact that the little boats offer the Promethean myth on a manageable and personal scale. Here a solitary individual grasps fire and makes it do what he wants. He can build in his home shop the whole steam launch mechanism needed to enslave fire, and thenceforth he knows with certainty, from his own experience, that man is master of nature.

The psychologist could have listed other human impulses that find an outlet in steamboating, and I would include among these the appetite for sharing achievements, aspirations, insights, and problems in a community of like-minded fellows.

I began an article that ran in *Motor Boating,* in February 1945, with these words: "Like everyone else, I was born with a desire to do some particular thing. Some people do great things; some nothing of any importance. . . . It has always been my desire to own a steamboat." The article continued with a description of my first steam launch, *Lourick,* which I had assembled in 1943 and 1944. In a later article I wrote: "In 1944 I sincerely thought I was the only small steamboat operator in the world."

Here it is almost 40 years later, and the central theme of my life hasn't changed a bit. Since 1944, my steam launch activities have introduced me to many people, especially those from New England, who are as preoccupied with "fire boats" as I am, and I have found that the personalities one meets through steamboating are a good part of the pervasive magnetism of the pastime. Although most of them have never believed that their ideas, design innovations, or mechanical skills would alter the flow of history, these men have enriched my life, so if history begins on a personal level, they have, after all, had some small effect. Most of all, I value having been a close friend of George Eli Whitney during the latter years of his 101-year lifespan. I learned how a man who always loved steam power felt about steam launches in 1882 — and 1900, and 1948, and 1962.

GEORGE E. WHITNEY

When George Whitney came into the world, America was torn by the Civil War. Sleek clipper ships were contending with doggedly advancing steamers for commerce, and steel and machine tools were revolutionizing manufacturing. In those years the whistle of the "iron horse" was a harbinger of decline for the extreme and elegant high-pressure steamers of America's Western rivers. In England, John Thornycroft and Alfred Yarrow were building their first steam launches.

As Whitney grew through boyhood, the compound engine was writing a new chapter in the history of steam navigation. By the time he became a teenager, the Golden Age of steam launches was just beginning.

As an inveterate steam man myself, I've often envied Whitney's opportunity to grow into manhood squarely in the middle of the great steam age, when iron and coal turned the wheels that built America. George Whitney lived then, and he went on living until 1963, his mind sharp and active to the end. He saw the great expansion of industrial America and played a significant role in it. He made his mark on the development of his greatest interest, the steam launch, and lived long enough to see the warm and graceful craft revived in the post-World War II era.

A book could be written about Whitney's life, but much has been told already. When I first met him, in 1946, his wife had been dead for many years and Whit himself was 84 years old. It grieves me that our

Olive, *the first boat that Whitney built for his own use, is shown on Long Wharf in Boston in 1884* (Olive*'s boiler is seen on page 306). When the author first visited Whit's Bridgeport apartment, it was lined with large photographs of the boats and activities of a long career. Copying all of these was a group-financed effort. The originals have since been lost, but the copy negatives remain. (George Whitney collection)*

paths did not cross much earlier, yet I'm happy that they finally did, and that I had the chance to know him well over the short span of years that followed.

There were others, like Yarrow and Herreshoff, who were better known and did greater things in the world of steam launches — Whitney never employed many men or mass-produced his products. His fame, though perhaps fleeting, did not result from his steam launch activities alone. The horseless steam carriage he built in East Boston in 1896 generated a great deal of publicity and established Whit as an important pioneer in the development of the automobile in America. (At one point the Whitney Motor Wagon Company sued the Stanley Brothers for patent infringement.) Evers Burtner (who also appears in this chapter) made available to me a copy of the *East Boston Argus-Advocate* of May 1897, which describes in detail the carriage built by the Whitney Motor Wagon Company.

Altogether, Whit was credited with numerous inventions; copies of many of his patents and drawings are now in my possession. These include the steam carriage; a steam-powered machine for compressing asphalt paving blocks; another for compressing baled hay; a bottle-capping machine; a coal-furnace stoker; a hospital service tray; a bicycle brake; an audible low-water alarm for steam-auto tanks; a liquid gauge; a boiler safety device; a flash boiler; an exhaust-dissipating apparatus for motor vehicles; and a hydrocarbon burner.

Whit was born in Boston, but some of his early boyhood was spent in North Tunbridge, Vermont, where he lost his right eye while playing with an old flintlock rifle at the age of nine. Later, he studied mechanical arts at the Massachusetts Institute of Technology, forgoing completion of the course in favor of other pursuits. After valuable experience in his uncle's tool shop (which, according to Whit, later

George E. Whitney fashions some marine hardware in his East Boston shop, probably in the late 1880s. (George Whitney collection)

became Pratt & Whitney), he established his own shop at 32 New Street, East Boston, and went into production of marine engines, boilers, "propeller wheels," and auxiliary equipment for launches. Whit designed the hulls himself, then subcontracted their construction. Altogether, 95 Whitney-designed steam launches were built, ranging in size from 20 to 90 feet. These boats were equipped with vertical fire-tube boilers or with the Roberts-type water-tube boiler, of which George built 46. He turned out most types of steam engine at one time or another, and he never hesitated to throw in his own innovations. All in all, he did pretty much as he pleased and was successful nevertheless.

The completed boats, including six side-wheelers for the Currituck Sound Fishing and Hunting Club, were shipped by rail to many parts of the country. Ten identical steamers were shipped to the sugar plantations of Cuba, and Whit went along to instruct the Cubans in the art of steamboat engineering.

Perhaps the best known of all Whitney boats was his 94th, the *Ida F,* built in 1900 and named for his daughter. After the shakedown cruise to New York and back — four days each way — the *Ida F* was transported overland to Big Island Pond, in Derry, New Hampshire, where Whit had a summer camp and shop. There she served him faithfully for 48 years, patiently doing his bidding.

It was to nearby Chase's Grove that the huge textile mills of Lowell and Lawrence, Massachusetts, sent their employees for company picnics. Whit was quick to perceive an opportunity, and the *Ida F,* with her load capacity of 25 people, carried thousands of paid passengers over the summers. Whit once told me that at 25 cents per passenger, he paid for his steamer in two years.

The *Boston Post* of June 1, 1910, gives a vivid account of an incident involving four pirates and the *Ida F.* Those erstwhile villains coaxed the lovely steamer onto a dray pulled by six horses and took her to Haverhill, Massachusetts, where she was launched in the Merrimack River. There "the law" took up pursuit, and poor, noble *Ida F* carried her thankless cargo to the mouth of the Merrimack, the stoic heroine in a dramatic chase complete with gunplay and derring-do. Finally the scoundrels were taken at gunpoint and later brought to trial, and the steamer was returned once more to tranquil Big Island Pond.

Whit loved people and always welcomed them in Derry. A great storyteller and correspondent (who usually wrote in green ink), he enjoyed having his picture taken and regaling his guests with memories of the old times. I was aboard the *Ida F* many times but only had one ride with George. The piston whistle was a familiar sound around the pond; the tune "How Dry I Am" was wont to echo from the surrounding woodlands. Under the canopy top was stored a long table that could be set up with one end attached to the engine. Seated at this table, on summer afternoons of 40 years, Whit and his friends shared lobsters steamed by the boiler.

George spent much of his life in Bridgeport, Connecticut, where he built seven houses over the years. When I first knew him, he summered at Big Island Pond, where he did his engine work, and during the winter he drew plans and kept up with correspondence at his apartment in Bridgeport.

In 1947 he sold the *Ida F* to Arthur Bradstreet, and she burned one night in the following year in front of Whit's camp. Dick Mason of Wolfeboro, New Hampshire, bought the remains and resold the engine and boiler to Bill Willock, then of Long Island. Willock hired Whit to rebuild the power plant; the deck hardware and rudder were installed in my *River Queen.* By 1964 Willock had moved to Chestertown, Maryland, where he contracted to have the power plant installed in a 1902 hull that had been shipped down from Lake Champlain. Fire broke out in the Lloyds (Maryland) shipyard, however, and the hull was destroyed. Willock saved the power plant, but his motivation to have it installed in another hull was exhausted. It is now displayed in Bill's private museum in Chestertown.

My story of George Whitney occupies a place of honor in this book because, in his long and varied life, Whit spanned the years from the Golden Age of steam launches to the steam revival of our generation. To many of us, it seemed that we could look back into Whit's experience and see there the historical antecedents of our craft. He never deviated from his love of steam power, even to the day he died. If he could not have a steamboat, he

IDA F

IDA F

George Whitney in 1948.

Oscar D. York in 1950.

wanted no boat at all. As a very old man, he sat in his chair and carved wooden engines, "wheels," and crankshafts.

On June 10, 1962, Fred Merriam, Dr. Henry Stebbins, my son Gary, and I attended Whit's 100th birthday party; letters came from New Hampshire Governor Wesley Powell and President John F. Kennedy. The last of over 300 letters in green ink that found their way into my mailbox was dated January 31, 1963. On December 3 of that year, George E. Whitney died in Manchester, New Hampshire, at the age of 101. Because he lived so long, the once-substantial resources he had accumulated during his career were used up by the time he died, so a group of eight of his friends financed his grave marker.

OSCAR D. YORK

Captain of the mail steamer *Columbia,* "O.D.," as his friends knew him, never became the chairman of the board, nor did he ever wish to. He was smart enough to have done almost anything he set his mind to, but as he liked to say, he had at one time held a steady job for two weeks and never made that mistake again. Oscar York of Wolfeboro, New Hampshire, by Lake Winnipesaukee, was the kind of person who did just exactly what he wanted to do, no more and no less. His legend will live longer around the big lake than if he had sat at the head of "the varnished table."

The appearance of his property was not aesthetically pleasing to his lakeside neighbors even on their good days. O.D. and his wife Karin lived in an old board camp that was never completed, while a cobblestone foundation for a house that was never begun stood nearby. The remainder of the property consisted of an abandoned, "booby-trap" sawmill, old boilers and engines, an Owens electric automobile, a Stanley steamer that he had converted to burn wood, a Terraplane car, and his "shivvy" truck. Construction of a massive steel barge had come to a fitful halt when neighbors complained of O.D.'s late-night riveting, which could be heard all over Wolfeboro Bay. Oscar usually slept late in the

Columbia, *shown beginning her days as a U.S. mail and passenger boat on Lake Winnipesaukee (1905). Built by George F. Lawley in South Boston,* Columbia *had a Ward water-tube boiler and a Fore River compound engine, 7.5'' & 15'' x 8''. In Oscar York's ownership, she served the islands for 25 years.*

U. S. MAIL

STR. "COLUMBIA"

THE ISLAND TRIP FROM

WOLFEBORO

ON

LAKE WINNEPESAUKEE

OSCAR D. YORK, Mgr.

TWO TRIPS DAILY (EXCEPT SUNDAY)
DURING JULY AND AUGUST

Fare, One Way, - -	35 cents
Round Trip, - -	50 cents
Fare to Weirs or Lakeport,	
One Way, - -	70 cents
Round Trip, - - -	$1.00

The "Columbia" may be hired for excursions on Sundays and evenings. For further information apply to the captain.

Morning Trip discontinued after Sept. 1.

morning and passed the remainder of the day getting ready to do something or perhaps just talking to the many people from every walk of life who beat a path to his door to listen, gaze, admire, or snicker. Sometimes along toward evening, if the wind was right, he might settle down to do a little work. Natives who liked him used to remind the "flatlanders" that, after all, Oscar was there first.

In 1905 O.D. bought the steam yacht *Columbia* for $5,000 and had her brought to Wolfeboro from the ocean on a flat car, and for many summers he plied the waters in his only persistent trade, carrying passengers and delivering mail to some of the 274 habitable islands that dot the 44,586-acre lake. But time and change are inevitable, as much in New Hampshire as in the streets of New York, and in 1930 the aging *Columbia* was hauled for the last time. Years hurried after years, and the harsh New England winters beat the once-proud mail steamer into submission. Oscar bathed in the lake, lounged in his underwear in the summer shade, and talked of all the things he was going to do — while his property rotted and rusted away around him. When he died, in 1960, Frank Coughlin bought the old *Columbia*'s engine; Nat Goodhue, of Goodhue & Hawkins Navy Yard, was appointed administrator of the estate and sold most of O.D.'s prodigious collection of scrap iron. Proceeds were used to renovate the old camp so that O.D.'s widow could live out the remainder of her life in respectability.

Sewell Road is just not the same, for gone are Chet Hawkins, Nat Goodhue, the old *Columbia, Swallow,* and Oscar. No longer can we say, "Let's go pay Oscar a visit."

PROFESSOR EVERS BURTNER

Professor Evers Burtner in 1954. (Courtesy Donald Fellows)

Evers Burtner's stamp is deep upon this book, for during more than 30 years many people, myself among them, have profited from his expertise in naval architecture and marine engineering.

The son of a Congregational minister, he was born in Hagerstown, Maryland, in 1893. Because his grandfather made many tools in his blacksmith forge, Evers developed an early fascination for tools and their uses, and this grew to embrace marine steam when, as a boy, he traveled back and forth with his parents from Baltimore to Boston on the Merchants & Miners steamships. Sometimes he made the trip from New York to Fall River, Massachusetts, on the steamers of the old Fall River Line. Childhood impressions etched themselves deeply in the boy's mind.

For a few years after 1910 the Burtner family spent many vacations at the New Hampshire lakes. Evers can recall getting off the Boston train one day at Alton Bay on Winnipesaukee and seeing at least a dozen small steam launches waiting to take guests to their summer cottages. He also recalls his rides on some of the old Woodsum steamers on Lake Sunapee. Perhaps it was his early exposure to steam-powered vessels that prompted the young man to enroll at the Massachusetts Institute of Technology, where he studied naval architecture and marine engineering.

After graduating in 1915 Evers began a teaching career at M.I.T., meanwhile gaining practical experience in marine engineering at the Portsmouth Naval Shipyard and with a Boston firm. He became Associate Professor of Naval Architecture and Marine Engineering at M.I.T., a position he held for over 40 years before retiring in 1963. He measured ocean-racing yachts for the Eastern, Corinthian, and Boston yacht clubs for more than half a century, from 1916 to 1968; he was named an honorary measurer for the Boston Yacht Club and one of a select few honorary members of the Eastern Yacht Club. He is also an honorary member of the Society of Naval Architects and Marine Engineers.

For many years, first at his home in Wakefield, Massachusetts, and later in Kingston, New Hampshire, where he now lives, Evers has carried out experiments with small water-tube boilers, various engines that he has built and rebuilt, and his two launches, *Ala I* and *Ala II.* These boats embodied an unusual design *(Ala II* is still in commission) — each consisted of two smaller boats that were clamped together. One section was kept on shore and could be paddled out to the other section, which held the steam plant and was moored offshore. *Ala I* was built of plywood, but the main section of *Ala II* is derived from a commercial aluminum hull.

Although he does not drive as far or as readily as he once did, Evers has remained faithful to the hobby and to his friends, attending steam launch events around New England. He still operates (and no doubt experiments with) steamers and sailboats on the lake behind his house, and he keeps his advice and encouragement flowing through a heavy load of correspondence.

Like many others, Professor Burtner has made some invaluable contributions to steamboating through his writings. To the examination of an 1890s engine he brings the technological perspective of a 20th-century marine engineer. The following articles appeared in *Light Steam Power:*

"Yacht Engines by Fore River Engine Company," Nov.-Dec. 1959; "The Stickney Steam Launch Engine," Mar.-Apr. 1960; "The Lawley Steam Engines," Nov.-Dec. 1960; "Seabury Launch and Yacht Engines," May-June 1962; "Engines by Murray & Tregurtha's," May-June 1963; "Advanced Boiler Designs by Herreshoff," July-Aug. 1963; "Steam Engines by Herreshoff," May-June 1964.

Also, in *Steamboats and Modern Steam Launches:*

"Herreshoff Three-Drum Launch Boiler," Jan.-Feb. 1961.

Fred Semple, steam launch power-plant manufacturer, in a photo taken a number of years ago.

FREDERICK H. SEMPLE

No other engine or boiler builder in modern times has contributed more to the steam launch hobby than Fred Semple has, and very few who are alive today were steamboating, just for the fun of it, earlier than he. It was in 1937, at his summer home on Kezar Lake in Lovell, Maine, that Fred "assembled" his first steamer. The components of this launch included an old 18-foot dory that was resurrected from a dooryard grave and a 2-h.p. Shipman engine (Rochester model) with one cylinder, two pistons, and a Stephenson-reverse, rocking cylindrical valve. Fred *thinks* the water-tube boiler came from a White steam automobile. No doubt more than one old-timer in that staid summer community was rocked gently by a wave of nostalgia raised by the little steamer underway.

His next launch used a 24-foot by 5.5-foot motorboat hull of 1915 vintage (Fay & Bowen, Geneva, New York). Fred built the 3-h.p., horizontal return-tubular boiler and the double-simple, 4" x 3.5" engine. When the whole business was strapped together and the moment of truth arrived, it was discovered that the engine was too big for the boiler, but a man of his ingenuity was not to be so easily dismayed; he bushed one cylinder down to two inches and "compounded" the simple engine. The resultant creation fulfilled its intended function with aplomb, while Fred's experience continued to accumulate.

In 1953 he unveiled the newest fruits of his spare hours, a 5-h.p., vertical fire-tube boiler and a 3" x 4" simple engine that together formed a package much like the marine unit now in production. Fred used this plant in his steamer, the *City of Lovell,* for seven years before replacing the engine with a Semple-built 10-h.p. "V"-compound. Professor Burtner was along on one of the steamer's many excursions during that seven-year span and was so impressed that he convinced Fred to build the power plant commercially. Thus, from the suggestion of a perceptive engineer, the Semple Engine Company was born in St. Louis, where Fred made his winter home.

Burtner, his interest aroused, guided an M.I.T. student into a thesis on the performance and efficiency of the single-cylinder engine. Several improvements in Semple-engine efficiency resulted from experiments that Jonathan Leiby carried out with Burtner's guidance. The lessons learned were also applied to the 10-h.p. compound, 3'' & 5'' x 4'', that was brought into production in 1961.

Semple power plants are now in use, both in boats and as stationary units, in the United States, Canada, and many other countries. Fred Jr. is active in the business, and it is to be hoped that new Semple engines will be appearing on the market for a long time to come.

Cliff Blackstaffe

CLIFF BLACKSTAFFE

Whenever the term ''native genius'' comes up, people who have known Cliff Blackstaffe think of him first. Possessing a ninth-grade education, Mr. Blackstaffe has been the primary consultant for scores of steam launch projects, whether proposed by millionaire yachtsmen or hopeful teenage enthusiasts. Many discussions of technical steam launch questions have terminated with the easy answer: ''Let's call Cliff and find out.''

Cliff Blackstaffe spent most of his 43 working years at hard, brutal labor in Vancouver Island sawmills, then became engineer for the Empress Hotel, Victoria, British Columbia, and finally first engineer of the S.S. *William J. Stewart,* a Canadian survey steamer, before retirement. Long experience made him equally at home at his lathe, in front of a large industrial boiler, at the throttle of a steam locomotive, or in a ship's engine room. Through those 43 years, and to the present day, he has been widely known as a steam-model engineer and writer with an intuitive grasp of everything that happens inside a boiler or a steam cylinder and everything that should be considered in designing an engine frame or a woodburning firebox. Between 1961 and 1971 Cliff and his partner, Eric Wood, found time to build 36 steam plants for small launches, including vertical and horizontal water-tube boilers and simple, fore-and-aft compound, and steeple compound engines. These power plants sold themselves so well that the partners never had to advertise. The classic of them all was a little triple that Cliff built for his own launch, *Sparrowhawk.* A thrifty craft, *Sparrowhawk* consumes just one gallon of fuel per 12 nautical miles.

His ever-fertile brain has created more innovations than many noted inventors — his letters often include a brand-new engine design — but Cliff never pursued false hopes that his inventiveness should entitle him to wealth and fame. He remains the steamboat hobbyist's mentor.

FRANK FUEGEMAN

Master Machinist Frank Fuegeman has been blessed most of his life with the quiet, frequently renewed satisfaction that comes from reaching toward perfection through one's work. Sixty or more

Frank Fuegeman

years ago Frank began to learn to shape beautiful and functional mechanisms from shapeless metal, and indications are that he has never wanted to stop. How many of us can claim to find the fullest expressions of our temperaments while earning our wages?

Born in the Harz Mountains of Germany in 1905, Fuegeman was apprenticed to learn a machinist's art at the age of 15, and for five years he existed on long hours, no pay, and no room or board. In 1928, his apprenticeship and three additional years as a journeyman behind him, he emigrated to the United States, eventually opening his own shop in White Plains, New York.

Even as a boy the "appropriateness" and detail of small steam engines caught Frank's eye, and years later, when time finally permitted, he started to build them himself (including many from the castings of the well-known model engine company Stuart Turner, Ltd.). His engines began to appear at gatherings from New Hampshire to Florida, and increasing numbers of steam launch and engine builders began to benefit from his expertise.

In 1969 he retired and with his wife, Ada, moved to Berlin, Maryland. There he again set up his shop, began to take orders for the customary variety of jobs, and with his happy and productive dawn-to-dark routine, he settled into a retirement that is to all outward appearances identical to his working years. Now in his late 70s, Frank will still help a steam enthusiast with any problem if he can, though much of his time these days is devoted to small engine construction.

There is another drain on Frank's time that can't be overlooked — a 25-foot fantail launch that was once pushed by a naphtha engine. Over the years Frank has built a dry-back Clyde boiler and con-

Hull restoration or machine work, Frank's standards never lapse. Frank's launch Ada. *(Courtesy Frank Fuegeman)*

verted a stationary simple engine to marine use. This exceptionally handsome power plant will turn an old-time steam wheel with square-tipped blades, hung from a lovingly rebuilt hull that could be a centerpiece in any living room. After Frank's launch finally feels the water, maybe he'll spend some time steamboating — but that would mean stealing time from the shop, where the real fun is.

JOHN S. CLEMENT

What is John Clement? Well, he's a fellow New Englander, hailing from Franklin, New Hampshire, where he operates a hydroelectric plant for his buddy Ted Larter. And he's an electrician, a carpenter, a plumber, a steam fitter, a welder, and a machinist, among other things. In fact, an easier question might be, "What isn't John Clement?" He's not bored, I guess.

If you're a steam buff (Can you spot a bronze injector in a pile of scrap metal from a distance of 50 paces?) and you live in New England yourself, you probably know John already. If you live in Albany, San Diego, Olympia, or any other American town, you may know him. If you don't, you know someone like him. Someone who can be joking or serious without changing his expression or his tone of voice. Someone who on occasion seems to answer a question before it is asked. ("How does a steam engine work?" a stranger once asked, at a very inopportune time. "Damn good," the reply shot back.) Someone who might once have come across a derailed freight train beside the New York State Thruway late at night, and might have slapped on a railroad cap, bluffed his way past an officer of the law, and freely inspected the cordoned-off wreck before driving on.

John has had two steam launches on Winnipesaukee and is putting together a third. At one time he assembled a 50-foot steam tug with a large three-cylinder compound engine (the engine presently powers the new Boston Harbor excursion boat *Calliope).* He helped to restore the passenger boat *Sabino* (now operated by Mystic Seaport) and after sitting for his chief engineer's ticket, he served as her engineer on the Merrimack River. But John will not be found all decked out with cocktail in hand, loudly

John Clement

ruminating on past accomplishments. He will be found in his work clothes, muttering around the stem of his pipe, up to his elbows in grease.

FREDERIC C. MERRIAM

Sometimes a person is spurred by a bond of friendship, a shared interest, or simply an unselfish kindness to give freely of his time, energy, and knowledge. It is no great revelation that time is usually a dearer gift than money, and that a person who gives his time frequently and generously is a rare person indeed. Fred Merriam is, in the eyes of many steam launch owners, such a person.

A brilliant machinist (and distant cousin to George Whitney), Fred has been in and around

Fred Merriam on board Richard Hovey's Skookum Jack *in 1974.*

steam launches for years. During the 1950s and 1960s his skills and counsel were of immeasurable assistance to Bill Viden and Bob Bracchi, co-owners of the *Staurus* on Lake Winnipesaukee. Later he served in a similar capacity, friend and adviser, for Dr. Henry Stebbins, on the launch *Zephyr* out of Marblehead, Massachusetts. Fred was always aboard, keeping operations smooth, when the *Zephyr* served as the starting boat for the Jeffrey's Ledge yacht races.

In more recent years he found time to do extensive boiler and engine work for his friend Percy Stewart of Hinsdale, New Hampshire, who owned the launch *Gemini II.* It is gratifying to know that before Percy passed away in 1978, Fred, who had contributed to so many steam launches but had never owned his own, became a co-owner of *Gemini II.*

ARTHUR E. HUGHES

Knowledge is one thing you can always give away yet still retain, and Art Hughes of East Braintree, Massachusetts, has an abundant fund of knowledge and gives it away all the time. He is one of the professionals who have guided the efforts of the amateurs, frequently saving us from the frustrating pitfalls of a technology that is isolated from the marketplace mainstream.

Art Hughes

Born in the "golden era" of steam, Art vividly recalls the Merrimack River of his childhood, where he used to watch the steam tugs with vertical boilers, fancy steam yachts racing in from the sea, and the stern-wheel passenger boat *Merrimack,* which plied the river between Newburyport and Haverhill.

His father worked in the yard of the Gas Engine & Power Company and Charles L. Seabury & Company, Consolidated, in New York, so it was no surprise when Art chose a career in marine engineering. Beginning in 1916, he served his apprenticeship in the Quincy, Massachusetts, yard of the Fore River Ship & Engine Company, and he later saw the world as engineer on a succession of freighters, tankers, and liners. While his shipmates were frequenting the beer halls of exotic ports, Art would often study engineering or mathematics texts or visit the engine rooms of foreign ships. He was a passenger many times on German tugs in Hamburg, and he remembers to this day the details of their engine rooms, even to the sounds of the pumps.

Competence, ability, and attention to detail advanced Arthur steadily in his career, and he became chief engineer for the liner *City of Hamburg,* holding an unlimited license for any United States ship, steam or diesel. He returned to the Fore River yard before full retirement in 1963.

Art's interests are broad, and he is as much at home at a model steam locomotive meet as he is at a steam launch gathering. The 4-6-2, ¾-scale locomotive that he recently completed had been in progress through all the many years of our acquaintance. Exquisite in detail and workmanship, it could only have been a product of his unique temperament. During its construction he would labor patiently over a single intricate, beautiful piece — and when it was done, he would store it in a little wooden box, lined inside with green felt and covered outside with several coats of paint or varnish, that he built solely for that purpose. Each piece, each perfectly crafted little assembly was stored in its own felt-lined box like a precious gem, until the time came to add it to the emerging locomotive. The boxes seemed appropriate in Art's shop, glistening in the lights beside his immaculate machines and orderly rows of tools, his chief's ticket proudly displayed overhead.

Bob Thompson and the author's son pose beside Rum Hound *in 1962.*

ROBERT W. THOMPSON

The woods and lakes of New Hampshire form the backdrop to Bob Thompson's life. A native son, he was born on the shore of Squam Lake; he saw the last of the commercial steamers in the lakes region; and he was the last paid fireman and engineer on the old steam yacht *Swallow*.

Bob listens to his own counsel, like it or be damned, and he moves when he wants to, at the speed he cares to make. When I first met him in 1943, he was living in Wolfeboro and working at the Goodhue & Hawkins yard. I remember quite clearly that one morning when he was installing deck hardware on a refinished speedboat, the boat's owner appeared and with many "helpful" suggestions endeavored to tell Bob how to do the job. On the lunch hour Bob went home — for the day. Soon after one o'clock the owner was back and asked the boss, Nat Goodhue, where "the man" was. Patiently Nat explained, "In the city you can tell a man what to do and how to do it, but up here in New Hampshire you ask him if he'll do a thing, and you trust him to do it right." As Bob told it later, he was "up to the moving picture theater with the family and a big box of popcorn," and the wealthy owner was shorebound.

Bob's love of steamers made him crave one, his limited means dictated the terms, and his ingenuity made it possible. He rebuilt a 30-foot Truscott hull for steam and named his craft the *Rum Hound* — she has survived to this day, first on Winnipesaukee, later on Squam Lake, and most recently on Hawkins Pond in Ashland, where Bob now lives. How Bob has been able to keep the contraption steaming all these years challenges the imagination.

To ride on the *Rum Hound* is to touch base with the basics of steamboating. She is a pretty boat, though not prettily decked out, and she carries with her a certain earthy, pleasant sensation. She may be cluttered, but not with "extras" or useless paraphernalia. Boiler water is supplied by an injector near the floor — the pipes are a plumber's nightmare. Paint

Bob in one of his good moods — most of them are.

and varnish, except on the outside of the hull, are unnecessary luxuries; the bilge is ankle-deep in oily water, wood chips, and ashes.

Before going for a ride in *Rum Hound,* branches for the fire are gathered in the nearby woods and broken against the side of the boiler by the owner's boots. A sharp jackknife shaves a pine stick for kindling, and soon the smoke is billowing from the rusty stovepipe that projects from the tar-paper canopy at a jaunty angle. The owner throws some oil on the two-cylinder, high-pressure engine — one of many that have powered the boat over the years — and when the pressure starts to climb, he kicks over the flywheel with his foot. We begin to ghost away from shore.

Now the owner/captain sits to one side in the stern, where he can steer by pulling on the tiller rope. Covering his head is an old felt hat spattered with sawdust; a piece of rope holds up his shapeless trousers. He is surrounded by his steamer, and as we watch we realize that his boating attire is in keeping with the unbuttoned spirit of the *Rum Hound;* man and boat seem inseparable, and the lack of spit and polish of the whole is a welcome expression of the owner's independent ways. We are comfortable, for this is a boat that one can have fun in.

From the lower end of Little Squam Lake a small river flows to the old Ashland dam, a distance of perhaps three miles. This little waterway was once navigable for a shoal-draft boat; now it has silted in extensively and is very shallow in places. In 1961 Frank Coughlin, John Clement, my son Gary, and I spent a carefree day with Bob on the *Rum Hound,* steaming on Big Squam and Little Squam lakes. When someone suggested we try to navigate the river to the dam, it seemed like the natural thing to do. With a great deal of laughter, whistle blowing, and poling we finally did make it down and back, but many times we had to climb overboard and shove her off the mud. If we grounded once that afternoon, we grounded 30 times. In *Rum Hound* you can do that sort of thing.

The afternoon shadows cast by the hills around the lake are lengthening, and we are putting in again to the creek from whence we started, our laughter and Bob's stories of the old days noisier by far than the steamer that carries us. Captain Thompson douses the fire with a bucket of water, and another layer of ash dust settles over hull and cabin. He closes the draft with an old license plate and pushes the ashes up against it with his foot. We are ashore once more; *Rum Hound* waits comfortably in the gathering darkness.

4

The Naval Architecture of Steam Launches

by Bill Durham

Much of the popular appeal of modern steam launches originates in the fact that they are unpredictably and unevenly varied, reflecting all levels of taste and skill. Each is the expression of an individual's private vision, strongly colored by his experience and by the practical need to improvise and adapt. Each, in its own way, is unique.

What is wanting is a feeling for abstract ideals of excellence, with rigorous examination of basic principles and design objectives. Hundreds of small steamers of the 1880s and 1890s exhibited high degrees of balance and appropriateness in their designs; the only steam launches of this century to demonstrate such qualities as clearly have been a few powerful naval and commercial boats, and these attained their most advanced stage of development at the moment of extinction. No late-20th-century pleasure steamers have yet approached the high, even perfectionist standards that have become common since 1960 in other hobbies and mechanical pursuits.

I believe that this failure has occurred because designing a good steam launch requires expertise in more fields than one person can master — unless he is a naval architect with a touch of genius. (Showpieces of wooden boatbuilding or old car restoration are commonplace because of the narrower focus of the pursuits.) For this reason, steam launching remains a fresh and immature avocation still in its salad days. All of the best steam launches have yet to be built. Since the 20th century's Thornycrofts and Herreshoffs are not devoting their manifold abilities to steam launches (they're probably in banking or video games), the knowledge and experience of several people must be distilled in a steam launch design. This is most cheaply and efficiently accomplished by reading other people's published opinions on the several subjects. The boats that modern texts on naval architecture concern themselves with do not have many of the characteristics of steam launches, but all of the principles of the science relate to steam launches in one way or another.

The source that I have found most instructive on

hull design is a readable, not overly mathematical text, *The Naval Architecture of Small Craft,* by Douglas Phillips-Birt. Weston Farmer's plentiful writings on boat design are full of seasoned wisdom, original insights, and design shortcuts. His work is scattered through dozens of periodicals, over a 60-year span. There is a concentration of some of the best in issues of *National Fisherman* published during the 1970s, and in his book *From My Old Boat Shop* (International Marine Publishing Company, 1979).

The most easily available technical source on some of the principles of steam power (including nuclear) is the U.S. Navy's training manual, *Principles of Naval Engineering,* available from any government-publications source for $10. A nontechnical, uneven record of the steam launch experiences of many amateurs is in *Steamboats and Modern Steam Launches* (Howell-North Books, 1981).

POWERING AND HULL DESIGN

Post-1950 amateur designs (and amateurish professional designs) have suffered from fuzzy thinking about the essential nature of steam power and recreational steamers. In small craft, steam power imposes restrictions, which, if they are understood, can be exploited to design a successful and beautiful boat.

Gasoline power is extremely light, cheap, and convenient, a prime mover without peer for dynamic-lift boats of very high speeds and powering. For people who want to enjoy the benefits of abundant mechanical power without having to appreciate the power source itself, internal combustion is the choice.

Steam power excels in silence, simplicity, adaptability, and mechanical interest, and a steam engine's torque characteristics are ideally suited for screw propulsion of vessels of modest speed and power. In small craft, steam is five to ten times more fuel-wasteful than gasoline power, and the combined weights of power plant and fuel would be 20 to 100 times greater if the same amount of power were wanted. The installed cost of a steam horsepower will be at least 10 times greater than a gas horsepower (although fuel costs may be nil).

These attributes make two or three horsepower per ton the most suitable level of powering for a small steamer. A steam launch needs to be lean, clean, and graceful to make good use of this powering and to be in harmony with the warmth and gentle, slow-moving strength of the power plant.

To plan a steam launch with hull lines well adapted to its powering (or vice versa), it is necessary to employ the concept of "speed/length ratio." A vessel's speed (in knots) divided by the square root of its waterline length (in feet) is its speed/length ratio at that speed. A 16-foot boat going 4 knots will make waves in the same pattern, and require the same powering per ton of displacement, as a 36-foot boat at 6 knots or a 100-footer at 10 knots. These are examples of a speed/length ratio of 1.0, which is attainable with around one horsepower per ton.

Most of the world's half-billion tons of commercial shipping crosses the oceans with less than one horsepower per ton, at speed/length ratios below 1.0. Recreational motorboats commonly carry 40 or more horsepower per ton, permitting them to plane over the water at speed/length ratios above 2.0. Scientifically, these are closer relatives of airplanes than of ships. Steam launches belong in the realm of "overpowered small craft," that most difficult and interesting area of boat design in which speed/length ratios climb from 1.0 to 1.6, while the required power rises from one to nearly ten installed horsepower per ton of displacement. Within this range, power requirements increase exponentially, and fundamentally different hull forms are optimal for design speeds that differ by as little as 1½ or 2 knots.

Given the ravenous fuel appetite of a steamer, most pleasure steam launches will spend the greater part of their time using one to three horsepower per ton to cruise at speed/length ratios between 1.0 and 1.2. These speeds call for a hull with a low prismatic coefficient (that is, bow and stern fine, with the displacement concentrated amidships), a narrow entrance, and an easy run. The longer and lighter the boat, the more power-efficient. A prismatic coefficient (Cp) under 0.55 will impose some penalty at speed/length ratios above 1.3, but this penalty will be less than what a high Cp (above 0.60) would impose at the steamer's normal cruising speed.

All steam launch builders are optimistic about

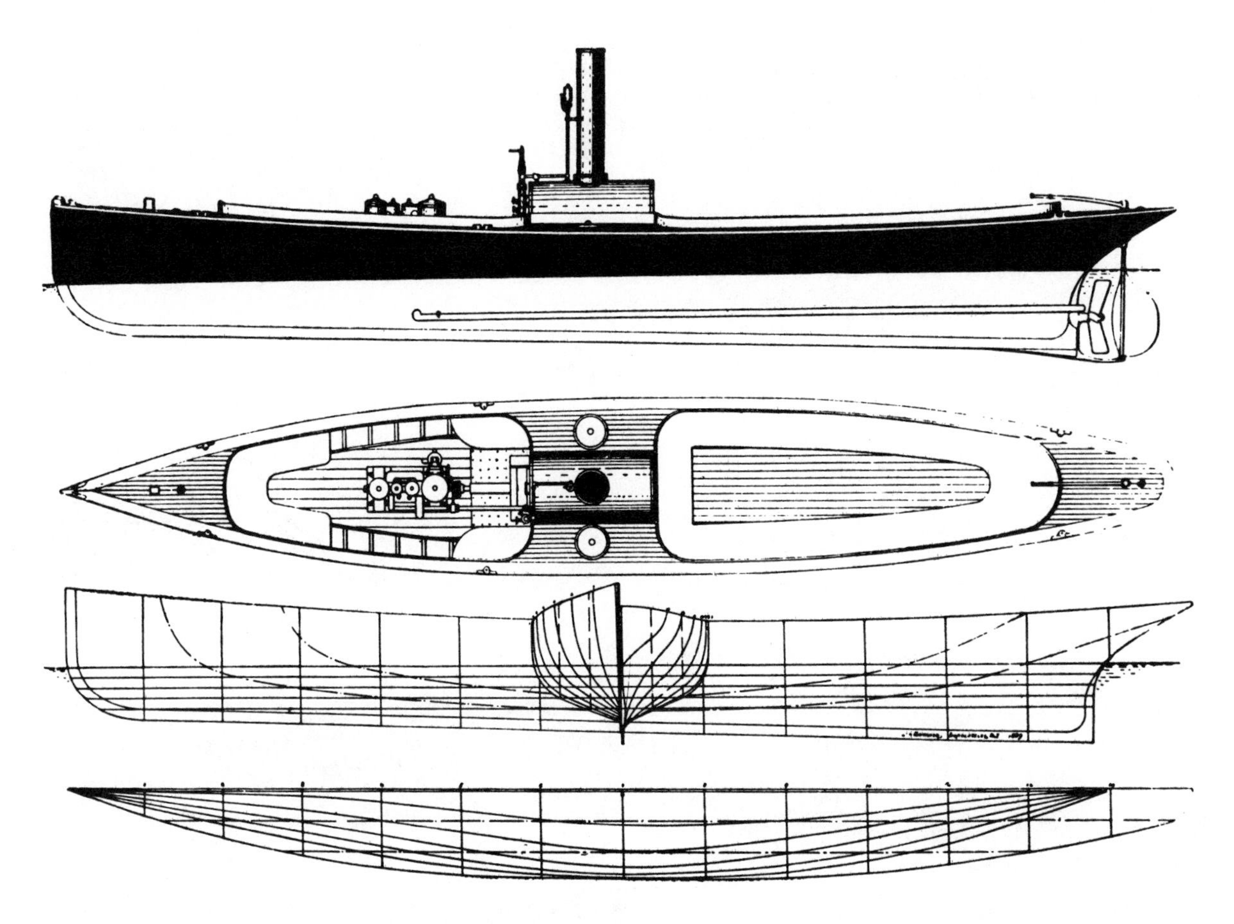

This 28' x 4.7' boat carried a 3" & 5" x 4" forward-mounted engine and a water-tube boiler that could be fired with coal or oil. In 1889, five-beams-to-length proportions were considered appropriate for cruising launches. Faster or more easily driven boats needed to be narrower. (From C.P. Kunhardt, Steam Yachts and Launches, *1891)*

their boats' anticipated speeds, and many have chosen quite unsuitable sources for their design ideas — tugboats, heavy gas boats, or naval steam launches, for example. (Naval steam boats were built short and heavy to withstand rough usage and for hoisting and stowage on deck, and they were given high powering. Tugs are floating power plants.) As a result, the lines of many modern steam launches are appropriate for a level of powering more than twice as high as that which the boats will employ in normal service. If a boat is designed for a speed/length ratio of 1.4 or 1.5 — typical "trawler yacht" or naval steam launch speeds — she may require twice as much fuel to operate at a speed of 1.2 as she would need with proper pleasure-launch lines.

When some sturdy, elaborately designed steamers of the 1970s were given both great beam and hull lines that swept broadly from end to end (Cp's above 0.60), one suspected that the owners planned to drop 90-horse diesels into their boats as soon as the novelty of 10-horsepower steam wore off — and the image of steam launches as heavy, beamy, and underpowered was further reinforced.

A powerfully rigged modern sailboat may have about the same Cp and design speed as a steamer. Studying sailboat hull design can be instructive, but wind power imposes so many special requirements that a sailboat hull is rarely suitable for steam propulsion. A keel sailboat is normally heavier and much bulkier than a steam launch of the same length, because of the generous beam, ballast, and freeboard needed for carrying sail.

Three-beams-to-length proportions are advantageous for the modern boats, sailing and powered,

The hull lines of the U.S. Navy 40-foot and 50-foot steam cutters of 1915 are a striking illustration of the use of a transom stern and flattened run to accommodate high power, although the 10-horsepower-per-ton 50-footer shown here would be too big and powerful to suit a hobbyist's needs. Today's steam launches with transom sterns frequently look like fast boats going slow.

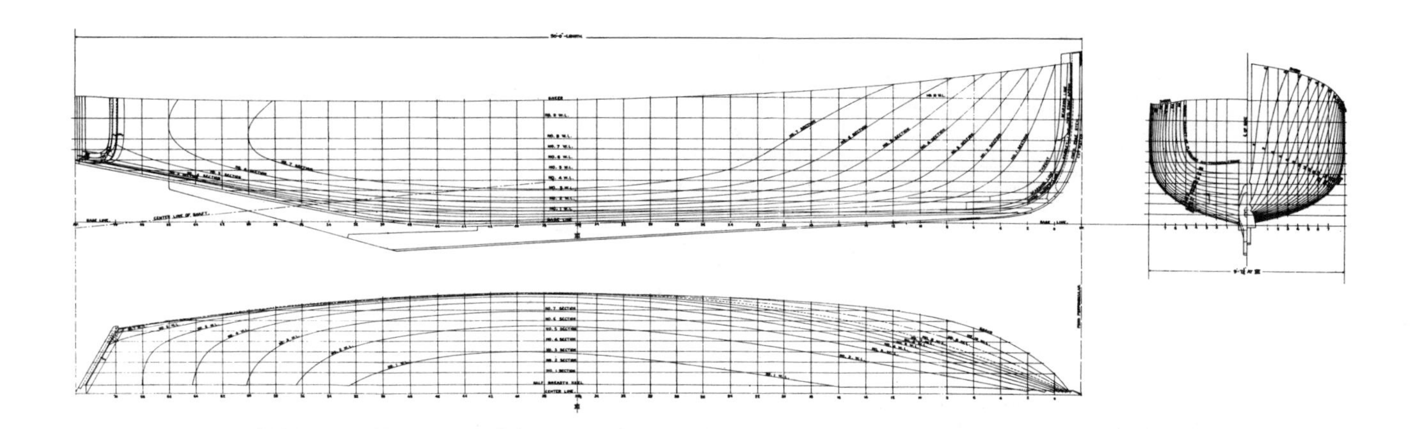

that millions see at boat shows, but these craft have impaired our collective ability to envision the shape of a "proper boat" as a dynamic thing, dependent on powering and service. There is also a common tendency to see boats and ships as much beamier than they really are.

On a North Atlantic crossing in 1942, I whittled out a block model of my ship, the C-1 freighter *Exiria.* My shipmates laughed and laughed at the model, because they could see that I had somehow got the beam only half as great as it should be. My model was 14 inches long by 2 inches broad, while any messboy could look out on deck and see that the ship *appeared* to be only three or four times as long as she was broad. My shipmates were all mistaken. The *Exiria* had the seven beams to her length that is normal for ocean freighters.

A reasonably efficient "trawler yacht" (10 horsepower per ton, Cp around 0.65) can be designed with 15 feet of beam on a 45-foot overall length. It is scarcely possible to draw a good two-horsepower-per-ton hull (Cp around 0.53) to these proportions. If the length is four or five times the beam, the hull of a steam launch will fall naturally into easy and efficient lines. All boat design is compromise, however, and a short, chunky hull will be a good deal more cost-effective than a long, graceful hull if the boat is underway only the few dozen hours each year that is said to be typical of recreational craft.

Evers Burtner once reminded me that most modern pleasure steamers are rather small, many steamboat boilers are top-heavy, and few enthusiasts employ a naval architect to check their boat's stability. We should first of all encourage the building of safe, stable, and commodious steam launches. I heeded Professor Burtner's advice and kept quiet about proper steam launch design while the first generation of modern pleasure steamers was appearing, from 1950 to 1970. I hope that some of the

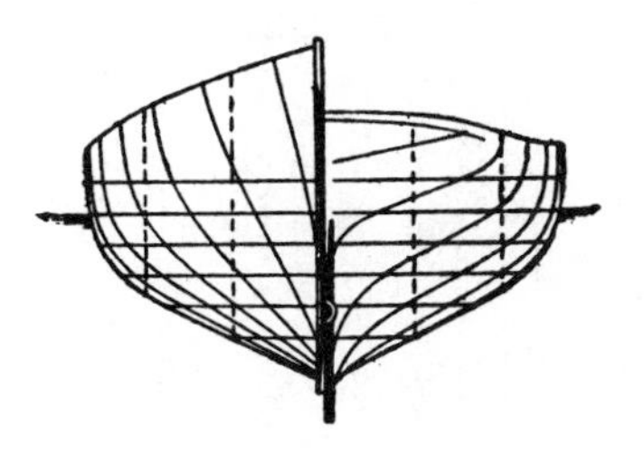

The lines of this 40' x 8' clipper-bowed launch reveal that the seemingly low and light hulls of the classics often incorporated deep and powerful underwater lines. When the boat was running under full power with that 36-inch screw there would have been only inches of freeboard aft, but what a vision of steam power! (From C.P. Kunhardt, Steam Yachts and Launches, *1891)*

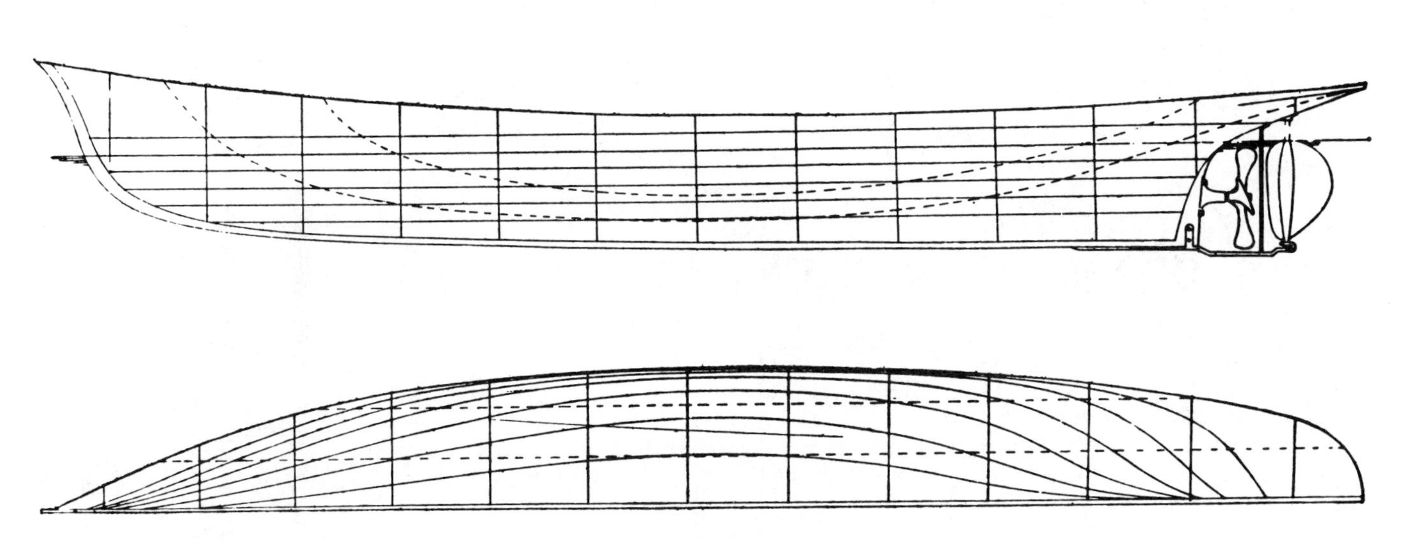

second-generation boats of the 1980s and 1990s will hew more closely to the imperatives of good steam launch design.

Everyone admires the grace and natural motion of a seakindly hull. When a boat's power density is of the same order of magnitude as that of a trout or a mallard, its contours and movements are not dissimilar from those of living things. Many steam launches, old and new, have seakindly hulls that will experience less stress in rough water than the hulls of modern motorboats, but seakindliness should not be mistaken for seaworthiness, which in powered small craft is as much a function of power as of form. These same launches may be inadequately seaworthy because they have low freeboard, tall houses, and large, open cockpits, and they lack sufficient power to get out of a storm's way quickly or to make headway through it.

If a steam launch design is based on fundamentals of power, speed, engine characteristics and hull proportions, its lesser details can be varied almost without limit. A raked bow can provide more flare forward than the traditional plumb stem, leading to a drier boat. A canoe or cruiser stern gives good results with a relatively small, fast-turning propeller. A cut-off transom stern can be used with low power and a narrow canoe-form underbody, or with the flattened run and broad stern required by high powering.

Quantities of talent and money have been invested in improving the hull designs of sailboats and planing motorboats. Very little — if any — modern research has been devoted to small powerboats in the speed and powering range of steam launches. Model testing could be especially fruitful in improving steam launch designs, since wave-making resistance reaches its maximum at high steam launch speeds. The builder of a small steamer is not free to add unlimited quantities of light, cheap power, so modern steam launches may revive an interest in efficient and graceful pleasure-boat design. "Energy-efficient motorboats," with powering several times higher than good steam launch design, cannot provide much guidance in this direction.

Deciding on the best hull form, powering, and propeller constitutes only part of the naval architecture of steam launches. The structure must be

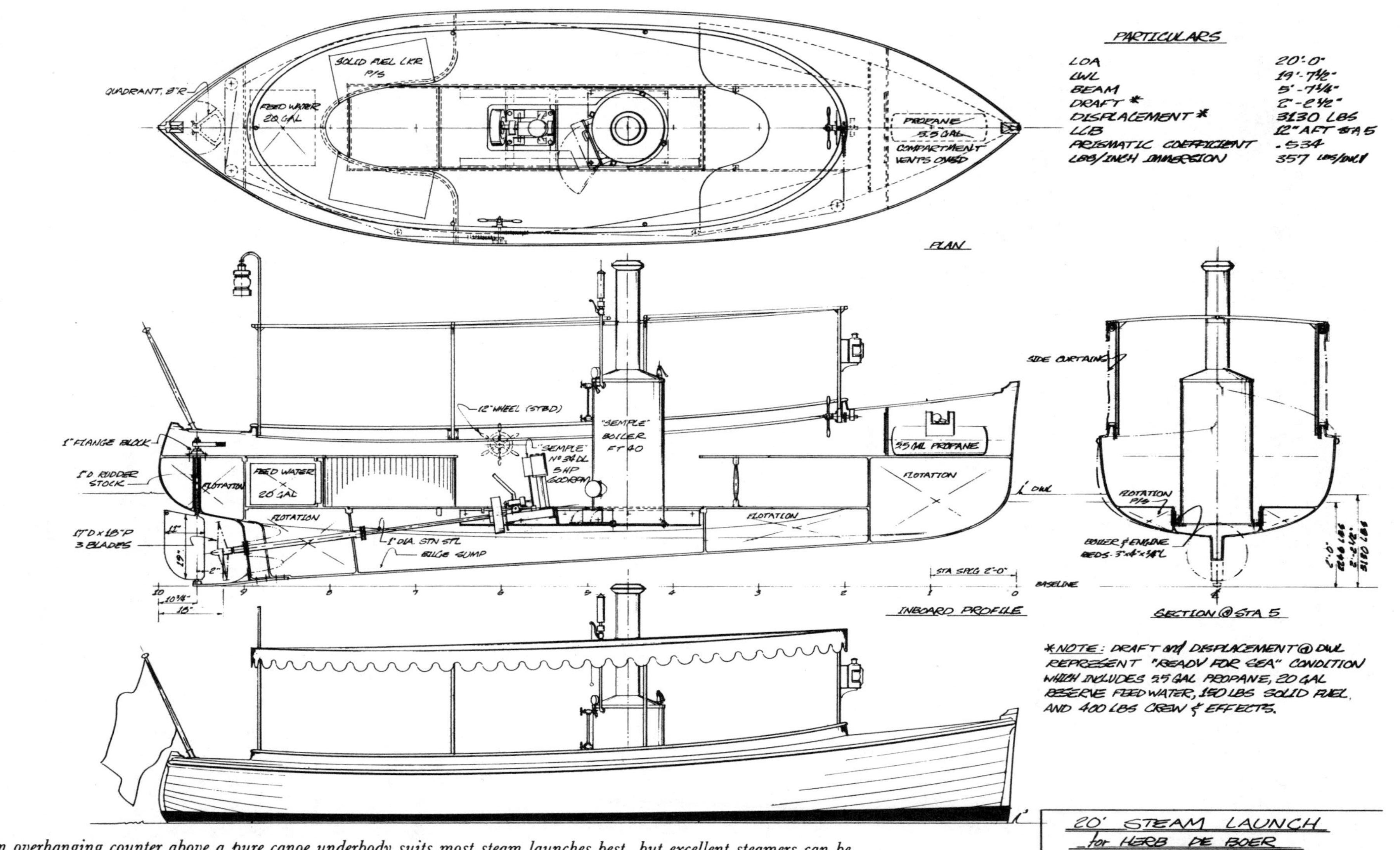

An overhanging counter above a pure canoe underbody suits most steam launches best, but excellent steamers can be designed with canoe ("cruiser") or transom sterns, and these designs permit higher powering and greater capacity on the waterline length. Tim Nolan, of Seattle, is one of the few naval architects interested in designing powerboats adaptable to steam propulsion. His Ajax design of 1976 has a prismatic coefficient under 0.55 — typical of keel sailboats — and the narrow entrance, shallow body, and easy lines that keep wave-making to a minimum. The hull is in production in fiberglass, and in 1980 the architect worked out a steam-powered version for Herb de Boer, of Bainbridge Island, Washington. It would be difficult to improve on the particular combination of qualities of this 20-footer — capacious, easily driven, and stable enough to carry a tall and heavy boiler.

planned, a major consideration often being the distribution of weight concentrated in the boiler. Accommodations and fittings must be detailed and the whole boat kept in harmonious proportions. Non-design — mere assembly of miscellaneous components — is an old tradition in the steam launch revival, and some newcomers receive the impression that this is how it should be done. If you can afford it, why not be different and employ a young designer, perhaps one with sailboat experience, in the spirit of contributing to his education while you are getting some help with structural design and stability calculations?

PROPELLERS

A steam launch engine works at a rotative speed that corresponds with the highest propeller efficiencies possible for displacement craft. An ideal steam propeller might conceivably achieve propulsive efficiencies between 70 and 80 percent, beyond the fondest dreams of designers of non-racing motorboats — and might cost more than the engine! The steam wheel screws its way slowly and powerfully through the water, converting what Weston Farmer called ". . . the long elastic shove of steam . . ." into forward motion. The transfer of ergs of energy from the mechanism to the medium is almost visible at 300 or 450 r.p.m.

All steam launch owners are aware of the importance of their propellers, and they know that their needs are not answered by gas-boat practice. This awareness has led to a "steam wheel" folklore that is as unscientific as folklore usually is. Single-minded emphasis on coarse pitch or large diameter may result in an inefficient propeller or an overburdened, underpowered engine.

A propeller may be given correct pitch for the boat and engine speeds, but with a diameter so small it can transmit only a fraction of the engine's power as effective thrust. In another boat a small engine is given a propeller big enough to lug it down to 150 or 200 r.p.m., and the engine can develop only a fraction of its rated power. Low-r.p.m. propellers are quite tolerant of deficiencies in trueness and balance but unforgiving of gross errors in selecting diameter and pitch.

When consulting a manufacturer's propeller-determination chart, keep in mind a few fundamentals of the science. Starting with accurate estimates of the engine's horsepower at the desired r.p.m. and the boat's speed with the available power, propeller diameter is determined first, then pitch. The slower the engine, or the slower the boat, the larger the diameter of the propeller. It is self-defeating to fudge on size; if the tables say you need a 26-inch propeller to transmit 10 horsepower at 400 r.p.m., don't try to get by with a 24-inch wheel. A 15" x 30" propeller is undoubtedly suitable for steam power, but its diameter permits it to transmit one horsepower at 400 r.p.m., not two or three.

There are numerous propeller-selection graphs and tables. One of the most useful to steam launch designers is Weston Farmer's 1941 chart of "torque-diameter propeller determinants," published in *Steamboats and Modern Steam Launches,* September-October 1962, in *National Fisherman* in July 1976, and in Farmer's book, *From My Old Boat Shop.* Books and boating magazines from the era of low-speed gasoline engines, approximately 1895 to 1915, also contain a great deal of propeller data useful to small steamers.

I will list here the approximate diameters of propellers to suit some typical steam launch engines in 6-knot boats.

Horsepower	Speed (r.p.m.)	Diameter	Speed (r.p.m.)	Diameter
1	600	12"	1000	9"
2	500	16"	750	13"
3	450	18"	600	15"
5	400	22"	600	17"
10	350	28"	500	23"
20	300	34"	400	30"

Choose a 10 percent larger wheel if it is two-bladed, or a size smaller if four-bladed.

The U.S. Navy employed four-bladed propellers in its steam cutters of 1900. (The engines' power ratings are at 350 r.p.m.)

Rated Horsepower	Speed (knots)	Diameter/Pitch
14.28	6½	26"/42"
18.65	7½	30"/48"
48.87	9	32"/54"

A small increase in propeller diameter absorbs more power than a large increase in pitch or blade area. Insufficient diameter can be compensated for by abnormally high pitch and/or great blade area, but only at a major loss in efficiency. Boats that are so compromised might run better with ('orrible thought!) smaller propellers driven through step-up sprockets à la 1845.

The pitch of the propeller gears the rotary motion of the engine to the linear movement of the boat and should not be required to do anything else. One should base propeller pitch on a realistic estimate of the boat's expected speed — multiplying the square root of the waterline length by 1.2 will do. Convert the anticipated speed into feet per minute and divide by engine r.p.m. to obtain the nominal pitch in feet. Add 15 to 25 percent to this figure to allow for propeller slip.

The beautiful 2.0 or 2.5 pitch-to-diameter propellers we all admire are not likely to suit any hobbyist's steam launch. A 20'' x 50'' wheel would be fine in a boat running 15 m.p.h. with 4 h.p. at 300 r.p.m., but our boats don't really go that fast. Most launches need propellers with a pitch-to-diameter ratio of 1.0 to 1.5 — right-hand only, of course.

When owners fabricate their own propellers — to save money or to create a unique design — it is usual to employ the hub of a scrapped propeller, with new blades (midwesterners say ''flukes'') brazed or riveted to it. Weston Farmer did not fail to explain propeller design and manufacture for amateurs (is there any boating subject he did not add to?) in his article in *Steamboats and Modern Steam Launches,* September-October 1962. He recommended constructing a ''false sheet-metal wheel'' first, to learn what diameter and pitch will best balance engine power and speed. It may be learned that engine r.p.m. must be increased or decreased slightly, to avoid critical periods of vibration or rattlety-bang couples among the engine parts. This is best discovered before investing in a refined, permanent wheel.

If you can find a gas-boat propeller reasonably close to your diameter and pitch requirements, it is not difficult to shape the blades to a ''steam wheel'' profile with a saber saw and disc sander. The main purpose is to reduce the blade area, for higher efficiency at low r.p.m. and powering. The blades can

Opposite: *The drawings and photograph of the 1934 ''Canal Grande'' class of Venetian steam water buses were kindly contributed by Dr.-Ing. Artú Chiggiato. Dr. Chiggiato has been an attentive observer of these boats since their inception, and he provides some insight on their design and performance:*

> *Steam* vaporetti *were chosen over diesel-electric because of the enormous ''ahead'' and ''astern'' requirements of the service. There were 20 stops in each 56-minute run, and electrical/mechanical reversing gear could not survive these stresses for long.*
>
> *The hull lines were chosen from the best of three models, all of which were tank-tested thoroughly to find the form that would cause the least resistance and waves in shallow (17-foot minimum depth) water. I love to watch these boats running past at 8 or 9 m.p.h., the water closing in astern with scarcely a ripple.*

The 75' x 13.8' x 5.9' boats were given 140 indicated horsepower, to make 10.25 knots. Displacement was 32 tons light, 45 tons with the addition of 13 tons of passengers. It is interesting to note that the original testing showed the boats required more than twice as much horsepower to make 9 knots in 17 feet of water as was necessary under the same conditions in water of 33-foot depth:

Knots	*6*	*7*	*8*	*9*	*10*
Horsepower required in water 17 feet deep	*8.6*	*14.6*	*25.6*	*64*	*120.5*
Horsepower required in water 33 feet deep	—	*11.4*	*18*	*27*	—

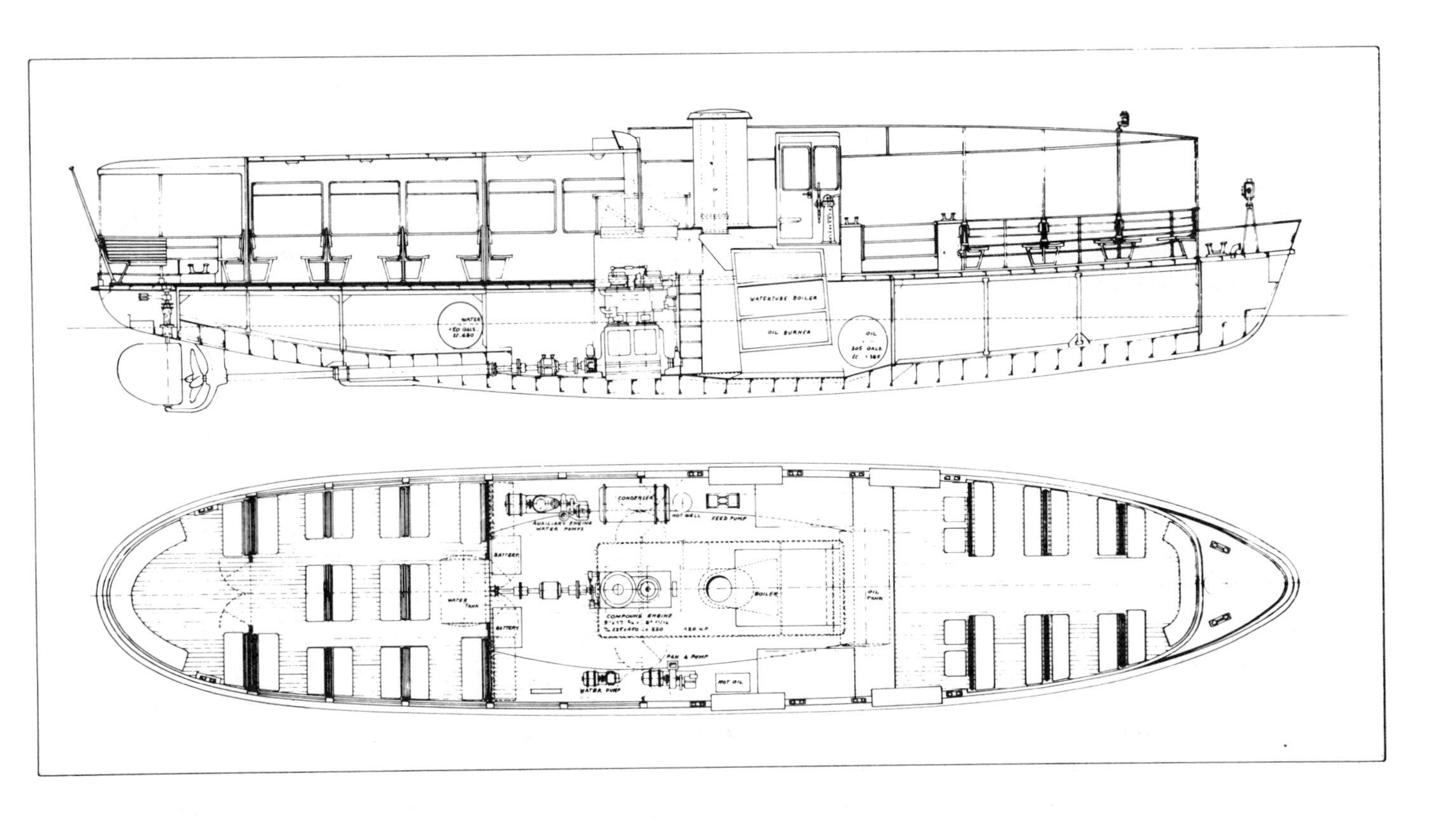
WATERTUBE BOILER
OIL BURNER
CONDENSER
BOILER
BATTERY

56

be cut to a fanlike "common propeller" profile, which was standard in Victorian times, to one of the voluptuously curvaceous designs popular early in this century, or to the narrow ellipse that is technically preferable for most installations.

A sound propeller can be re-pitched upward as much as 50 percent, but this may not be possible with some of the most appealing antique propellers. Most old propellers have markings to indicate that they have been re-pitched downward one, two, or three times (as gas-engine r.p.m. increased, from 1910 to 1940), and these may not stand another re-pitching.

A successful steam launch will probably have about the same power density as a powerfully rigged keel sailboat under the best sailing conditions or under maximum auxiliary power. Therefore, the steamer's hull form should be closer kin to a sailboat than a motorboat (although this is not evident to everyone, because a sailboat's form is modified by the special requirements of wind power). The technical data in *Theory and Practice of Propellers for Auxiliary Sailboats,* by John R. Stanton (Cornell Maritime Press, 1975), serves steamboats better than similar studies of motorboat propellers.

MATERIALS

The design of a boat is more important than the choice of structural material, but at some point the builder must decide what substance he will employ to realize his ideas. At this juncture in small-craft history it would be foolish to recommend any common boatbuilding material as best for steam launches. Serviceable steamers have been built with all of the materials presently available, but superior materials and techniques are long overdue. A creative designer keeps a weather eye on materials developments.

If the qualities of an ideal boatbuilding material were listed alongside those of fiberglass, there would be very little congruence, but fiberglass has been so cheap, adaptable, and well suited to series production and mass marketing that it has pushed alternatives into the background. The "fiberglass successor," when it arrives, will be fireproof, shatterproof, less dense than water, a good insulator, easily repaired, and practically free of hull-manufacturing labor costs. While fiberglass dominates, more steam launch builders should take advantage of its ability to assume any form and to provide any concentration of localized strength that is wanted.

A steam launch's hull in fiberglass should be considerably lighter than a motorboat's. There are no flat surfaces requiring massive thickness and, since all surfaces are compound curves, no supporting frames except those for machinery installation. Stresses from powering and wave action are much less in the slower, low-powered type of boat.

A steel- or alloy-built boat, with its small scantlings and 100-percent "fastenings," can be made more trim and taut, more beautiful and apt, than a wooden boat. Iron and steel were the preferred materials of industrial steam launches for half a century, especially in Britain. This choice was economic for the time and place, since the working launches were built in shipyards possessing plate-rolling equipment and plentiful metal-working skills. Steel would be equally desirable today, especially for working boats (just try to insure a wooden steam launch carrying passengers for hire!), if plate-forming skills and equipment were still widespread, but these have become uneconomic, superseded by cheap welding and fiberglass techniques.

Small steel hulls today are "developed" from flat sheets, just as plywood hulls are. Hulls with hard chines and flat or developed surfaces can be made to be thoroughly suitable for low-power steam launches; the distribution of weight and volume on length are of more importance in steam launch design than surface fairness or curved beauty. Several of the powerboat hulls designed for plywood by Pete Culler (a catalog of Culler's designs, compiled by John Burke, will be published by International Marine Publishing Company in 1983) and by Harry Sucher *(Simplified Boatbuilding: The V-Bottom,* W.W. Norton, 1974) are suggestive of ways to achieve efficient and sufficiently beautiful metal steam launch hulls. I wish there were more interest in building proper steam launches simply and cheaply, in plywood or metal plate, for then we would have a greater number of good launches sooner.

Light alloys have several well-known advantages over ferrous materials, and a few drawbacks. They

Donola, *representative of the river launches of Great Britain, is the most distinguished steam launch preserved under cover. Here she is making a last run on the Thames before being beached in the small-craft wing of the National Maritime Museum. Built at Teddington in 1893, the 58' x 7.8' steel-hulled launch served 76 years afloat, requiring re-boilering in 1923 and renewal of some hull plates in 1947. The 7'' & 11'' x 6.5'' engine turns a 32'' x 42'' propeller. The boiler is a side-fired locomotive type, 82 inches long overall. The coalburning firebox is of 7/16'' copper; the boiler has 94 1 1/4 '' o.d., 12-gauge brass tubes. (National Maritime Museum, London)*

necessitate especially close attention to cathodic protection and the control of stray electric currents, but aluminum alloys will generally not require any organic coating, as steel does. Long ago, a few steam launches were built in phosphor bronze or brass (for fresh water). If the cost of copper continues to decline relative to aluminum and steel, these materials will again appeal to a few builders. There are several excellent books on the design, manufacture, and maintenance of metal small craft. These include: *Boatbuilding with Steel,* by Gilbert Klingel (International Marine Publishing Company, 1973. This book also contains a chapter on building in aluminum by Thomas Colvin); *Small Steel Craft,* by Ian Nicolson (Adlard Coles Limited, 1971); *Own a Steel Boat,* by Mike Pratt (International Marine Publishing Company, 1979); and *Metal Corrosion in Boats,* by Nigel Warren (International Marine Publishing Company, 1980).

Wooden boats are beautiful and nostalgic — and flammable and labor-intensively expensive. Since the time of the first experimental steam launches, 200 years ago, wood has been the most common steam launch building material. Some genres of steam launches still demand wood — for its naturalness and warmth, or because the association of mahogany, brass, teak, and oily steel is so time-hallowed now that it should not be deranged. Lithe

Jim Webster's 30-foot Cloud Nine, *shown here at the 1980 Kingston, Ontario, steam launch regatta, is a good example of the use of developed surfaces and sheet materials. Webster built the aluminum hull himself. The boat is now powered by a Roberts-type WT boiler and a Navy Type B compound engine* (*see page 244*).

river launches with gleaming brasswork and ladies lounging on the cushions in their summer organdy are unthinkable in anything but natural wood — but for heaven's sake, put enough asbestos between the firebox and the teak and organdy.

Most private steam launch builders choose wood because it is the easiest and most pleasant material for them to manipulate and live with while the building of "the steam launch" dominates their lives for a few months or years. Good steam launch designs for construction in wood should be sought in the boats of 80 or 100 years ago, before gasoline profoundly altered boat design (and, more importantly, altered how naval architects and laymen *perceive* powerboat design). There are many good 20th-century books on wooden-boat materials and construction, but what these books have to say about hull lines, scantlings, or mechanical installations should be regarded with great skepticism — they are not about steam launches.

ACCOMMODATIONS

The accommodations in a steam launch that will cruise at 6 knots should be very different from those of a runabout or motor cruiser capable of 15 or 25 knots. Passengers in a slow boat require more flexibility in sitting and moving around, more vantage points or things to do to while away long spans of time devoid of the excitement and tension of high-speed travel. The automobile-like seating desirable for banging across the wave tops at 15 knots seems static at 6 knots on smooth water.

The extensive top-hamper on some of the old cabin boats was not top-heavy. It was, however, light to the point of flimsiness, and dangerously flammable. Building superstructures of the old-fashioned spaciousness to modern standards of strength and fire resistance can lead to a dangerously unstable boat. Think aluminum, thin fireproof paneling, and light foam insulation.

STEAM LAUNCH AESTHETICS

It is risky to say what a steam launch should look like. There is widespread agreement on what makes a good-looking American locomotive of 1890, a superior classic car of 1930, or an appealing country house of 1750. Steam boats are a subject less easily grasped. Their range of uses and length of history are greater than those of cars or locomotives, the functional bases of their design less widely understood than those of a house.

The most important aesthetic considerations for steam launch design are that the hull and propeller should conform to the natural laws of hydrodynamics and that the power plant should exploit and exhibit the best qualities of steam. Other design decisions are superficial by comparison.

A modern pleasure boat must be saleable to people who may be more accustomed to shopping for automobiles, refrigerators, or stock powered playthings. Some of the inherent grace of traditional steam launches comes from the fact that they matured before mass-marketing techniques greatly altered the goals of product design.

I was shocked when I first encountered the virulently anti-democratic bias of some naval architects — primarily yacht designers — of a generation or two ago. I am more understanding now, having witnessed a continuing erosion of taste in boat design, as pleasure boats have increasingly become mass-market commodities that must appeal to very large numbers of people — symbols of success, danger, or sex — or recreational tools that are only secondarily boats. Individuality, for better or for worse, remains the norm in steam launches, which don't sell in very large quantities.

Steam launches originated when warm colors predominated in popular taste and machines were designed to reflect masculine artistic values. There are no good reasons to revise these emphases now. The ground color of most classic American steam launches was white lead; the hulls of many British launches were black, relieved by bright teak and mahogany from overseas. Good colors for steamer surfaces or detailing include all shades of earth reds and yellows, live or dead browns, magenta, and olive and bottle green. Other acceptable colors are most buffs, ivory, nile green, dove gray — and sky

It would be hard to design a working steamer more plain, honest, and functional than this Skagit River logger. Timber-country utility boats were often paddle-propelled, to make effective use of whatever sawmill engines were at hand and to work log tows in shallow water. (Courtesy Everett Arnes)

Lady Elizabeth, *used by George and Ruth Pattinson's children on Lake Windermere, is as perfect of its kind as the larger steam launches in the Pattinson fleet (now on display at the Windermere Steamboat Museum). A plain white funnel is a convention of many British pleasure steam launches. (Courtesy George H. Pattinson)*

blue, royal blue, carmine, and gold leaf used sparingly.

The main colors to avoid in a steam launch are the garish reds and greens favored by collectors of old farm machinery and the cool pastels that came in with interior decorators and Madison Avenue influence. The cold, harsh grays engraved on the memories of ex-navy men by several modern wars have no place in a pleasure steamer. "Navy gray" belongs with high-pressure turbines, diesels, and electrified ships. (Some civilian grays are warm as toast and suit a steam launch as well as buff or brown.)

Steam vessels have been built in all sizes, for 200 years. Much of the accumulated equipment installed in modern steam launches was designed for other uses, and many of the components that an owner wants to incorporate in his new steamer are a little — or a lot — too large to suit. The clumsy appearance of most modern cabin launches, compared to boats of a century ago, often results from faults of scale and proportion. A smokestack or wheelhouse suitable for a 60-footer may be dumped on a 30-footer; the steering wheel, whistle, or name board may be greatly out of scale. The first essential is to prepare drawings of a proposed boat, to see how things go together.

On the other hand, copying large-steamer elements, reduced in direct proportion to fit a much smaller boat, is not recommended either. A miniature, model-like quality results.

Variables such as speed, power, resistance, stress, and strength are more or less precisely quantifiable, but a boat's beauty is finally defined by personal taste and prejudice, modified by history. It depends first of all on the fitness of the design for a certain service, or level of powering, or socio-historical context, or owner's viewpoint. Much of a steamer's natural beauty derives from its speed and powering, which require the most beautiful of hull forms.

FAST STEAM LAUNCHES

A speed/length ratio of 1.34 is of considerable historical and technical — and aesthetic — interest. This is the speed at which the distance between the crests of the waves generated by the boat's passage is equal to the waterline length of the hull. Above this speed, power requirements become enormous, and shaping the hull to achieve dynamic lift becomes increasingly necessary.

Historically, a speed/length ratio of 1.34 corresponds to the highest recorded speeds of sailing ships under a towering cloud of canvas, and long seemed a natural limit to speed through water. A speed approaching 1.34 also leads to the sweet poetic flow line that clings to and mirrors the curve of a hull and is much admired in keel sailboats and tugboats.

It can be argued that steam launches have no business fooling with speed/length ratios above 1.34 (8 knots on 36 feet) and that higher speeds should be left to boats with lighter and more efficient power plants. I am sure, though, that some steam launch fanciers will be tempted to plan faster vessels, and they need to have a realistic understanding of the power requirements for speeds above "hull speed." The torpedo-boat destroyers of World War I offer a good model for people planning a very fast steamer. The common 310' x 30.5', 1,050-ton destroyers (the four-stackers) attained 30 to 31 knots with 16,100 shaft horsepower — with 15.3 s.h.p. per displacement ton, the "tin cans" reached a speed/length ratio of 1.75. (They could make 8 knots with 150 horsepower — less than one one-hundredth of the powering needed for 31 knots.)

The U.S. Navy 40-foot steam cutters of 1900, with a 37.4-foot waterline length, made a claimed 9 knots with 48.87 indicated horsepower on 10 tons light displacement — about five horsepower per ton gave them a speed/length ratio of 1.45.

Seventy-five to ninety years ago, some steam launches attained, at great cost, speed/length ratios around 4.0. *Norwood,* 63' x 7.25', made 30.5 m.p.h., while *Dixie II,* 40' x 5', reached 36.6 m.p.h. Long, light, and powerful is all it takes, and I hope that someone with idle time and money (mainly the latter) will build one more toothpick steam launch for our modern delectation.

5

The Steam Launch Power Plant

The half-billion or so internal-combustion engines that provide most of the world's portable power are all substantially identical. Contrary advertising notwithstanding, the rotary engine in an imported sedan, the big V-8 in your old Chevy, and the snarling unit in the middle of a Porsche have the same processes going on internally, at approximately the same pressures, temperatures, r.p.m., and fuel efficiency. The diesel in a truck or bus is only slightly different from these.

By comparison, steam power is notable for the very wide range of conditions and mechanisms through which it can work. During the past 200 years steamboats have operated successfully with steam pressures ranging from zero to 2,000 pounds per square inch, with steam temperatures between 150 and 1,500 degrees Fahrenheit, and with engine speeds of 10 to tens of thousands of revolutions per minute. Commercial steam plants have earned a living with fuels as various as cow dung, green sagebrush, and fissionable uranium. Fuel consumption in different steam launches may range from one pound of coal (or equivalent) to 10 or more pounds of coal per horsepower-hour.

When a driver turns the key in his car, he rarely thinks of the energy content of his fuel, pressure differences in manifold and carburetor, the timing of the spark, or the heat lost in the exhaust. The events taking place in the coil, the combustion chamber, and the automatic transmission are remote, mysterious, and discouragingly complex.

The operator of a steam launch, on the other hand, is on familiar terms with all of the thermodynamic and mechanical transactions in his power plant. Most of them occur in slow motion, where he can observe them. He has a clear idea of the weight of firewood he will have to carry on board for an hour's run, the intensity of fire he needs for full-speed operation, and the way energy is transferred from the firebox to the boiler water, the

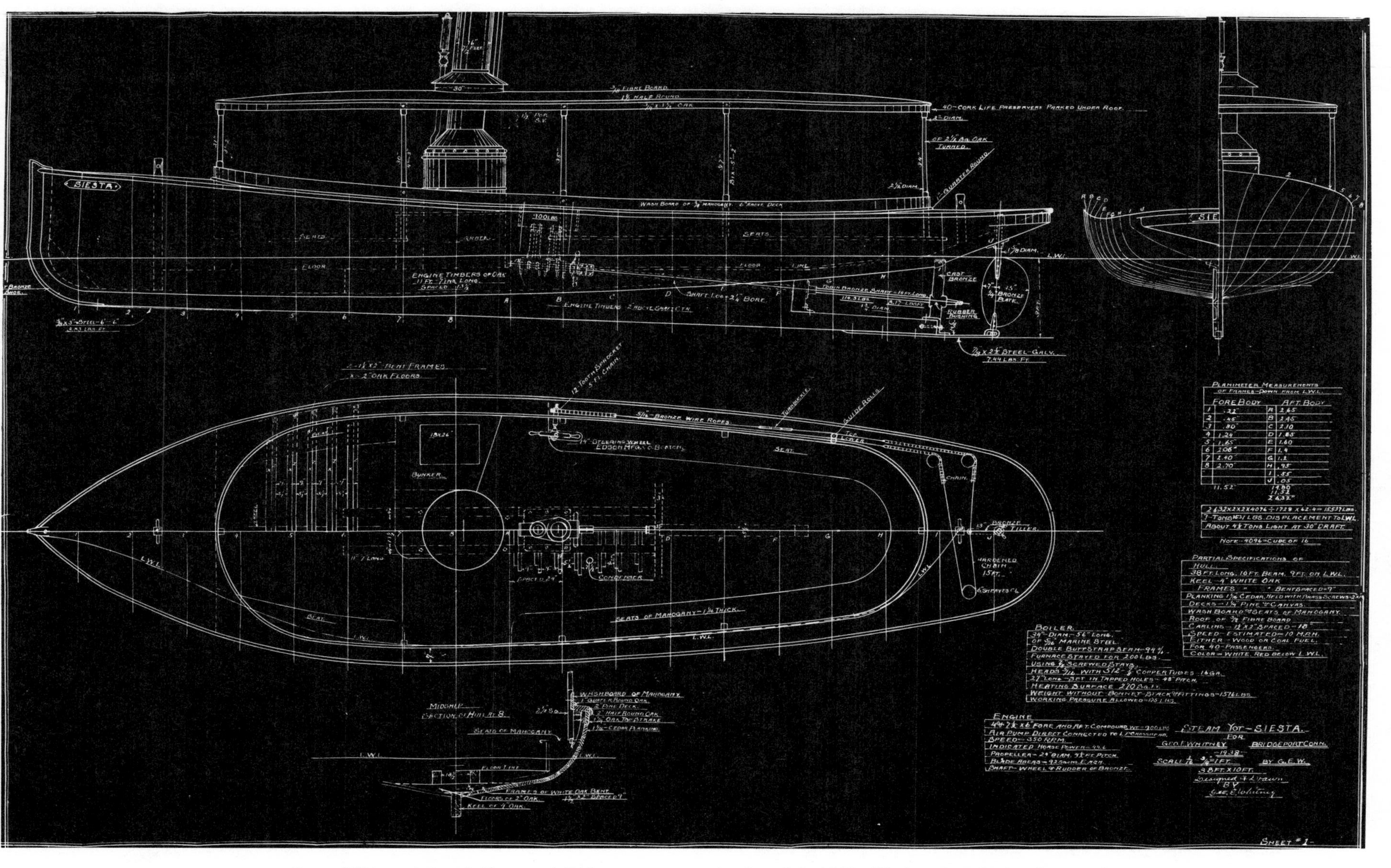

George Whitney designed this powerful 40-passenger excursion launch in 1938. The legend on the drawing reveals a sure professional hand, with planimeter measurements of hull sections at 18 stations and a boiler rather closely calculated for strength. Siesta's *design has a few singularities that set it a little apart. For example, the 24'' x 54'' propeller is extreme enough to thrill any admirer of ''steam wheels,'' but it is evident that Whitney achieved shallow draft and proper engine speed with this wheel at the cost of high propeller slip. The 4'' & 7.5'' x 6'' compound engine weighs 900 pounds. The VFT boiler has a 5/16'' steel shell and a working pressure of 175 p.s.i. At 350 r.p.m., the plant is designed to develop 44.6 indicated horsepower, almost 10 i.h.p. per ton light displacement. (George Whitney collection)*

engine, and the shaft and propeller. His power plant consists of simple forms of common metals — usually cast iron, mild steel, and bearing bronze. There are no hardened parts, precision-cut gears, electrical devices, carburetors, "computers," or plastic.

Late-19th-century practice in marine steam engineering embraced pressures of 60 p.s.i. to 200 p.s.i., one cylinder to three, and engine speeds of 100 r.p.m. on up, depending on the size and speed of the boat. Steam launch men see no need for radical change, and they avoid debate with steam car fans or power engineers, who live in another part of the woods.

The working components of a steam launch power plant resolve themselves into easily recognizable functions. The fire, the boiler, and the engine constitute the heart of the mechanism. Heat is liberated by combustion processes in the firebox and is transferred through the boiler heating surfaces to water, creating steam. The heat in the steam is converted to mechanical energy by the engine. Other power plant components supply the boiler with water and regulate the rates of steam creation and consumption.

A steam engine is of the piston, connecting rod, and crankshaft variety (reciprocating, or "up-and-down") that is familiar to nearly everyone on earth by now. If it is a double-acting steam engine, as most are, it has this structural difference from a lawn-mower engine: The connecting rod is coupled to a crosshead, not directly to the piston, and the crosshead is then linked to the piston by a piston rod, which passes through a packing gland in the lower cylinder head. The piston rod and crosshead are

A certain cockiness and pride went with the launch engineer's ability to make 100 horsepower do his bidding. Although Cyrene *had been on the Seattle-Mercer Island run nearly 20 years when this picture was taken, her compound engine shows not a drip or stain. Students of period detail should examine* Cyrene's *patterned engine-room floor. (Courtesy Captain Robert E. Matson)*

Left: *Cochran's, of Birkenhead, England, sent this workboat engine and boiler to the 1882 North-East Coast exhibition of marine machinery. Of a size for canal boats or powered lighters, the plant is clearly noncondensing, with no provision for exhausting the spent steam anywhere but up the funnel. A smokebox door left off the boiler reveals the arrangement of cross fire-tubes. (From* The Engineer, *September 22, 1882)*

Below: *The little Canadian yacht* Niska *was conventionally sleek outside but decidedly homey in her engine room. The 8" & 12" x 14" engine was built by the Doty Engine Works, Goderich, Ontario, in 1900. The boat served on Lake of Bays until Cameron Peck added her to his collection at Muskoka Lakes. (Photo by Edward O. Clark)*

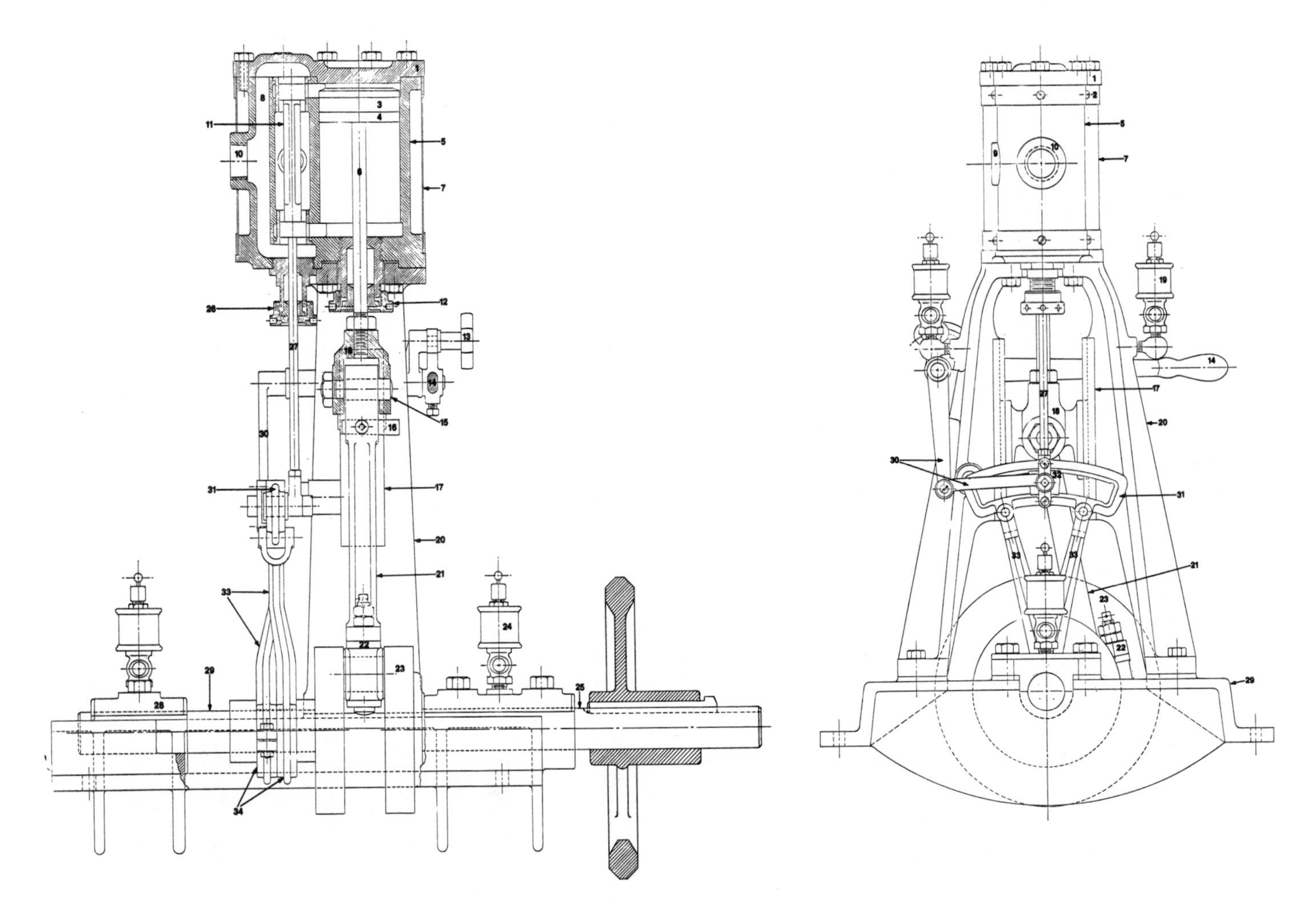

1 Cylinder head
2 Brass trim band
3 Piston ring
4 Piston
5 Cylinder
6 Piston rod
7 Cylinder jacket
8 Steam chest
9 Steam inlet
10 Exhaust steam outlet
11 Piston valve
12 Piston rod packing gland
13 Hand screw for reversing lever
14 Reversing lever
15 Wrist pin
16 Crosshead bearing key
17 Crosshead guide
18 Crosshead
19 Oil cup for crosshead guide
20 Frame
21 Connecting rod
22 Connecting rod bearing
23 Crank disc
24 Oil cup for crankshaft bearing
25 Crankshaft
26 Valve rod packing gland
27 Valve rod
28 Crankshaft bearing cap
29 Base
30 Connecting rod arm (between reversing lever and link)
31 Link (for throwing eccentrics in or out of gear)
32 Link block (travels freely in link)
33 Eccentric rods
34 Eccentrics

The Shipman Engine Company, of Boston and Rochester, was a major steam launch engine manufacturer in the late 19th century, and this 2-h.p. simple engine was one of the company's most popular offerings. A few of these engines, including the one (page 276) that powered the author's River Queen*, are still around. Arthur Knudsen, a steam-engine archivist in Alberta, made these fine drawings from catalogs, photographs, books, and measurements off existing engines. (Courtesy Arthur Knudsen)*

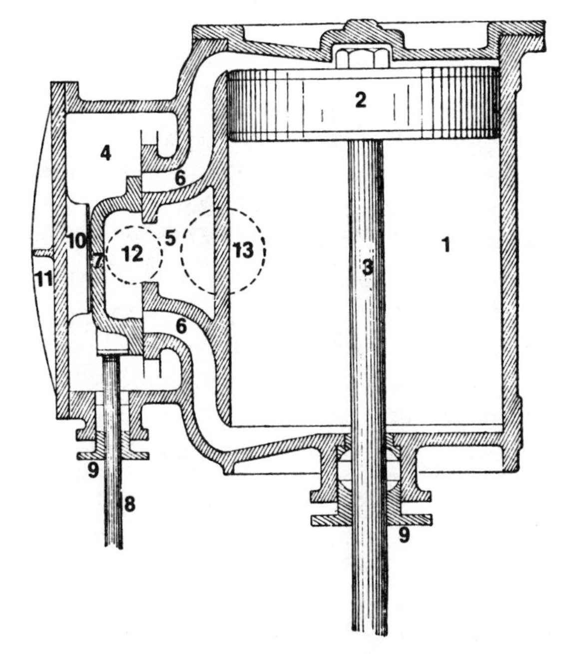

1. Cylinder
2. Piston
3. Piston rod
4. Valve chest
5. Exhaust port
6. Steam ports
7. Slide valve
8. Valve stem
9. Glands
10. Guide to relieve valve of pressure
11. Ribs on steam chest cover
12. Position of steam pipes
13. Position of exhaust pipe

In this section of a cylinder and valve chest, the piston is just at the end of its upward stroke, but due to the angular advance of the eccentric, the D-valve is already traveling downward. Live steam is beginning to enter through the upper steam port, and spent steam is exhausting below. When it travels upward again, the valve will cut off steam admission before terminating communication between the lower steam port and the exhaust pipe. Steam's expansion will then bring the piston's downward stroke to completion. (From C.P. Kunhardt, Steam Yachts and Launches, *1891)*

held in rigid alignment with the cylinder by crosshead guides, while the connecting rod pivots to follow the crank. This arrangement allows each side of the piston to receive a push from the working fluid (steam) during one revolution of the crankshaft (the "double-acting" principle). A single-cylinder, double-acting steam launch engine produces the same power per revolution as a four-cylinder, four-stroke gasoline engine of the same dimensions and cylinder pressure.

A steam engine is different from a gas engine principally in its torque characteristics. Everyone is familiar with the image (from the movies, probably) of a steam locomotive starting off at very low throttle — a heavy touch on the throttle would cause the wheels to spin. We all know that when starting off in a car we have to rev the motor to keep it from stalling. A gas engine has to be spinning to breathe in fuel and air; since a steam engine's working vapor is already charged with heat and pressure when it enters the cylinder, it will push the piston with its full force when the engine is turning at very low speed or not at all.

A steam engine cylinder is provided with an adjoining chest that receives "live" steam from the boiler at a rate determined by the throttle setting. The valve that governs the admission of steam from the steam chest to the top and bottom of the cylinder (i.e., to alternate sides of the piston) may be flat, sliding up and down across the ports in the side of the chest (slide- or D-valve), or a little piston covering and uncovering ports in a cylindrical chamber (piston valve). Valve motion is usually timed by an eccentric on the crankshaft; the eccentric corresponds to a gas engine's camshaft. Steam is not admitted into the engine during the entire stroke of the piston, since most of the heat energy would then be lost in the exhaust. Rather, the steam is "cut off," perhaps halfway through the piston stroke, then expands to do continuing work during the remainder of the stroke and exhausts at low pressure and temperature. In a condensing engine the exhaust steam is condensed, and the water eventually reused. A noncondensing plant exhausts to the atmosphere.

FUEL AND FIRE

Anything that will burn can fuel a steam launch. The few limits are imposed by such things as the total inefficiency of firing with water-soaked driftwood (most of the heat goes to evaporating the water in the fuel, and the moisture-heavy smoke will scarcely rise in the stack) or the complications created by choosing such fuels as straw or heavy crude oil.

Fuels of all kinds contain known quantities of potential heat. Petroleum fuels have around 19,000 B.t.u.'s per pound; good coal, 13,000; dry wood, 6,000; and uranium, about 25,000,000,000. A cord of heavy wood may equal the heat value of a ton of coal or 200 gallons of oil. The heat available from petrochemical wastes — old tires, scrap plastic, crankcase drainings — is near the high heat value of petroleum itself. The heat potential in all kinds of dry biomass is comparable with wood, but waste wood is so abundant and inexpensive that steam launch operators rarely exploit other renewable fuels.

To fire a boiler successfully, large quantities of air must be intermingled with the fuel's hot surfaces and volatile gases. This need sets a lower limit to the size of the firebox, dictates whether grates are required, and determines the amount of draft necessary. Adequate firing capacity (fuel supply, combustion space, and draft) is often the culprit when amateur designs are unsuccessful. (A fire that releases 40,000 B.t.u.'s per horsepower-hour is about right for a good, average engine.)

Some of the fire's heat, 15 to 30 percent, typically, will be lost up the stack. The height and temperature of the column of heated gases in the stack will determine the amount of "natural draft." Steam launches never have too much natural draft; they rarely have enough, since their stacks are no taller than the chimney in a one-story house. Launches with relatively tall stacks and ample firebox and boiler capacity may cruise on natural draft alone, requiring "induced" or "forced" draft only when starting up, racing, or steaming on poor fuel. Other launches, though well designed, may need some artificial draft most of the time.

Artificial draft is frequently induced — when needed — by directing jets of live steam (as opposed to exhaust steam) up the stack — a method especially suited to a solid-fueled boat that can normally cruise on natural draft alone. A small, battery-powered blower may be worth having for firing up from cold a solid-fueled boat with a short stack.

The fierce and noisy draft characteristic of a steam locomotive never occurs in a steam launch. A draft of one-half inch of water (as measured by a U-tube manometer) and the burning of 15 pounds of coal per square foot of grate per hour represent a firing rate high enough for a steam launch.

Wood is the easiest fuel to use, needing little more than a spacious firebox and a controllable air supply, but most woodburners would benefit from a bit more care and forethought. A few minutes spent cutting wood to the right size for the desired rate of combustion, keeping it dry, and mixing poor fuel with good can save hours of frustration with a slow fire and a slow boat. Hardwoods are clean burning and hold a steady fire.

Successful firing with coal is one of the more prosaic arts lost in this generation. A 12-year-old managing the household furnace in 1920 knew more about hand-firing small boilers with coal than the average mechanical engineer knows today. There are substantial differences among coals from different sources, and a preferred grate design and firing technique for each. Soft coal needs a lighter fuel bed, a more even thickness, and more frequent stok-

Two hundred pounds of "anthracite beans" will steam A.E. Moulton's Alice *200 miles on the River Avon, in Wiltshire. Her 2-h.p. Blackstaffe-Wood plant, built in 1969, is extremely light weight and finely tuned. The compound engine is 1.5" & 3" x 2.5", and the water-tube boiler has 10 square feet of heating surface and an effective grate area of 100 square inches.* Alice *can be seen in her entirety on page 133. (Courtesy A.E. Moulton)*

ing than hard coal. The Babcock and Wilcox volume on *Steam* (available in many editions, most libraries) presents this information in a technical manner. *Fuels and Fuel Burners,* by Kalman Steiner (1946), is a gem of a book, describing state-of-the-art coal heating just before cheap oil put coalburning furnaces out of business.

The burning of coal necessitates iron grates in the firebox, strong forced or induced draft (most of the draft is required to push air through the fire bed), and the ability to handle large volumes of coal and ashes in a neat, efficient way. Coal is apt to mean a dirty boat and a dirty boiler, but it's a more concentrated fuel than wood, and coal smoke is the most significant and evocative odor of the past 200 years — the very essence of industrial wealth.

Petroleum fuels have become popular in post-1950 launches because of their concentrated energy, convenience, and availability. There are numerous ways to burn oil efficiently. The gun-type burner from a home furnace can make steam for a three-horsepower boat, and with larger nozzles and more draft, this type of burner can be adapted to much higher powers. For those who prefer burners not requiring electric power, steam- or pressure-atomizing equipment is a good alternative, suitable for any size launch and especially adaptable to the burning of crankcase drainings and other waste oil. Vaporizing burners are effective with clean, light oil.

BOILERS

The boiler is the simplest power plant component, in principle no more than a vessel to contain pressure and allow part of the fire's heat to be transferred to water. Heat is transferred across the metal surfaces intervening between combustion gases and the boiler water, and there must be ample surface area for this transfer to be efficient. A boiler's duties lead to cylindrical and tubular parts, which increase surface areas and are self-supporting against pressure.

Heat transfer is easier through copper than steel, and much easier through clean surfaces than through soot or scale. Copper heating surfaces are especially satisfying for men who are sensitive to the gradations of temperature and pressure within their power plants, for heat fairly leaps through the copper to the boiler water. Copper heating surfaces are chosen if they are not excluded by considerations of strength, temperature, manufacturing, or cost, as they usually are. Copper boiler parts are considered to be in the soft, annealed state, and they are allowed 30,000 pounds tensile strength — less than one-half the strength of steel. Most hobbyists limit copper boilers to 100 p.s.i. — 150 at most.

Boiler efficiency is attained more through design than by choice of materials. Most boilers are fairly efficient (thermal losses are much greater in the engine), but they are bulky, heavy, and expensive, so boat boilers have usually been chosen for their low price or ease of manufacture. The steam launch revival in America has depended on boilers that are poorly suited to the service but were more easily available and less expensive than others.

Boilers are categorized as fire-tube (commonly referred to as "FT"), water-tube ("WT"), or flash. In the first, hot gases pass through tubes surrounded by the boiler water, while in WT boilers the water circulates within tubes, and the gases flow around them. Flash boilers are of the water-tube type, but in these the water is pumped into the hot tubes only as steam is required by the engine — there is no residual water in the boiler. Flash boilers have some advantages in steam cars, but they are not used in boats.

Fire-tube boilers are crude and heavy tanks containing a large volume of water relative to their duty. These qualities are not necessarily undesirable in a steam launch — the large water reservoir makes a fire-tube boiler a steady steamer, relatively immune to wild and sudden oscillations. If the feed pump check hangs up or the fire goes out, the operator has time to get things squared away.

Vertical fire-tube boilers (VFTs) are the commonest type in service, being most easily obtained. These are sturdy, and the vertical flow of gases permits the use of small tubes with good heating surface, but a VFT can make a boat top-heavy, and its firebox volume is small.

Horizontal FT boilers have a low center of gravity well suited to boats, and some types have good firebox volume. Steeped in tradition, they were the common steamship boilers of 50 to 120 years ago and were often installed in steam launches. Horizontal marine boilers are "direct" or "return-tube."

In most modern VFT boilers, the tubes (or flues) run up into the head of the shell, but a century ago, many donkey boilers had tubes completely submerged, as required by U.S. law governing marine uses. Charles P. Willard and Company stocked several sizes of submerged-tube vertical boilers and would custom-build any size wanted.

SUBMERGED TUBE VERTICAL MARINE BOILERS.— Where boats are required for use in waters that are under United States Marine supervision it will be necessary to have boilers that conform in every respect with the United States Marine laws. These require, among other things, that Vertical Boilers used for marine purposes should have submerged tubes. The fact that a boiler has passed Government inspection is the strongest guarantee of the quality of the material and workmanship, as the requirements are extremely rigid.

We furnish these Boilers, with all the fixtures complete, consisting of safety valve, check and blow-off valves, steam and water-gauges, and three compression gauge cocks, at the following prices:

Horse Power, Marine Rating.	SHELL.		FLUES.			FIRE BOX.			Depth Cone.	Top Cone.	Square feet Heat'g Surf'ce	Weight.	Price.
	Diameter.	Height over Base Plate.	No.	Diameter.	Length	Height.	Diameter.	Height of AshPit.					
2	22	36	33	1½	12	15	19	3	6	8	20	330	$150
4	26	48	43	2	18	18	22	4	8	14	43	580	200
6	30	48	68	2	16	18	26	4	10	16	60	1040	225
9	36	50	87	2	18	18	32	5	12	16	92	1190	250
10	40	53	99	2	18	18	36	6	14	16	100	1530	300

This Kingdon boiler was sectioned after six years' service by its builder, Simpson, Strickland and Company, to show its condition. Patented in 1881, the Kingdon VFT was recommended in the company's promotional literature because it had a larger grate area with more compact dimensions than a horizontal boiler, but the company did not recommend it "for hard driving, [or] for use with inferior qualities of coal." The squat design enlarged grate area while keeping the center of gravity low. A "water leg" (lower left), interrupted only by the firebox door (lower right), surrounded the fire, increasing heating surface and obviating the need for insulating firebrick. Other models had submerged tubes. In a modern boiler, the rivets are likely to be replaced with welds. (Crown Copyright. Science Museum, London)

The former was often called a "gunboat" boiler; the latter is the "Scotch" boiler. Both are extremely heavy, bulky, and expensive to build for the power supplied and are rarely seen in steam launches today.

There are innumerable varieties of water-tube boilers — amateurs can invent their own, and often do. They may be as compact as a VFT and are normally lighter. They would be expensive if ordered from a boiler works, but much of their fabrication can be done by an amateur in his home shop. Each WT-type has its benefits and drawbacks, its ardent proponents and its detractors. Some basic forms are straight-tube, bent-tube, pipe (threaded joints), porcupine, coil, header . . . . The illustrations in Part Two show these best. Most are designed with one or more "steam drums" on top, one or more "water drums" below, and connecting tubes. (Alternatively, a central standpipe and radiating tubes, as in the porcupine type, are used.)

Water-tube boiler designers pay close attention to the temperature in various parts of the boiler and to circulation through the tubes. A froth of water and steam bubbles should discharge into the steam drum; there the gas and liquid phases are separated and steam is drawn off to the engine. In most WT boilers, cool "downcomers" ensure the positive circulation of solid water from the steam drum to the tubes next to the fire. Feedwater is pumped into the boiler at the location where it will cause the least thermal shock and contribute to the desired pattern of circulation.

The Roberts, or Almy, type of water-tube boiler was emblematic of American skills and values in the late 19th century. The firebox is spacious, ideal for wood, the American fuel. The boiler's parts are individually small and capable of being hauled to remote locations by packhorse or canoe. Manufacture requires only a single artisan with hand tools — hacksaw, pipe-threading dies, and pipe wrenches, principally. The Chinese-puzzle quality of the generating-tube nest was much admired 90 years ago. ". . . You see, the tubes are threaded left-hand/right-hand. You pull the boiler together with the wrench." The Steamboat Inspection Service was less enchanted by threaded-joint boilers and prohibited them in some services. (Courtesy Everett Arnes)

The fire's heat boils water and produces steam, an admirable fluid that contains a great deal of energy, is expansible — "elastic," they used to say — and is no more polluting or corrosive than water in its other forms. The output of the boiler is measured in pounds of steam per hour. (In testing a boiler's performance the steam is weighed as feedwater entering the boiler or as condensed water after use — it's hard to hold a pound of live steam on the scales.)

Forty pounds' weight of steam fed to a good, average steam launch engine should yield one horsepower for one hour. Three or four pounds of fuel oil (a half gallon), five or six pounds of coal, or 10 to 15 pounds of dry wood (a small armload) should evaporate 40 pounds of water. Think of boiling a five-gallon kettle or 10 two-quart saucepans of water to steam in an hour, on an open fire, and you will begin to sense the quantities of fuel and water employed in a one-horsepower steam launch.

The water that leaves the boiler as steam must be replaced by "boiler feedwater." (In a condensing boat, this is steam that has done its work in the engine, then returned to the liquid state.) The boiler feed pump must be rugged and utterly reliable. It has to overcome boiler pressures two to five times higher than household water pressure, and if it fails, the boat will soon be dead in the water. Preferred feed pumps are positive-displacement ram types, either a commercial "Hy-Pro" design or one homemade from lengths of bronze rod and pipe and two check (nonreturn) valves. The main feed pump should be driven off the engine, but there should be a hand-feed pump for emergency and other uses.

ENGINES

Stock two-horsepower launch engines of 1881 had the same cylinder dimensions as advertised five-horsepower engines of 1981, a difference derived from three sources. In an 1881 launch, steam was often cut off rather early in the piston stroke, and this resulted in a slightly higher fuel efficiency but considerably less power from an engine of a given size. Steam engines had no competition then, so engine design was based on good steam principles, with no attempt made to be as lightweight or compact as gas engines.

In 1881, a 60- to 75-pound boiler pressure was usual in small launches. Today, higher boiler pressures are favored.

The launch engines of 100 years ago turned 250 to 400 r.p.m. — 500 to 600 is popular today. The power of an engine is proportional to its average cylinder pressure and to the number of power strokes per minute.

While a steam launch power plant is rarely more than one-fourth as efficient as a gasoline engine, even this level of attainment requires good design and practice, but the finer degrees of engine efficiency need not concern steam launch owners. What a steam launch operator typically desires from his engine is simplicity, ruggedness, and reliability. The vertical-inverted (cylinders above the crankshaft), direct-acting (connecting rods joined directly to the

FOR FINE PLEASURE YACHTS

This is the most

Durable, Simple and Economical

Engine in the Market. It is Well Made,

HANDSOMELY FINISHED AND VERY LIGHT.

TABLE OF DIMENSIONS AND PRICE LIST.

HORSE POWER.	CYLINDER.		REVOLUTIONS PER MINUTE.	PRICE.	
	DIAMETER.	STROKE.		ENGINE AND PUMP.	WITH SNYDER'S PAT. LITTLE GIANT BOILER.
2	3	5	300	$150	$250
4	4	6	250	200	350
6	5	7	210	275	450
8	6	8	180	325	550

Terms same as for Snyder's Little Giant Engines.

The Boiler, mentioned in the above table, is exactly suited for pleasure yachts, a full description of which may be found in Snyder's Little Giant Engine Circular.

Steam Pleasure Yachts.

PRICE, COMPLETE, WITHOUT MACHINERY.

Full Length 20 feet,	$350
" 25 "	450
" 30 "	550
" 35 "	650
" 40 "	800

CABINS, UPHOLSTERY, ETC., EXTRA.

The above Yachts are of a handsome model, well made, light and graceful, and are built to run at a speed from 6 to 20 miles an hour, as required.

ADDRESS,

WARD B. SNYDER,

84 Fulton Street, NEW YORK.

The Semple Engine Company offers this 3" x 4", 5-h.p. launch engine with 20th-century refinements. A century ago, engines of this size delivered two horsepower. (Compare the sizes, power ratings, and speeds of the Ward B. Snyder engines in the accompanying 19th-century advertisement, courtesy John S. Clement.) The Semple engine employs a single "slip" eccentric. Reversing is accomplished by stopping the engine and, with a lever or wheel, rotating the eccentric (relative to the crankshaft) to its alternate position. The "wheel reverse" engine can be seen on page 254. (Courtesy Semple Engine Company)

This old friend of the author bespeaks ruggedness and integrity. When private steam launches were at a low ebb, Bob Thompson swapped a day's labor, worth four dollars then, for the 3'' x 4'' Shipman engine, rated at 2 h.p. The author installed it in River Queen *in 1957, and it suited both the boat and the owner for many years.*

Above: *Messrs. W. Sisson and Company, Ltd., built this non-condensing compound launch engine, 5.5'' & 8'' x 5.5'', in 1907. The engine, which could develop 30 h.p. at 150 p.s.i. and 480 r.p.m., has an unusual valve gear necessitating inclined valve rods. The scale in the foreground is 24 inches long. (Crown Copyright. Science Museum, London)* Below: *Triple-expansion engines carry the principle of compounding one step further, and triples, too, have been built in launch size. Many launch engines have been cut away to serve as teaching aids in maritime training schools; the principles of steam flow and valve motion in a multiple-expansion engine can be learned with a few minutes' study of one of these. This small triple has the high-pressure cylinder on the center crank throw (it is usually at the forward end of the engine). (Crown Copyright. Science Museum, London)*

crankshaft), single-expansion (''simple''), double-acting, reversible steam launch engine has seen few changes in over a century because it combines those qualities so well.

Compound engines are theoretically more efficient than simple engines. In a compound engine the steam is expanded in two stages, with the first stage occurring in a smaller, high-pressure cylinder. From the high-pressure cylinder, steam flows at lower pressure (and, consequently, lower temperature) into a larger cylinder, where it undergoes further expansion, performing more work in the process. Dimensions and steam pressures in both high- and low-pressure cylinders are adjusted so that the power output of each is approximately equal. Compound engines are more efficient because the metal surfaces of individual pistons and cylinders experience a narrower range of temperatures through each cycle, and the energy wasted in reheating these metal components at the start of a cycle is thereby reduced. In practice, mechanical complications and added radiating surface may cancel out the theoretical benefits of compounding — but the engines are lovely.

In a condensing engine, steam is exhausted to a vacuum created by the condensation of previously exhausted steam. The vacuum is equivalent to a 10- or 12-pound increase in steam pressure, at no cost. Maintaining the vacuum requires an "air pump" to remove noncondensable vapors from the condenser.

ENGINES AS COLLECTIBLES

People who have been connoisseurs of Bugatti roadsters or Colt's revolvers for years are sometimes surprised to learn that antique steam launch engines are prime collectibles. Why not? Each of the old engines was the invention of a creative mind. Many were custom-built by small shops, and are forever unique. Each part was handled by artisans during manufacture, fitting, and adjusting. The form and finish of many engine components is aesthetically pleasing in itself.

What principally distinguishes the collecting of steam launch engines from other antique-collecting hobbies is that nearly all of the collectors regard their engines as practical working machines, which they will install in a boat "as soon as I find time."

The assembled machine, producing work out of fire, had a magical appeal in its time that can scarcely be imagined by people who have lived with pushbutton electricity and gasoline engines for three generations.

The style and detailing of antique steam launch engines closely mirror the time and place in which they were built. The rough materials and craftsmanship of early engines gradually gave way to polished surfaces and carefully engineered designs, carried out in increasingly reliable materials. Advances in design accelerated steadily from mid-century iron stalwarts to the lean, taut, steel-framed compound engines of the 1890s — then ended.

Many steam launch engines, like this one from the Sumner Iron Works in Washington Territory, were custom-built by small shops and are forever unique. (Courtesy Everett Arnes)

A RIDE ON A STEAM LAUNCH

A steam launch power plant is perhaps best envisioned in operation. One need only select well-matched power plant components and, in his mind's eye, install them in a suitable hull and watch them function. Imagine a steam launch like the one in the accompanying drawing, with a compound engine and a wood-fired VFT boiler, and picture the procedure you might follow to prepare for an afternoon's run.

Since your boiler is fitted with a water glass and three vertically separated try-cocks, you can readily ascertain the boiler water level. If this is normal, and the ash build-up in the firebox is not excessive (too much ash could either cause the grate to warp or choke the draft, which enters the firebox around the ashpit door), you are ready to kindle the fire.

Small sticks of hardwood, such as the trimmings from a furniture factory, suit the normally modest firebox dimensions of a VFT. One bushel might be required to raise steam, a second to cruise five miles or so. You might allow 15 or 20 minutes for the

Steam Launch Power Plant

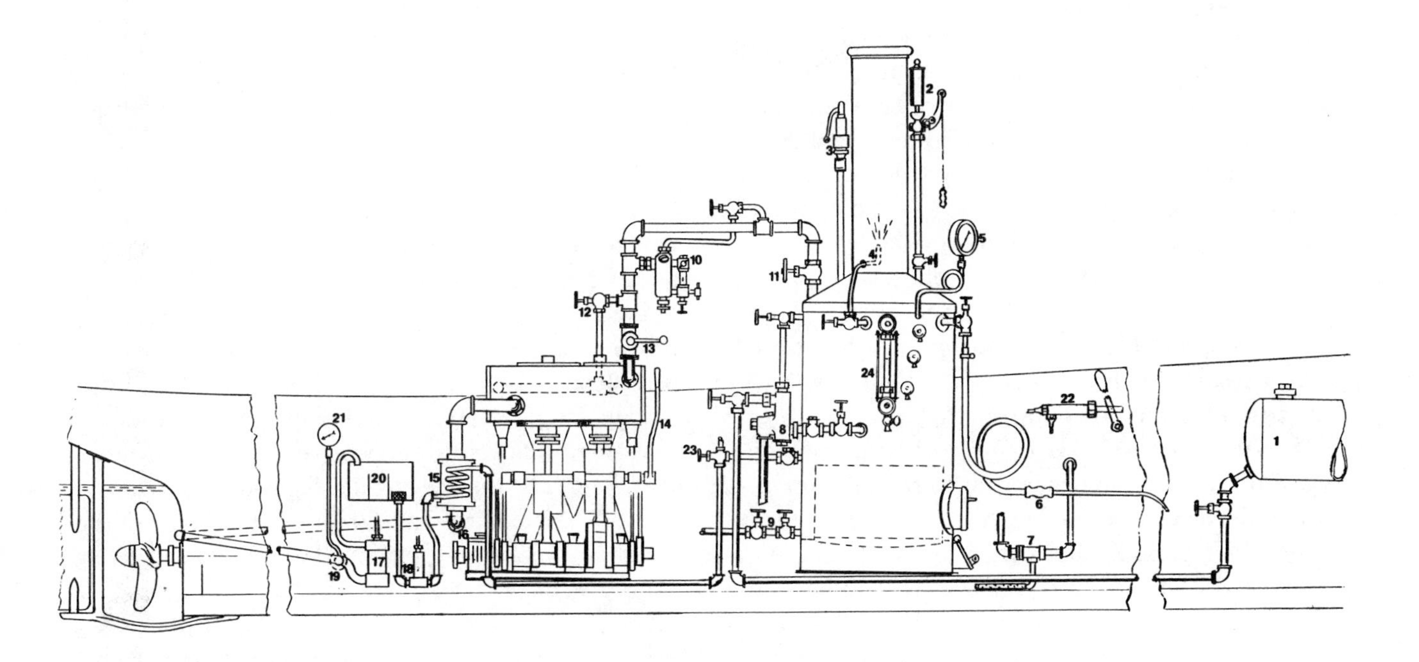

Man has always found visual pleasure and emotional security in systems arrayed in an orderly, purposeful way. Farmers have their straight rows and stored bales, carpenters their symmetrical patterns of studs, headers, and trusses. Steam launches are similarly disciplined and orderly, and for the most part they are expressed in pipes and pressures, temperatures and controlling valves.

Much of the appeal of a steam launch lies in the frank and open expression of its technical nature. The author's concept of a steam launch engine room (drawn by W. U. Shaw) reflects his personal experience and values. Many historical and modern steam launches have gotten along with simpler layouts, but all the components illustrated here are standard equipment, of positive value to the boat. Some steam men would omit the injector or add a steam dynamo and storage batteries for lighting — it's a free country. (Courtesy W. U. Shaw)

1 Boiler feedwater tank
2 Whistle
3 Safety valve
4 Stack blower for induced draft
5 Boiler steam pressure gauge
6 Steam lance for blowing tubes, cleaning, or cooking
7 Bilge siphon
8 Injector (requires stop valve and check valve to boiler)
9 Boiler blowdown (discharges overboard through two valves)
10 Lubricator
11 Main steam valve
12 Bypass valve
13 Throttle
14 Reversing lever
15 Exhaust feedwater heater
16 Thrust bearing
17 Air-and-condensate pump (removes water and noncondensable vapors from condenser)
18 Boiler feed pump (picks up filtered condensate and forces it through feed heater to boiler)
19 Through-hull connection to external "keel" condenser
20 Hotwell
21 Vacuum gauge
22 Auxiliary hand feed pump
23 Feedwater bypass back to hotwell (regulates rate of boiler feed)
24 Boiler water-level gauge glass and three try-cocks

steam to approach working pressure, since the shell and tubes of an FT boiler must be heated slowly to avoid possible stress from uneven heating.

While steam is rising, you can dispense with some small but necessary tasks. For example, when the steam pressure climbs sufficiently (the boiler steam pressure gauge will tell you this), a "blowdown" valve or cock at the base of the boiler may be opened, allowing the boiler to blow down — no more than a few inches, or until clean water appears in the overboard discharge. The valve is then closed, and, if steam pressure is adequate, the boiler water can be replenished with an injector.

If you have decided to lubricate the cylinders you'll fill the lubricator, as well as the oil cups that supply the crosshead guides. A full oil can should be on hand for periodic lubrication of engine bearings.

Once working steam pressure is attained, the cylinder drains are opened, the throttle is opened a bit, and the engine is "rocked," first forward, then in reverse, while its metallic surfaces are heated toward working temperatures. (This process can be speeded up by temporarily opening an engine bypass valve to admit high-pressure steam into the low-pressure cylinder.) If your engine employs a Stephenson-link reversing mechanism, adjusting a link by means of a lever causes either a forward or an astern crankshaft eccentric to be engaged; switching from one to the other changes the timing of valve events and the sense of crankshaft rotation.

When the engine begins to make complete revolutions, the cylinder drains are closed, and the excursion can begin.

While the boat is underway, the boiler pressure will depend on the rates of steam creation and consumption, which in turn depend on the throttle setting, the fuel supply, and the intensity of draft. Natural draft is regulated manually by adjusting the opening of the ashpit door; supplemental draft, such as stack blast, can also be adjusted manually. For steady cruising at fixed throttle without the necessity for constant adjustments, the fuel supply and draft should be manipulated to obtain a steady steam pressure and a clean fire.

While operating at cruising speeds, the engine is likely to be "hooked up" somewhat on the link, which shortens the valve travel and cuts off steam admission partway through the piston stroke. For that occasional impromptu race, you can "link out" the engine to increase its power output (and decrease its efficiency).

Exhaust steam from the low-pressure cylinder of your engine flows through the jacket of a feedwater heater, giving up heat to the feedwater in the coil. Partially condensed, the exhaust is then routed outboard to a tapering pipe, the keel condenser, where condensation is completed. An air pump, run off the engine, pulls the condensate back inboard to a hotwell, and there cylinder oil is removed from the oily condensate. The boiler feed pump draws up clean condensate and forces it through the feedwater heating coil and a bypass valve to the boiler. Only as much feedwater as is needed to balance steam consumption is admitted to the boiler — any excess is returned to the hotwell.

Steam is lost through the whistle, the stack blast, leaky joints . . .; "make-up water," to replace what is lost, comes from a tank or from the "sea" (as long as the "sea" is fresh water).

If the excursion were to include a brief layover, you might wish to stop firing the boiler 15 minutes or so in advance, leaving enough steam to maneuver into a berth, but not enough to lift the safety valve with the engine at rest. After berthing, you might dampen the draft and leave just enough fuel in the firebox to keep the pressure up; to get underway again, you would simply open the draft and stoke the fire.

Part Two

A Steam Launch Album

Commercial Launches

Working machines have a kind of hearty integrity that is lacking in mechanical playthings. A dump truck has a more durable style than a Porsche roadster. Tugboats have a stronger definition than yachts.

Nearly all early steam launches were commercial, since they existed at a time when machines were built only for war or profit. A little later in history, many workaday steam launches were very similar to privately owned recreational boats of the time. In America the same boat was often used both for private pleasure and for money-making enterprises. Pleasure steamers died out early in this century, but hundreds of small commercial steamers, mainly in faraway places, continued to earn money through the 1910-to-1950 hiatus in private steam launch use.

Although commercial steam launches were not numerous in the 19th century, they were relatively more important as passenger carriers than any small boats have been since then. Before 1900, wherever a horse and wagon couldn't go, a steam launch was the only alternative for fast, sure transport. Today, wherever a car can't go, an all-terrain vehicle, helicopter, or floatplane surely can. In America the smallest, prettiest commercial steamers flourished where parties of adventurous people needed to be run from railheads or scheduled steamboat landings to resort hotels and logging or hunting camps far back in the woods.

In England, paying steam launches served large urban populations. The numbers of warm bodies that could fit into a small space determined English passenger-launch designs, and most of these were pretty drab. The British Empire was another matter, with thousands of powerful and distinctive steam launches helping to bring trade and order to the jungle rivers and tidal creeks of three continents.

Empire steam launches not only dominated the backwaters of the world from the 1830s until the ascendancy of diesel power 80 years later, but they continued to pour out of English and Scottish yards long after the internal-combustion-engined boats had taken over European and North American waters. The Empire launches were of all-steel construction, with steam machinery rugged enough to cast doubt on the frailties of early gas boats. They were designed for operation by native labor in remote parts of the world, where a defunct magneto or fuel injector might take a gas boat out of service for months. A steam launch's bent piston rod or cracked boiler could be blacksmithed back into service on the riverbank.

The industrial steam launches of Victorian and Edwardian times were by nature more durable than other small craft built before or since. In fresh water, their hull plates were good for 50 years, and after that, new steel could be riveted in place where needed. Since the engines turned one-tenth as fast as modern diesels, they were expected to steam 35 years — maybe 100,000 hours and a million miles — before requiring refitting. Fuel economy was mediocre from the beginning, so a little loss of efficiency after 15 or 20 years was disregarded.

Since 1950 American airplanes and boats with German or Japanese diesel engines have supplanted most of the British steam launches, but a few durable coalburners out of Scotland or London still furrow the Hooghly, the Huang-p'u, the Godavari, the Congo, the Paraguay . . . .

In the good old days, when business financing was primitive and sometimes chancy, most steam launch operators had to show a good profit in a hurry. Monopoly conditions were devoutly desired, and rates on a new service displacing ox-teams or keelboats were set sky-high. Within a few years, the natural forces of competition, costs, and experience would lead to a stable rate structure that might last for years, though perhaps adjusted temporarily to take advantage of a land boom or the burning of a competitor's boat.

In the 1860s, steam launch passage from Old Town to Lincoln, Maine, was $2.00, and the freight charge was $4.50 per ton. Those prices would be equivalent to $1.50 per passenger-mile and $3.50 per ton-mile in modern terms. A rail line ended the monopoly pricing in 1869, and fares declined abruptly.

The "cheap" fueling of woodburning steamers depended on a good deal of sweated labor and human misery. In Maine in the 1860s, a contract called for split four-foot hemlock at $1.00 a cord, exploiting men who could not find other employment during the winter. As recently as 1910, by one account, crews of skid-row drunks were recruited in Seattle to hand-cut steamboat fuel in the San Juan Islands for $1.00 a cord ($2.00 to the middleman).

Commercial steam launches declined and died out in the industrialized nations within a generation after the private steamers were gone. Murmurs about reviving money-making steamers began to be heard by 1960, but these usually involved a fun-fair or excursion-boat concept.

During the late 1960s and the 1970s there was a flurry of market-feasibility studies and design proposals for making money with "nostalgic" steamboats. These were all paddle-wheelers, and all the proposals eventually settled on petroleum fuel as the designers learned about the stark impossibility of firing 100- or 500-horsepower boats with solid fuel, given modern smoke-abatement requirements and workmen's salaries. Most of the potential money-making uses of small steamers in the modern world have not yet been explored.

Above: The Puget Sound country is extensively broken by old glacial channels, now water-filled, and its settlement was entirely dependent on water transport. *Mocking Bird* served the waterfront hamlets that were springing up around the fringes of the ''big timber'' in the 1880s. (Joe Williamson photo)

Right: Myra, on Highland Lake, New Hampshire, seated her passengers on portable chairs so that the decks could be cleared when she turned to her other duty of log-towing. In this postcard, her boiler appears to be from a threshing machine. (Courtesy Louis Pollard)

Nearly every part of the world once had its own variety of steam launch. How could your average central-Oregon steam launch enthusiast improve on *Mazama,* the first twin-screw vessel on Klamath Lake? *Mazama,* 50' x 12', cost $7,000 in 1908, including the boiler and two 12-h.p. engines (which cost $3,500). After four years of hauling hay, she was torn down and rebuilt, and she then pioneered in freighting sawn lumber (instead of logs) to "motor trucks" at the road head, and in transporting parties of tourists bound for Crater Lake. (Courtesy Klamath County Museum)

Acme, of 1899, earned her living on a 15-mile commuter run from Seattle to the suburb of Bothell, traversing both a large lake and five miles of the sluggish little river called Sammamish Slough. This graceful "water bus," seen in a quiet sylvan setting, is emblematic of all the values destroyed by gasoline vehicles. She provided precisely the same kinds of services that are performed today (for rather more people) by 10,000 cars, trucks, and buses roaring along Bothell Way. (Joe Williamson photo)

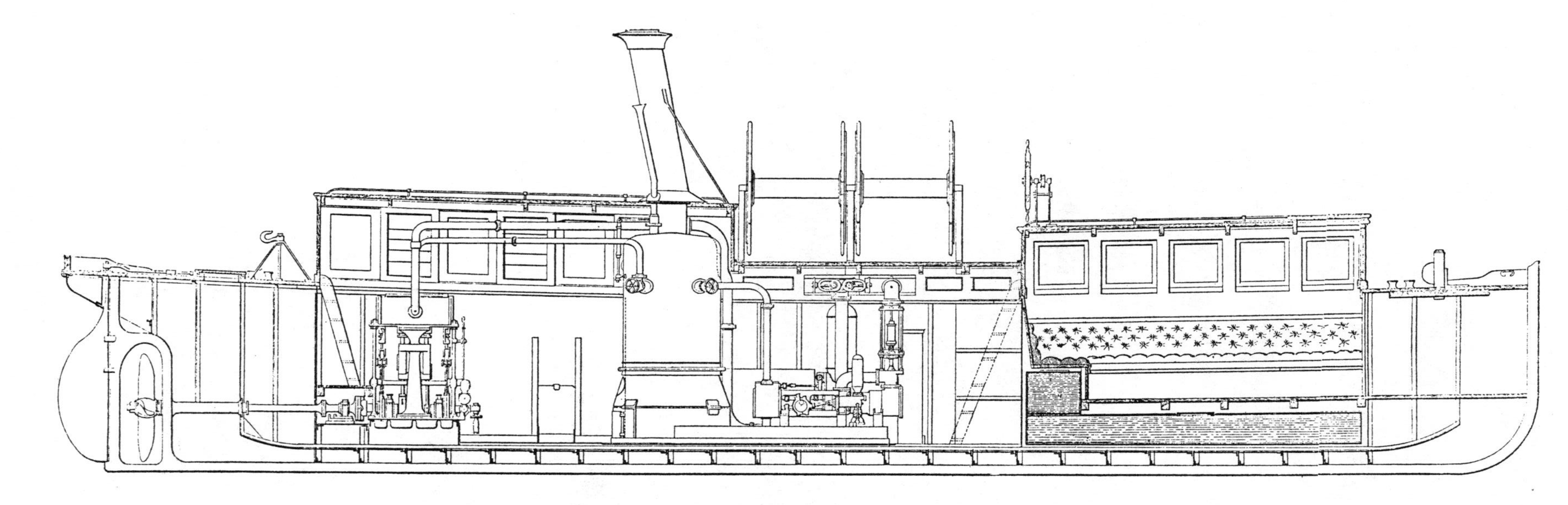

Shipboard or waterfront fires held terror in the days of wooden ships and bucket brigades. Fire-fighting launches were among the earliest and most numerous of industrial steam launches. The November 10, 1882, issue of *Engineer* magazine described the many refinements in a new London-built "fire float" for Brazil, contrasting it with an 1856 fireboat by the same firm (Shand, Mason and Company), which was still protecting the London docks.

This 51-foot steel launch could get up 100 pounds of steam in her water-tube boiler within 9 minutes, 31 seconds after light-off. With 160 pounds of steam on her triple-ram pump (forward of her boiler), she could maintain 220-pound water pressure and throw a stream 200 feet in the air through one 1¾", two 1¼", or four ⅞" nozzles. The high-pressure (double-simple) propulsion engine indicated 82 horsepower at 260 r.p.m., driving the boat at 10.5 m.p.h. There was 6,000 feet of sewn-canvas hose, tested to 400 p.s.i., carried on reels forward of the stack. The price, complete except hose, was £2,050.

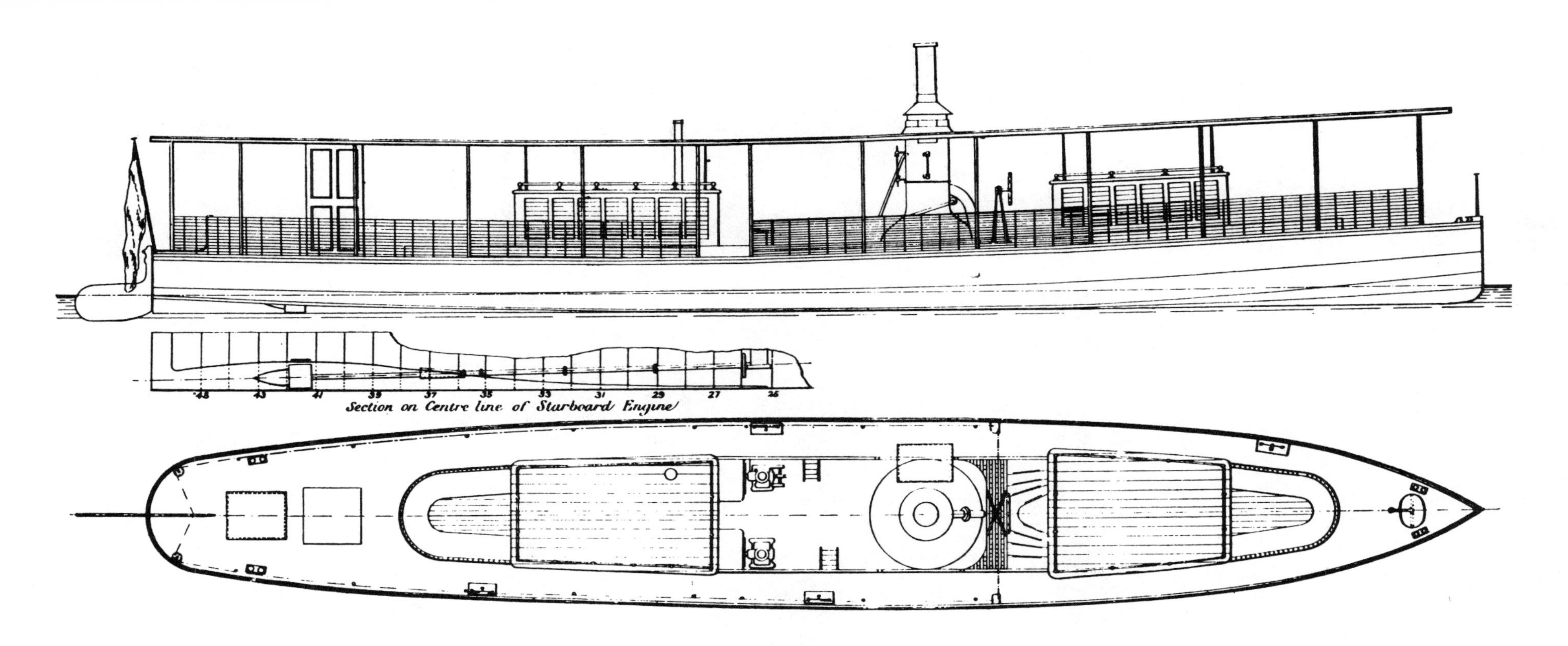

Peace was built by Thornycroft for the Baptist Congo Mission in 1882, only five years after Henry Stanley first followed the Congo River to the sea. She was designed to float 10 tons on a 12-inch draft, outrun war canoes at 12 m.p.h. on wood fuel, and break down to 64-pound bits for backpacking past the lower cataracts to the Stanley Pool.

Native packers charged a penny a pound to carry *Peace* 250 miles from tidewater to her erection site. The irreducible crankshafts weighed more than 64 pounds and were delayed while the packers haggled. Four erecting engineers were sent out from England in succession, and each of the four died of fever before he arrived on the job. To spare the lives of additional engineers, the directing missionary, George Grenfell, finally supervised the assembly.

The propeller tunnel-races were well designed, but the screws themselves were of an excessively "scientific" guide-vane type developed by Thornycroft. At the trailing edge these had a pitch of almost six feet (on 16 inches diameter), and their efficiency was only 57 percent — little better than a modern outboard motor.

The launch's Thornycroft water-tube boiler, first of the type, was a carefully reasoned improvement on the Herreshoff coil boiler, which had excited interest in British steam launch circles a few years earlier. A large grate area accommodated poor jungle fuel. To deal with the risks of small water capacity in a solid-fueled boiler, the entire top of the steam drum was spring-loaded, designed to lift as a giant safety valve. (From William H. Maw, *Recent Practice in Marine Engineering,* 1883)

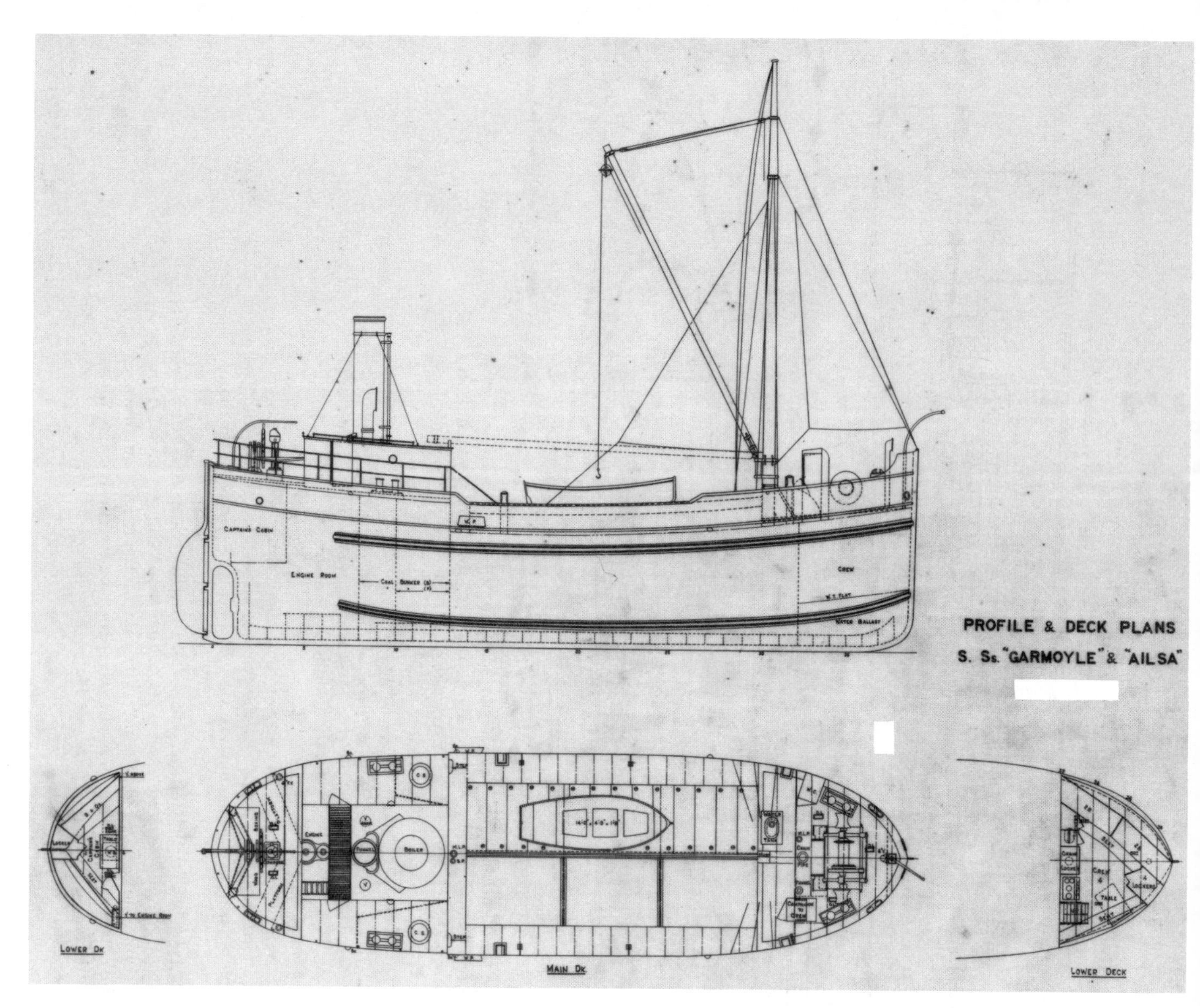

Above: West Highland (Scotland) coasters, capable of lifting 100 tons of cargo on a 66-foot waterline length, were the most familiar small steamers in that thrifty land. *Garmoyle* was delivered in 1904 for £2,000 (at a loss of £716 to the Scottish builder!). She had a 10'' & 20'' x 16'' engine and a vertical tubular boiler. (National Maritime Museum, London)

Right: Of course *African Queen* belongs in the ''Commercial Launches'' section. At the time of the fictional events, every steam launch in Africa was hauling goods or passengers for profit. After the filming of the movie, United Artists' mock-up of *African Queen* was brought to America and carried paying passengers at a resort near Bend, Oregon. Presently, the beloved, 30-foot *Queen* sits in front of the Key Largo Holiday Inn, near where Humphrey Bogart and Lauren Bacall shot scenes for the 1948 classic, *Key Largo*.

In a curious way that Humphrey Bogart and Katherine Hepburn probably never knew about, *African Queen* was a rallying point for steam launch men. In the late 1950s and 1960s, word would crackle out over the telephone lines— ''*African Queen*'s on Channel 7 tonight'' — and many private steam launch dreams would be refueled. There used to be hot debates on whether Bogart's friction-tape repair of the main steam line was really adequate to get him past the German fort. (Richard C. Powers photo; courtesy Jerry Heermans)

UNITED STATES

No. 29

Inspectors' License to Pilots.

—PILOTS—

This is to Certify that George E. Whitney has given satisfactory evidence to the undersigned Local Inspectors of Steam Vessels for the District of [illegible] Mass that he is a skillful Pilot of Steam Vessels and can be entrusted to perform such duties upon the Waters of [illegible] Harbor and Sound between [illegible] and Gay Head Mass on Pleasure Steamer of Fifty [illegible] and he is hereby Licensed to act as SECOND CLASS PILOT on Pleasure Steam Vessels of Fifty gross tons for the term of five years from this date, upon the above named route.

Given under our hands this 14th day of September 1906

11577

[illegible]
Inspector of Hulls.

Andrew J Savage
Inspector of Boilers.

The Maine State ferries, plying between coastal ports and offshore islands, were once powered by steam. The *Vinal Haven* ran between Rockland and the Penobscot Bay island from which the boat took its name, providing island residents with a year-round lifeline to the mainland in the same way that its diesel-powered descendant does today. (William Gribbel collection)

Steam Launch "Phantom."—The following is a general description or rough specification of a class of boat, such as the "*Phantom*," designed for passenger traffic across rivers, etc. Length over all 40 ft.; Beam outside 7 ft.; Depth inside of skin to gunwale 3 ft. 6 in. Carvel built, of pitch pine copper fastened, with keel, stem, stern post, and timbers, of oak; swan stem, and counter stern; lined with fir down to the benches. Cuddy deck at each end of the boat, with doors to form lockers; seats arranged fore and aft, with cross benches in stern sheets and forward of boiler. Coal bunkers amidships on each side of boiler. Rudder fitted outside of propeller, with galvanised iron tiller. Engines, a pair of vertical inverted cylinders direct acting high pressure 5¾ in. diameter, by 6 in. stroke, fitted with link motion reversing gear, and feed pump. Boiler, of the vertical cross tube type, 3 ft. 3 in diameter by 4 ft. 0 in. high, with nine tubes; tested by hydraulic pressure to 150 lbs. and designed to work at 60 lbs. per square inch. The boiler is felted and lagged with mahogany and brass bands, also fitted with all necessary furnace and steam mountings, spring balance safety valve, with escape pipe, steam pressure gauge with double dial facing fore and aft, double set of water gauges, blow-off cock, steam whistle, blower cock, and hinged chimney. Propeller of steel, 36 in. diameter; screw shaft of wrought iron, 2 in. diameter; stern tube bushed with gun metal, and fitted with thrust bearing and collar. Steam and feed pipe connections of copper, exhaust pipe of iron; steam starting valve, back pressure valve and suction cock fitted. A set of stores, tools, spanners, and oil cans supplied.

The vessel is copper sheathed to the water line, and has a cabin forward 12 ft. long, framed of teak, with framed and glazed sashes, seats and lockers, the doors open forward into a small cockpit and aft on to the gangway amidships, with companion and scuttle hatch. The after part of the cabin is divided off to form a w.c. on one side and a bath room on the other. The engine room is covered in with a framed teak cabin, with fixed round lights and scuttle hatch. Gangways are laid round the engine room, and a small open cockpit arranged aft. A No. 3 Korting's Universal Injector is fitted to the boiler with the necessary valves and pipes, also an ejector to discharge bilge water.

A galvanized Trotman's anchor is supplied with 15 fathoms chain, and fitted into a chain locker forward. A steering wheel is fitted forward, and a spare tiller to steer from stern sheets. A patent ship's closet is fitted in the starboard compartment of cabin, and a zinc sponge bath in the port compartment.

Phantom was a passenger ferry for a cold climate. (From James Donaldson, *Practical Guide to . . . Small Steamers,* London, 1885)

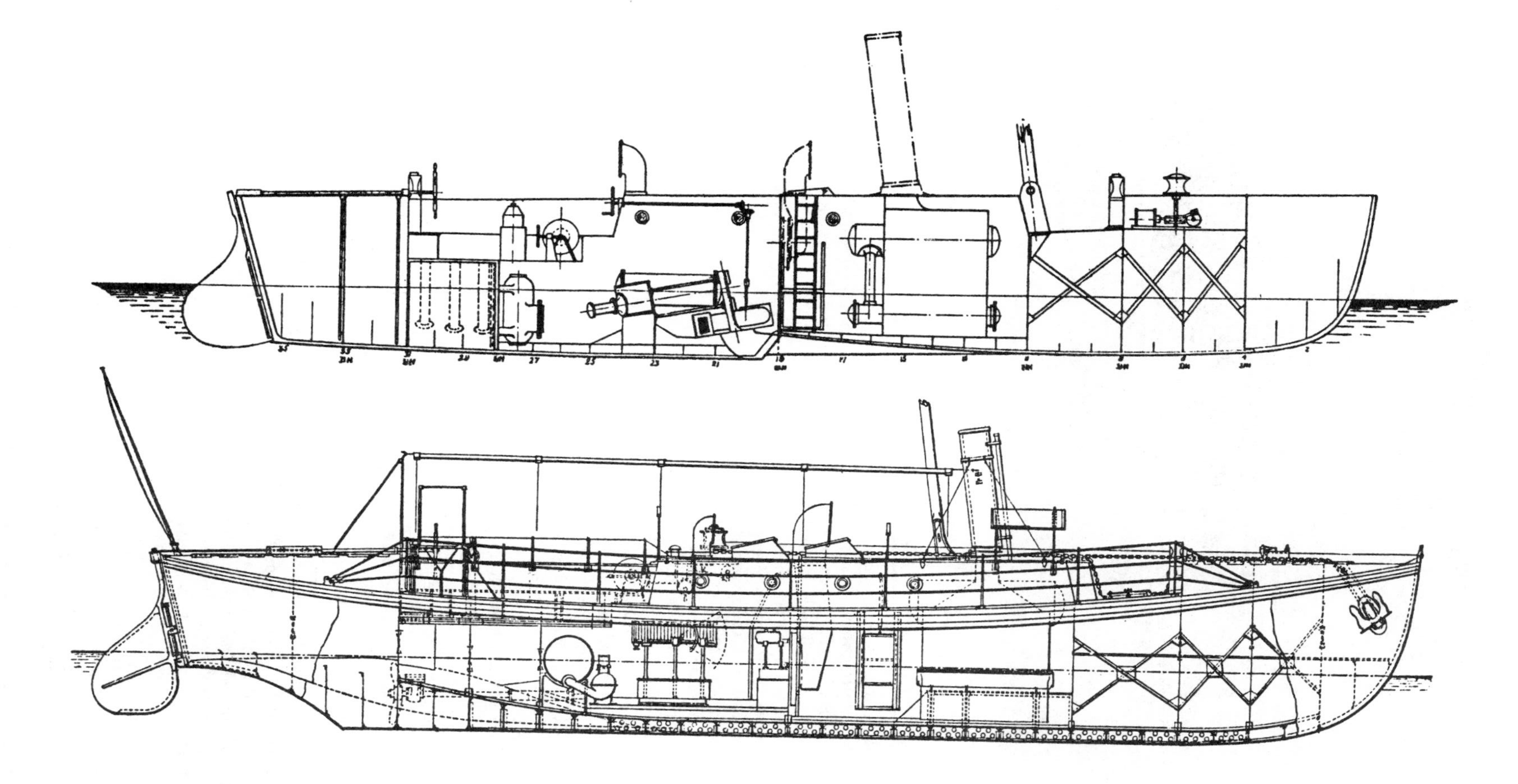

When powerful steam lifeboats came into demand, the propelling device was considered a major design problem. An exposed propeller might tangle with wreckage, mutilate swimmers, or suffer fatal damage if the boat grounded hard. Always ready to experiment, the Thornycroft Company chose "hydraulic" (jet) propulsion for 1895 and 1897 life-saving boats for Holland. The centrifugal pump expelled a ton of water per second through 9-inch nozzles.

Queen (top), of 1897, attained her design speed of 8.5 knots with 123 horsepower; with 246 horsepower, she made 9.29 knots. Being experimental, she had 100 percent excess power. *Queen* was equipped for both coal and oil fuel. Lacking advice from Russians or Americans with oil-burning experience, the Englishmen had to keep a coal fire going or the oil fire went out, and the boat smoked worse on oil than on coal.

Molesey (bottom), shipped in 1905 to Lagos, on the West African coast, had twin screws in semi-tunnels and reached 10.25 knots. Her galvanized-steel hull had 21 compartments, but none of the steam lifeboats was expected to survive a rollover, "like the old rowing lifeboats." (Inboard profile of *Molesey* from *Engineering,* September 22, 1905)

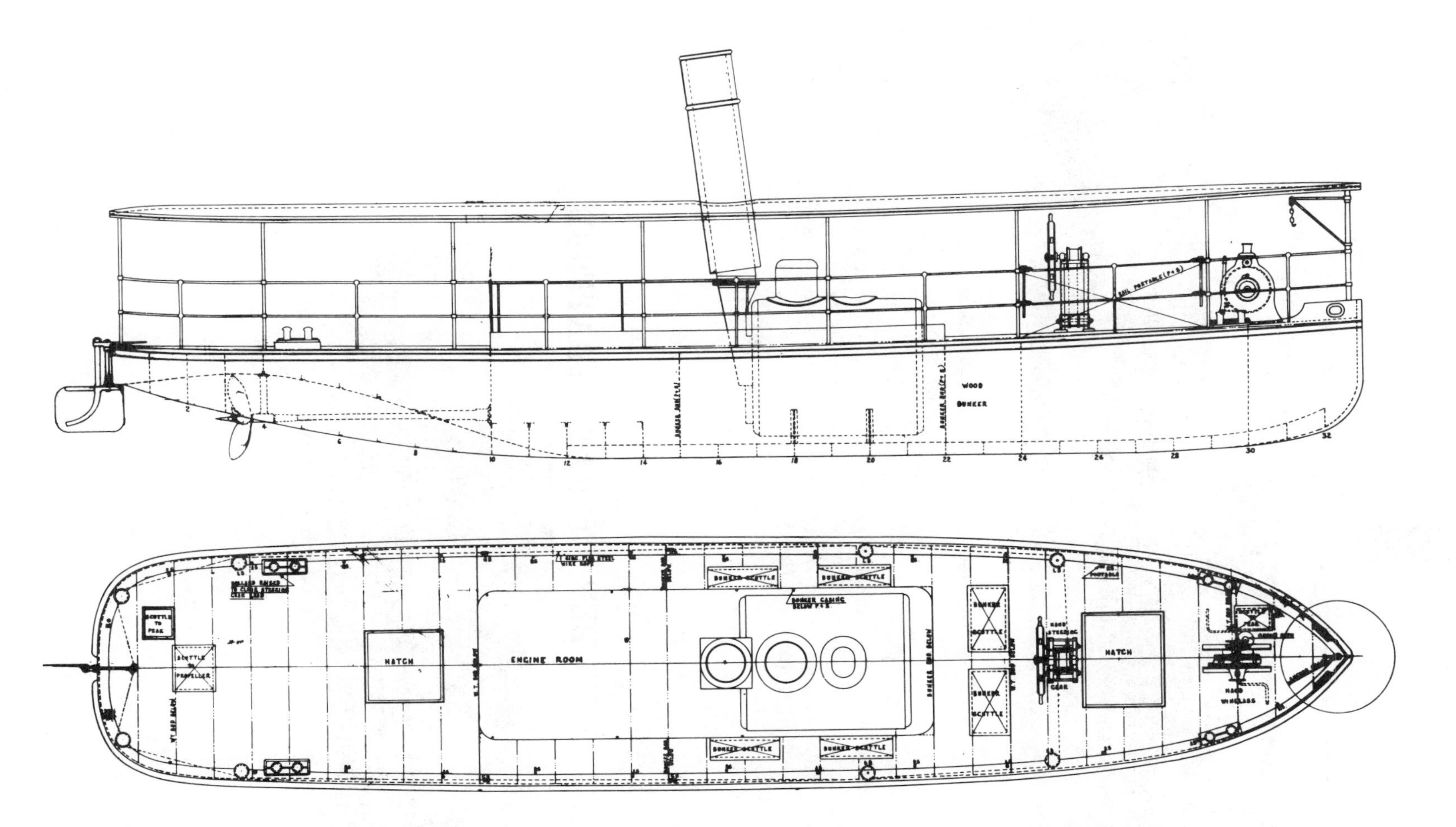

Kelvin was a no-nonsense, all-steel British steam launch of 1909, designed to maintain pilot services along a 60- or 80-mile "beat" of the Irrawaddy in Burma. With her tunnel stern, she could be ordered to the fast-flowing Chindwin or upper Irrawaddy, a thousand miles from Rangoon. The launch made 10.47 m.p.h. at 278 r.p.m. After working in Burma, from 1910 to 1916, *Kelvin* was sent to Basra (Iraq) for the Mesopotamian campaign, and did not return. (National Maritime Museum, London)

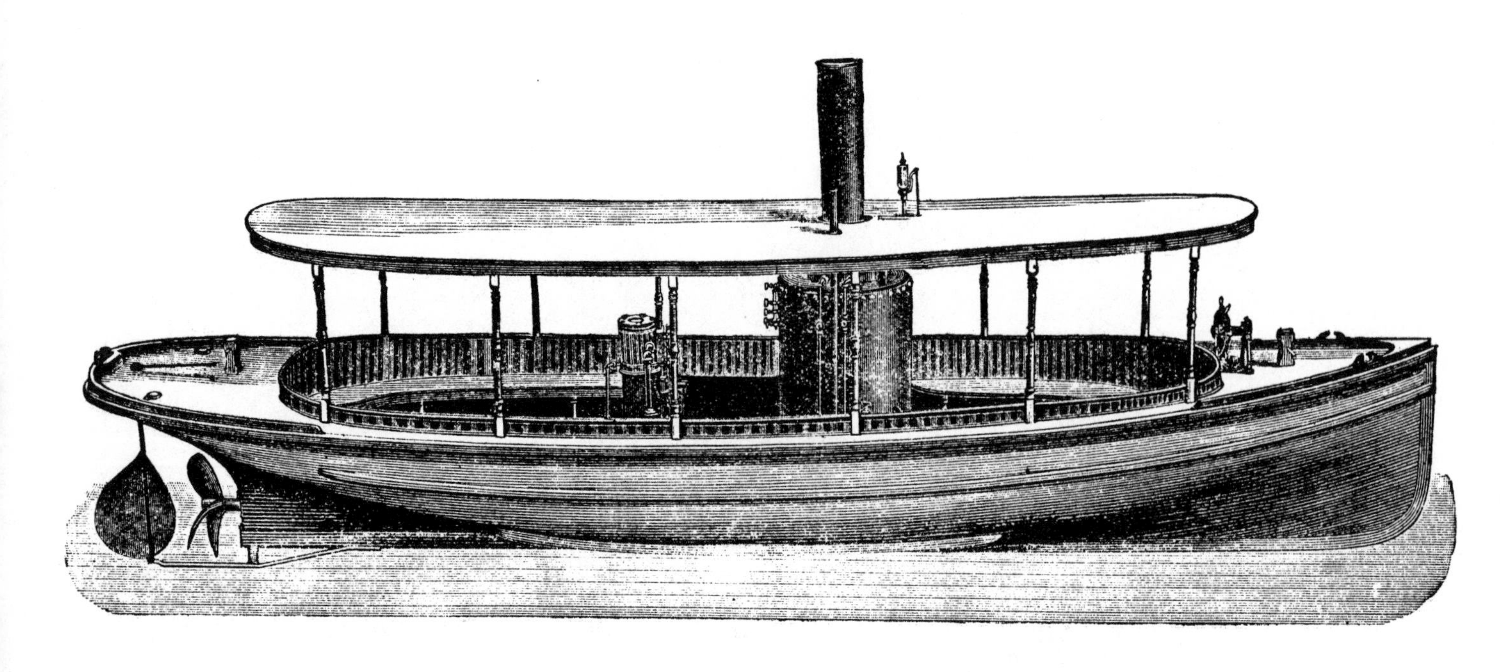

Large steam tugs fell into their natural format by 1870 and changed only by small degrees during the ensuing 50 or 60 years. Small tugs began as multipurpose steam launches, which were effective in towing because of the virtues peculiar to steam power. Eventually, some steam launches were designed as towboats, with strong hulls and oversize machinery. Merwin, Hulbert and Company's *Atlas,* 34' x 8', could be shipped anywhere by rail, to apply her 7'' x 8'' engine to towing logs, barges, or fish nets, or running cargo up swift rivers.

EDWARD HAYES, ENGINEER, STONEY STRATFORD.

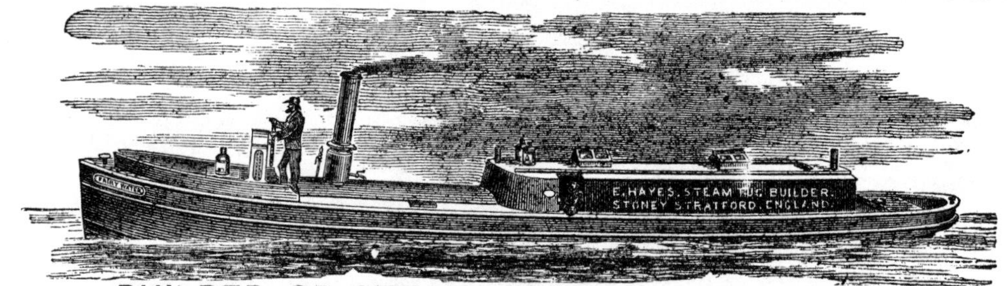

BUILDER OF STEAM TUGS AND LAUNCHES, 1579
ALSO, TUG AND LAUNCH ENGINES, &c. SUPPLIED TO BOAT BUILDERS AND OTHERS

A British advertisement of 1878. (Courtesy George Watkins)

When steam launches first flowered, they were a varied lot, and each was known individually to all the local people. These small commercial launches, members of a fleet of 10 on Squam Lake, New Hampshire, ran passengers and baggage from Ashland Dam to Sandwich. The larger launch on the left is the *Sandwich I,* owned by Alfonso (''Phon'') Smith.

Enterprise carried excursions on the 1½-mile-long expanse of Spofford Lake, New Hampshire, for many years. Crowding of small lakes by powerboats is more a function of speed than of size or numbers. At 5 m.p.h., *Enterprise* didn't crowd anybody. (Courtesy George Clapp)

These old postcards — passenger launches and mail steamers from New England and the Midwest — speak gently of an earlier time, drawing us back there and making us wonder, "Is this how it really was?" (Lake Winnipesaukee, New Hampshire, Crystal, Michigan, and Grand Rapids, Michigan.)

The dolt who moved, far right, was undoubtedly one of *Meteor*'s crewmen who had never had his picture taken before. All the rich people up from The Bay and The Valley knew how to hold a pose of easy affluence while the photographer took the cap off the lens. Early steam navigation on Lake Tahoe was devoted to logging operations supplying the Nevada silver mines. After the transcontinental rails passed nearby in 1868, increasing numbers of Californians rode the cars over the Sierra Nevada, to enjoy Tahoe's cool, 7,000-foot-high summers. (Courtesy Flora North)

Below: Tallac may be either berthing or getting underway, for no docking lines are visible. Whichever — no one seems overly concerned. (Courtesy Flora North)

Rough-and-ready steamboats had a natural affinity with sawmill and lumbering men. The same boilers, engines, and personnel were often used interchangeably in the woods, in the mill, or afloat — depending on the economic winds. A shingle-mill owner built *Jessie* at Rattlesnake Point, Klamath County, Oregon, in 1904, to fill a presumed need for a steamer capable of hauling 50 passengers on excursions or taking hunting parties up to Pelican Bay or the Indian Agency.

The " . . . nice roomy cabin was elegantly upholstered with red and dark green material, with substantial high-backed dining room chairs placed in the center." *Jessie* burned in her first season, so her machinery was installed in the planing mill at the Klamath Lake Lumber Company. (Courtesy Klamath County Museum)

Top: Startled Fawn, 71' x 9.5', was designed to win a boat race at the Philadelphia Centennial Exposition of 1876. It is claimed that she made 26 knots with the 70 horsepower given by her marine boiler and the two high-pressure engines on her shaft, but even 26 land miles per hour seems high for 1876. She was taken to Massachusetts in 1879 for service on the Merrimack River and to coastal ports between Boston and Portsmouth. The rise of electric trolley cars drove the *Fawn* out of service in 1896, although she could outrun the cars. Her tall, white stack blends with the background of this antique photo. (Courtesy Arthur E. Hughes) *Bottom:* The presence of several gas boats among the steamers identifies this Moosehead Lake (Maine) postcard scene as being after the turn of the century, but the ambience remains close to the preceding century, when mechanical power first began to make quiet, distant retreats accessible to urban populations. It is difficult to imagine anxious, nail-chewing passengers on any of these boats. (Courtesy Steamship Historical Society of America)

A hundred years ago, owning a 2-h.p. boat guaranteed a comfortable income — during the summer months, at least. *Wacantuck* served the Pine Grove Springs Hotel on Spofford Lake, New Hampshire. Entertaining hotel guests was her job, and her 4-m.p.h. speed was just right to stir a cooling breeze without ruffling the ladies' feathers. The ''cabin'' consisted of narrow booths along the centerline, probably enclosing only a toilet and the machinery space. (Courtesy Jon Knickerbocker)

These working steam launches are seen waiting at the outlet of Little Squam Lake for their next run up the lakes, taking guests and baggage to the resort hotels. The date is probably near 1900, since the boat on the left is gasoline-powered. (Courtesy Louis Francesco)

No small steamers carry passengers on scheduled runs in North America today, but a few of them, once among the workhorses of common carriers, survive as excursion or museum boats. *Sabino,* now at Mystic Seaport, holds special interest because of her small size and her long and varied career.

Built at East Boothbay, Maine, in 1908, as the *Tourist,* the boat worked out of Damariscotta until 1918, when she swamped. Subsequently raised, refurbished and renamed *Sabino,* she worked for a few years on the Kennebec, then operated on Casco Bay for 31 years. For work on the bay, she was sponsoned out to reduce rolling, and her main cabin was widened to the guards, giving her rather portly proportions — twenty-two feet three inches of overall beam on a fifty-seven-foot three-inch length. A coal-fired Almy water-tube boiler and a new stack were installed in the 1940s, but the original 75-h.p. Paine compound engine was left in service.

Sabino's first career ended in her 50th year, when she was sold to the Corbin family of Newburyport, Massachusetts. They spent six years " . . . in a labor of love, restoring her splendidly and fitting her out in grand style down to the velour curtains in the ladies' saloon." After a few years as an excursion boat, she was acquired by the Mystic Seaport Museum in 1973. She is now maintained in excellent order as a passenger-carrying museum exhibit. (Kenneth E. Mahler photo; Mystic Seaport, Mystic, Connecticut)

Commercial steam launches built in the 1960s and 1970s were aimed at catching the public's fancy, so paddle propulsion was usually chosen for its splash, flash, and frontier flavor. *Chautauqua Belle* is the most pleasing invention within this genre so far. The following statement, from *Chautauqua Belle*'s former owner, Jim Webster, of Webster, New York, can stand unadorned as a primer on how to get a boat built and steaming!

> I started thinking about building a steamboat back in 1972. Without any idea of where the money would come from, I just started making little pieces, like a whistle, etc. I had decided on a 65-foot sternwheeler. A contract for engines — 6'' bore, 24'' stroke — was let to Harry McBride of Fulton, Illinois. Through correspondence . . . I became acquainted with Alan Bates of Louisville, Kentucky, who consented to design the boat. By mid-1973, I began to realize the real size of my project, and I still had no idea where to get money.
>
> I finally decided to go for . . . a 100-passenger boat. Through a lot of tall talking and some simple arithmetic, money finally became available . . . . By December 15 [1975], the hull was decked over, the boiler had arrived, the boiler deck was finished, the stacks were up, the roof was framed, and the stern wheel was in place . . . . The boat was launched April 2, 1976.
>
> Steamfitting, electrical, plumbing, etc., were done after the launch.

The stern-wheeler has since been given to the Sea Lion Project, Ltd., of Mayville, New York, and she works for the museum, carrying passengers on Chautauqua Lake. (Courtesy Jim Webster)

The 75-foot, steel-hulled *Calliope* was envisioned by Peter Wensberg and his son Eric, designed by Halsey Herreshoff with design overview by William Baker of the Hart Nautical Museum at M.I.T., and constructed by Blount Marine of Warren, Rhode Island. Ron Fulmer, of Holderness, New Hampshire, built the mahogany deckhouse. The Davis engine, built in Ontario in 1906, is a three-cylinder compound, 6'' & 9.5'' & 9.5'' x 8.5'', and it can develop 80 horsepower at 250 r.p.m. and 180 p.s.i. The oil-burning, vertical, two-pass, tubeless boiler was manufactured by the Fulton Boiler Works, of Pulaski, New York.

The group of citizens behind the *Calliope* venture labeled themselves ''The Great Congress Street and Atlantic Steamship Company,'' a name that conjures mental associations with Boston's maritime history. With the aid and encouragement of Boston's Museum of Transportation, *Calliope* began Boston Harbor excursion trips in 1980. (Courtesy Peter Wensberg)

Paddle-Wheel Launches

Steam launches grew up with the screw propeller and have rarely departed from it. The same advances in marine engineering that enabled propellers to replace paddles in most services also made small steamers practical. When an owner today chooses paddle propulsion for his little steamer, there is always a strong special reason. In some cases, the builder is paying his respects to the heroic age of steam navigation, 1780 to 1840, when paddles were dominant everywhere; sometimes he has in mind the well-known paddle steamers that continued in certain services after 1840, such as those on America's rivers and sounds.

Occasionally, paddle launches are built because of a real need for one of the practical advantages of paddle propulsion. A paddle steamer can utilize more horsepower — especially low-r.p.m. horsepower — on shallower draft than a propeller boat. At low speed it can exert more bollard pull or move a heavier tow than a propeller. A paddler can work in shallow, weed-infested waters that would be impassable for a propeller, since the paddles tend to repel vegetation.

Historically, there were other reasons, besides those just named, for choosing a paddle over a propeller. The massive, slow-moving working-beam (later pronounced "walking-beam") engines once popular on America's East Coast had a deserved reputation for reliability and safety. (They used low-pressure steam, in contrast to the explosive, high-pressure Western river boats.) Some were still in service 100 years after technically superior designs became available. Side wheels provided maneuverability for tugs and coastal passenger vessels in windy British harbors.

Even after the American Civil War, when all British ocean steamships were iron propellers with inverted engines (cylinders above crankshaft), some American ships were given low-pressure engines and paddle wheels. American ships were still being built of wood, and it was feared that the vibrations of high-speed (50 r.p.m. and more) propeller machinery would loosen the fastenings.

Paddles, which break the water's surface, are offensive to some purists of marine propulsion. Not wholly of either the water or the air, their furor used to be likened to the splashy struggles of a wounded duck. The shuddering roar and splash of the paddle boats' ''perambulating waterfall'' was considered a distinct annoyance 130 years ago, and it was blessedly relieved by the screw propeller.

One century's public nuisance can become profitably packaged nostalgia in the following century. The following lavender prose described a 500-passenger excursion stern-wheeler proposed for a Western river in 1975:

> The stern-wheel steamer will captivate all who ride her, young and old, with her musical steam whistle and the powerful beat of her wheel churning the water to a cascading display astern. These and the hypnotic effect of her engines quietly and deliberately transmitting power to her pitmans at the stroking crossheads are the simplest manifestations explaining a mystical fascination these vessels have always exerted over those whose good fortune provided the experience.

Sport, 1881, was built at Newburgh, New York, for the president of the Lehigh Valley Railroad. He used her at his Sport Island estate on the St. Lawrence, where she fulfilled his private vision of a pleasure boat with her iron hull, oval windows, carved mahogany paddle boxes, and a miniature walking-beam engine.

During her 47 years as a family yacht, *Sport* provided employment for seven crewmen and enlivened the scene for a host of summer guests and local people. She was brightly lighted by electricity when all the houses along the river still relied on kerosene. One day each summer the boat was turned over to the 30 or 40 house servants for an outing that some remembered as the high point of their lives. *Sport* once got a stranded passenger steamer off a bank by tying up to a tree and directing the wash from her paddles under the steamer.

In 1926, the yacht was stripped of her upperworks and put to work as a car ferry on Lake Champlain at Plattsburgh, New York. She was broken up in 1942. (Courtesy Frederick G. Beach)

Operating the *Sport* for her owner was serious business, and this charming old photo *(inset)* reflects the prestige and responsibility that went with the job. The lad in the back row knows how his friends will react to this portrait, for few things could be more romantic than to be 17 years old, serving as an apprentice engineer and cruising on the rivers, lakes, and canals of Canada's interior. (Courtesy Frederick G. Beach)

Kiawanda was named for a headland in Tillamook County, Oregon. From 1890 to 1910, she carried freight and passengers on the Nestucca and Little Nestucca rivers, serving the towns of Cloverdale, Woods, Pacific City, Connely Hill, Ore Town, and Meda. Built by Charles Talbot and a partner, she was powered by a Kriebel oscillating engine. (Courtesy Western Engines)

Two carefully thought-out stern-wheelers with traditional machinery have been built in Oregon in recent years. Both have direct-connected, long-stroke engines, turning the 30 or 40 r.p.m. suitable for a paddle wheel.

The *Bonanza* (formerly *Beatrice*), shown above on the Yamhill River, employs 2.5'' x 20'' cylinders to turn a 54-inch-diameter wheel with twelve 32-inch floats. The woodburning "Thermite" boiler resembles a Roberts design. The boat has been accurately timed at 6.25 m.p.h. Tom Graves built her; Karl Carlson is the present owner.

Indian, 26 feet overall, is owned by Frank Fernandez, of Butte Falls. Her hull is 20' x 8.5'. She has an all-copper Bolsover-type boiler, 2.25'' x 16'' poppet-valve engines, and a five-foot wheel with feathering floats. Bill Edmondson and Ben Walch built the boat. (Photos courtesy Tom Graves and Bill Edmondson)

(From C.P. Kunhardt, *Steam Yachts and Launches,* 1891 edition)

These young men could paddle this canoe faster than the pin wheels can push it, but think of the fun they are having. *Sidewinder,* owned by Warren Hallum, is shown here on Eagle Lake in California. Power is provided by a Blackstaffe-Wood engine and boiler. (Courtesy Bill Edmondson)

Alice Austen was a pioneer American art photographer who often saw clearly into the heart of the matter. When she snapped this picture of *Mahopac* on a New York State lake at 11 a.m. of a fine August day in 1888, she may have had in mind a philosophical comment on man's work versus nature's. Steamboat men today just wish she'd taken the tarp off the miniature side-wheeler's engine and omitted some of that tiresome scenery! (Photo by Alice Austen. Reproduced by permission of the Staten Island Historical Society, Richmondtown, Staten Island, NY)

Turtle was a favorite name of George Eli Whitney (or his clients) for side-wheel steam launches. He built at least seven *Turtles;* six of them were 35-footers for the Currituck Sound Fishing and Hunting Club. This 70-foot *Turtle* was built earlier, in 1889, and steamed mainly around Boston's North Shore. (There may have been another, 90-foot *Turtle,* or else there was considerable confusion about the actual dimensions of a single big *Turtle* that was listed by the Eastern Yacht Club in 1900 and in a "for sale" ad in *Rudder* magazine a little later.) (Photo courtesy Muriel Vaughn and *Rudder)*

Barry Price, of Lockerby, Ontario, believes that his *Smokey Joe* is the smallest steam launch in Canada and the only one in northern Ontario. Whatever — it looks like a good way to pass the time if you're that far from Broadway. The driving mechanism defies comprehension, even with a magnifying glass. (Barry Price photos)

Little is known of the origin of this snapshot, but the boat's name, *Ooyoo,* is Burmese and the scene is 1890-ish.

The spectacular American paddle steamers of the 1840s and 1850s caught the world's eye, and they are still remembered as the epitome of inland paddle propulsion, but after 1860 most small paddle steamers were British built and were afloat in coastal waters at home or on the rivers of China, India, Burma, Colombia, Russia, Africa, and such places. (Courtesy Everett H. Fernald)

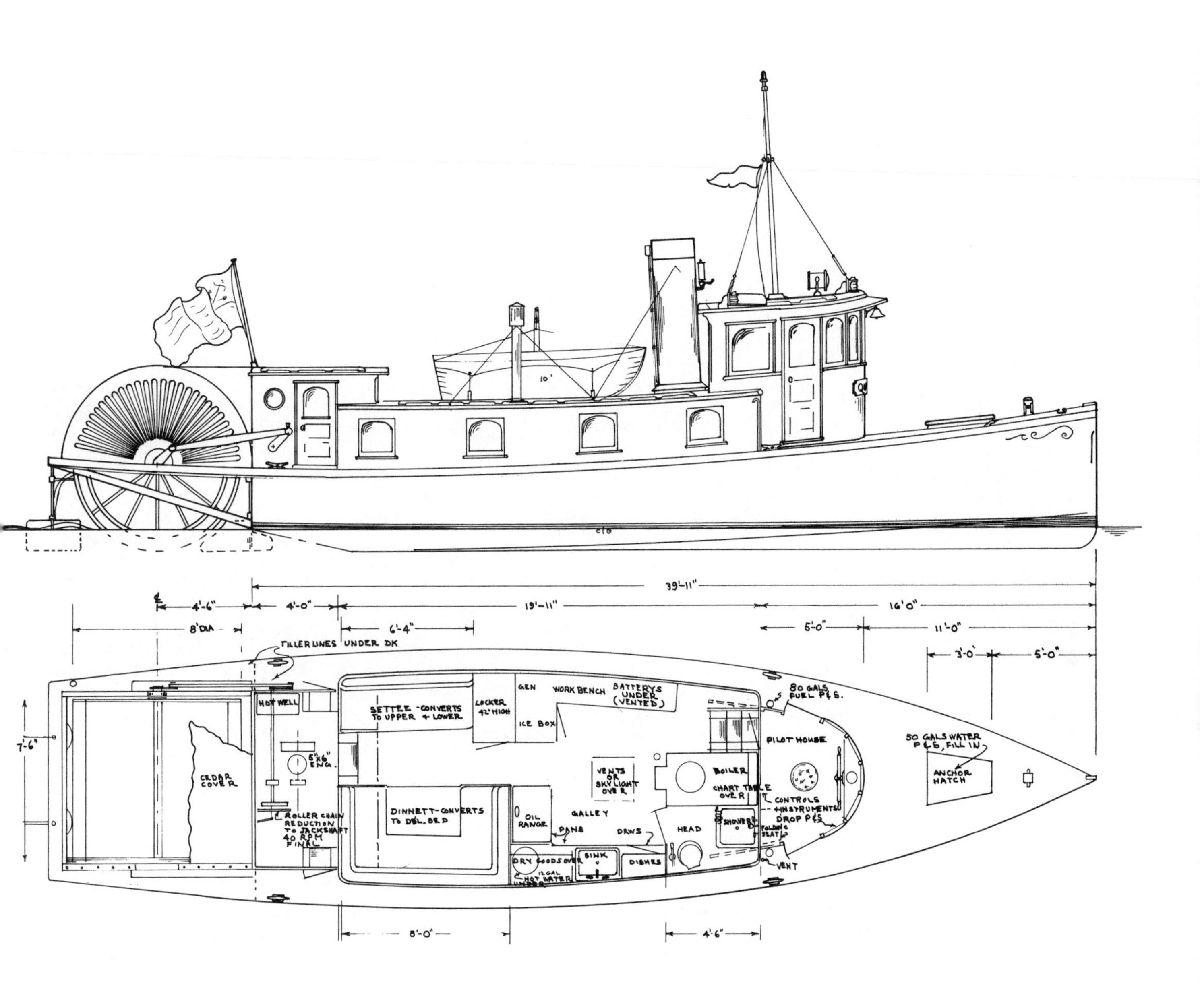

Gordon Sullivan's old propeller, *Quickstep* (see page 136), was a central figure in the steam launch revival in the Pacific Northwest during the 1950s. His new *Quickstep,* a 40-foot stern-wheeler, is emblematic of the refined, personalized designs available to enthusiasts who know what they want. The hull and deckhouse are sturdy, to meet winter weather on the "salt chuck" at latitude 49. Transmitting the industrial engine's motion through a roller-chain reduction to a jackshaft, then through pitmans to the wheel, is a practical choice. This arrangement escapes some of the manufacturing and operational drawbacks of traditional long-stroke stern-wheel engines. (Courtesy Gordon Sullivan)

Rippling Wave III was conceived by Robert Shapleigh as the Bicentennial Project for Dover-Foxcroft, Maine, near Sebec Lake. The colorful boat was designed with the assistance of a professor at the Maine Maritime Academy, her lines lofted on a basketball court.

All of Sebec Lake's *Rippling Waves* were side-wheelers. The first, of 1868, was a 90-footer capable of carrying 200 passengers. She had staterooms, a ladies' cabin, and a saloon where "fruits, confections, and cigars" were sold. *Rippling Wave II,* of 1888, was smaller, serving a hotel that the captain then owned. The 1976 *Wave* resembles this boat, especially in its decorative details.

The 39.5' x 13.6' hull was built of framing lumber and plywood in late autumn 1975. Further work was done by Shapleigh through a Maine winter, under a plastic cover. He fabricated the 4" x 20" engines of pipe, plate, cold-rolled shafting, and channel iron, employing eccentrics and reversing links from an old steam winch.

The boat was finished in such a rush, to get underway by July 4, 1976, that the rudder was not installed, " . . . but the boat handled so well with the two engines that we have steered . . . by means of two ropes, one from each reverse lever up to the pilot house. By pulling one rope or the other I could slow or reverse one engine or the other." During the first season, Peggy Mead, a "firewoman," was the only crew member capable of lifting the safety valve while the boat steamed at full speed. (Stephen Fazio photo)

In 1966, Henry Wiegert, of Kingston, New York, completed a 21.8' x 7.2' man-carrying model of *Norwich,* 1836. A machinist by trade, he designed and built a good 5'' x 15'' copy (see page 295) of the prototype's 40'' x 120'' engine.

Norwich became somewhat famous for working 85 years on Long Island Sound and (after 1843) the Hudson River. She was cut down to a tug in 1850, and for the next 71 years she was known as the ''Ice King,'' usually being the first boat through the ice in the spring and the last to tie up in the winter. She was the oldest boat in the Hudson-Fulton parade in 1909, and then worked another 12 years. Wiegert remembers that he began work in a machine shop at $9.00 per 55-hour week in 1923, the year *Norwich* was broken up. It seemed natural to use his accumulated skills to make a model of her engine 40 years later. (Collin Key photo)

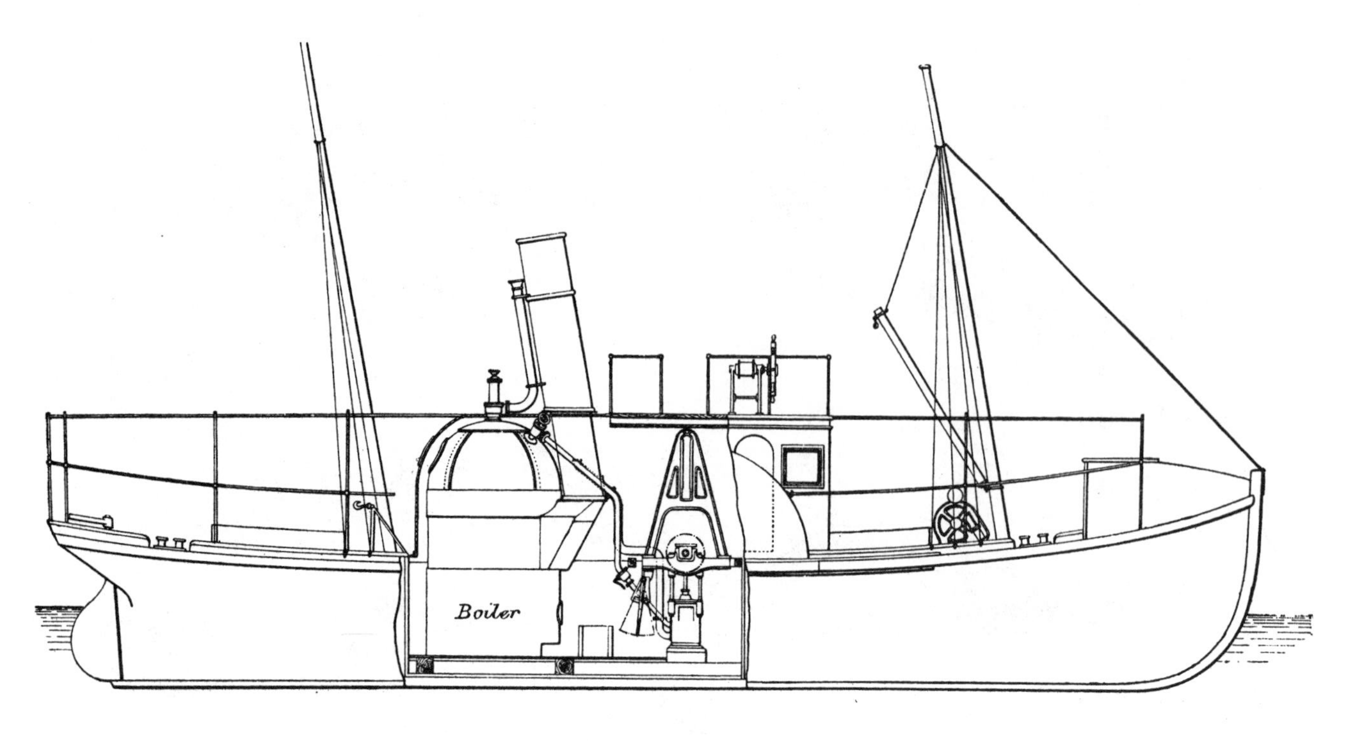

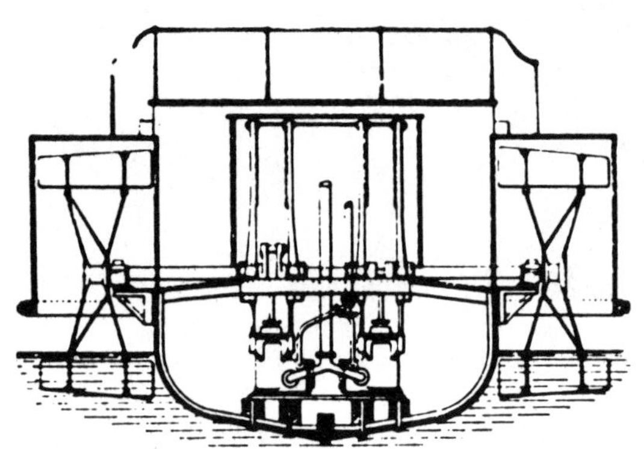

Mawuna, 55' x 12.5' x 3.2', was built about 1880 to carry oil on the Volta, a West African river. The little sidewheeler is atypical, with more resemblance to British paddle tugs than to usual colonial rivercraft.

The machinery was constructed by Cochran and Company of Birkenhead, England, the boiler being a Cochran cross-tube type widely popular at the time. The "steeple" engines, 12" x 20", employed 75-pound steam to turn the 8.7-foot wheels at 46 r.p.m. (From W.H. Maw, *Recent Practice in Marine Engineering,* 1883)

Above: Belle of Ottawa is representative of the low-powered river steamers that brought rapid transportation to countless mid-continent towns in North America, usually long before steel rails got there. A small steamer's smokestack often accurately reflects the horsepower of the power plant (because of considerations of draft through the fire, boiler tubes, and stack). This 60-footer had the power of 10 or 15 horses. (Courtesy C.B. Mitchell)

Below: Cliff Harris steams his 20-foot quarter-wheel boat on the Connecticut River in 1973. The launch was later sold to Frank Cook, of North Brookfield, Massachusetts. *French King Belle* is full of the personal whims and design eccentricities that abound in paddle launches. The hull is steel; the boiler is a home-built version of the British Merryweather type, burning wood for 4 m.p.h. The use of a walking-beam engine (see page 296) to drive quarter wheels is probably unique in all history. (Quarter wheels, common on British river steamers in Africa and Asia, were ordinarily driven by independent high-pressure engines. They provided both the maneuverability of side wheels and a stern wheel's protection from damage.)

Open Launches

Most open steam launches are a kind of equivalent of outboard runabouts or sportfishing boats. Their owners don't bother with creature comforts because the emphasis is on other recreational values. In the case of the open steam launch, this value is the pleasure of managing a small steam power plant.

Some small steamers that are a joy to behold ghosting past on a summer day — with the brasswork gleaming and heat-waves shimmering from the stack — are cramped and uncomfortable for passengers. The owner-designers are idealists, absorbed in the harmonious workings of a mechanism they well understand. They don't care if one foot is asleep in the bilge, the other half-fried next to the boiler, and the back aching from stoking wood in a cramped position.

Tending a small steam launch is at least as interesting and worthwhile as minding the sails on a sailboat, and a lot of people do that for fun, too.

Captain Nathan Young's *Mt. Maid,* originally named *Penacook,* was the first powerboat on Lake Sunapee, New Hampshire. A newspaper of 1877 advertised trips around the lake in *Mt. Maid* for 50 cents. (Courtesy *New Hampshire Profiles,* Charles Hill Collection)

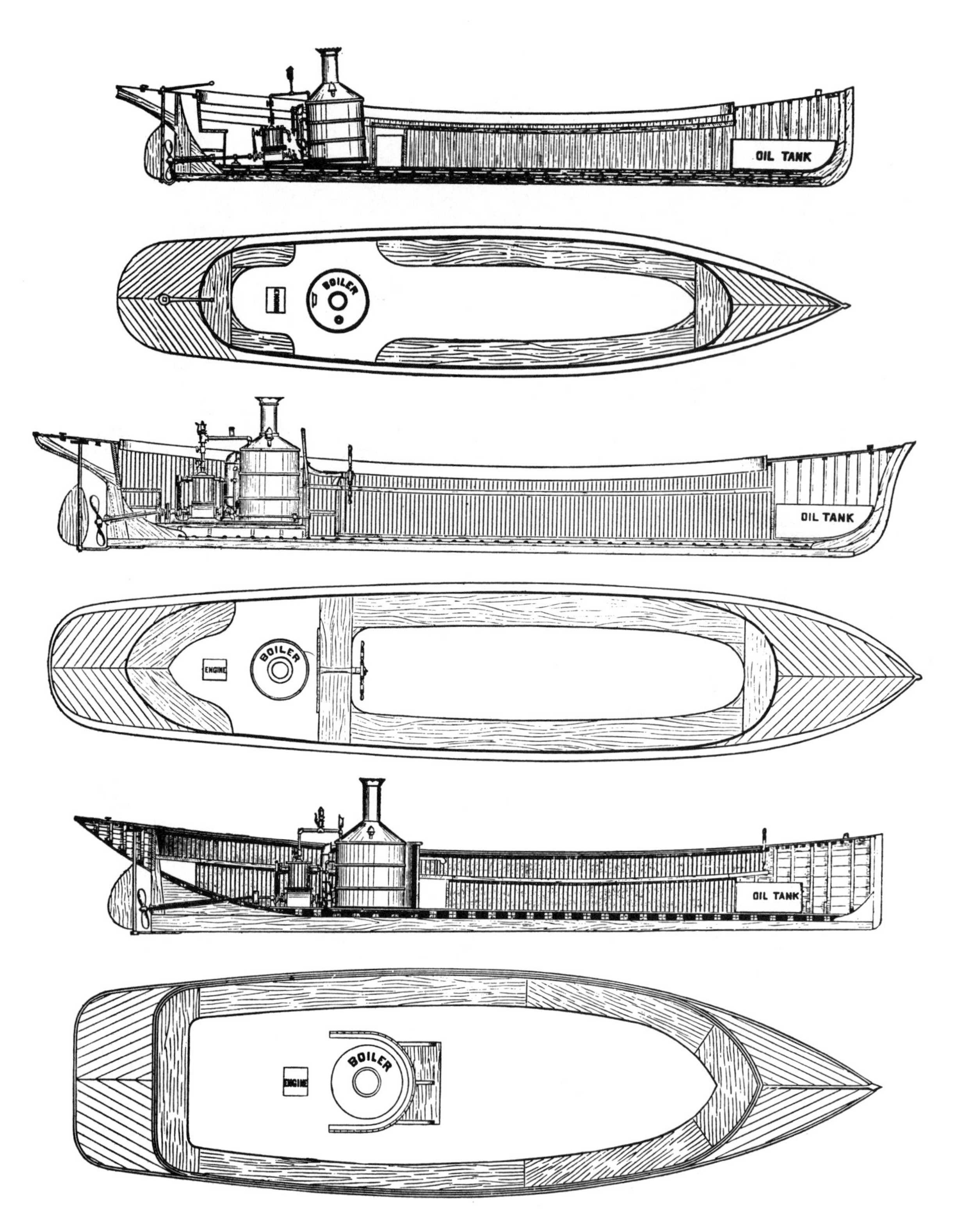

The Thomas Kane Company, of Chicago and Racine, Wisconsin, was a major manufacturer of church, school, and opera-house seating. Some say that the steam launches they also manufactured reflect more experience with church pews than with naval architecture. The boats were designed to float congenial groups (Methodist choirs?) around small Midwestern lakes. The use of petroleum fuel and closed-crankcase, single-acting engines suggests indifference to the established marine engineering practices of the day (when coal fuel and open engines were the norm).

Ted Larter's 23' x 6' *Jenny Lou* was built in 1910, but her engine is 40 years older, and the boiler, 60 years younger. The perfect harmony of the ensemble illustrates the timelessness of good steam launch design. The Davis & Smith engine was built in Dover, New Hampshire, in the 1870s; the VFT boiler was manufactured by Scanell, of Lowell, Massachusetts, about 1970. (Photo by W. Grishkot, Silver Bay, Lake George, New York)

Nip and tuck in Cambridge, England, in 1936. (Courtesy A.F. Leach)

Artú Chiggiato's little *Jela* surpasses "hull speed" as she cleaves Lake Maggiore, Italy. (Artú Chiggiato photo)

A.E. Moulton's *Alice* (see also page 78) is powered by a Blackstaffe-Wood plant built at Victoria, Canada. The 19.5-foot launch is based at Bradford-on-Avon, but here she demonstrates cup-winning grace at the Beaulieu Steam Boat Regatta in 1971. (Photo by Beken of Cowes; courtesy A.E. Moulton)

With her 14.5-foot overall length, ''Howie'' Bailey's steamer *Anne* may be the smallest ''classic fantail launch'' around. She was a prizewinner in 1974 at the annual Antique and Classic Boat Show, at Weir's Beach, New Hampshire.

The boiler is a ''tailor-shop VFT,'' a type that was plentiful in junkyards until recently, permitting many men to boiler their launches at negligible cost. The type is becoming rare now, and small launches will be a little less tippy in the future. Howie has since removed the boiler shown here and installed a low-profile boiler.

David Kyle's *Pussy Cat,* shown here off southern England, exemplifies a powerful yacht's launch of 1912. Her hull is not original, but the owner's knowledgeable design is true to the type and era and matches the authentic Simpson, Strickland machinery (see page 292). (Courtesy David Kyle)

While at M.I.T., F. Spaulding Dunbar was a student of Evers Burtner, one of the torch-bearers of the steam launch revival. Upon becoming a yacht designer and builder, Dunbar assembled his 22-foot steamer, *Psst,* about 1962. Since then, the handsome launch has had other owners, other engines, and other boilers. At present she is the *Mt. Rattlesnake,* operated by Peter Wensberg on Squam Lake, New Hampshire. (Courtesy F. Spaulding Dunbar)

Calm waters, and the trout were rising to the flies. The two gentlemen forward have hired a boat to "have a go." Reproduced from an 1885 stereopticon card.

Scott Nicoll, a retired mechanical engineer, served as chief engineer on the memorable 11,000-mile voyage of the paddle tug *Eppleton Hall.* For sheer pleasure, he prefers his 16-foot *Mikahala,* a quick, graceful boat with small parts and a small appetite for fuel. She is shown with a Blackstaffe-Wood boiler and steeple compound engine. *Mikahala* is now owned by David Dewey of Dunsmuir, California.

Gordon Sullivan and Kay Hogan's *Quickstep,* of 1957, was one of the best known of early "steam launch revival" boats. Her huffing and puffing and woodsmoke smells convinced a lot of Puget Sound yachtsmen that she was a survivor from the good old days. In fact, *Quickstep*'s fantail stern consisted of a mass of scrap 2 x 4s nailed onto the transom of a 1912 Navy dinghy. The boiler was retired from pants-pressing duty and the engine was a wheezy 1880s Kriebel oscillator that Sullivan found being used as a mooring anchor in an Idaho lake. Sometimes random happenings fall together just right, and *Quickstep* was such an instance. (Courtesy K.F. Hogan)

This 27-foot open mahogany launch, possibly a Simpson, Strickland and Company boat, was in service for its Venetian owners, about 1885, when this photo was taken. (Courtesy Artú Chiggiato)

Firecanoe has steamed almost 11,000 miles near Thomas G. Thompson's 30-acre island in western Washington, and frequently to the mainland. The six Thompson children have grown up to the rhythms of stoking driftwood in the firebox and to the tunes played on the 10-whistle calliope. Because of his decision to fuel the boat with saltwater driftwood, the owner has had to build three replacement boilers in addition to the original one he constructed while an engineering student in 1948. All firewood is used "as found," without sawing or splitting.

The Roberts-type boiler has 75 square feet of heating surface and an operating pressure of 125 p.s.i., and it supplies steam to a fore-and-aft compound engine, 3.5" & 6" x 3.5". "She'll steam 1½ miles on one armload of wood." *Firecanoe*'s engine is shown on page 283. (Thomas G. Thompson, Jr. photo)

Sarah has a checkered and colorful ancestry, as do many enthusiasts' steamers that have been developed and refined through long years — or a lifetime — of interest. The 17' x 6' molded-plywood hull once belonged to a Thistle sailboat. The damaged hull was repaired and converted to steam power for Jonathan Leiby while he was serving on board the ocean research vessel *Knorr*. The boiler consists of stainless-steel coils from a World War II smoke-screen generator (popular with builders of high-pressure steam cars). The 1954 Semple engine harks back to a time when Leiby was a student of Professor Evers Burtner at M.I.T. and helped Professor Burtner test the first production engine of the Semple Engine Company.

If you are handy with tools and machines, as Everett Smith of Homer, New York, is, putting *Panetelet* together is not such a big deal. Smitty was almost 70 years old when he started to build a little engine (see page 262) from a plan by Ray Hasbrook of New Paltz, New York. The 16-foot plywood hull, designed by Will Hardy, followed, and then came the assembly of all the components around a small, woodburning fire-tube boiler (which Smitty has since replaced with a WT boiler of his own make). Smitty has a great time running his little steamer at the Kingston, Ontario, and Winnipesaukee, New Hampshire, steam launch meets. His only regret — he should have started long ago.

It's a common thing in these "together" days for father, sons, and grandsons to enjoy their steamboating together, and sometimes the sons build their own steamers to outshine "the old man." When Clifton Hills assembled his 21-footer back in 1927, he was already continuing a family tradition. He used the same small engine (see page 263) his father had employed in a steam rowboat in 1905. Cliff's boat had a copper-tube Stanley steam-car boiler that was free steaming even with coal fuel instead of gasoline. Cliff removed the machinery and sold the hull in the 1930s. (Courtesy Clifton Hills)

George F. Brewster's launch *La Marotte* (French for hobby) is pictured about 1945 steaming on Lake Winnipesaukee. *La Marotte* had a homemade porcupine boiler and a compound engine that Brewster had piped in such a way that it could be operated as a high-pressure engine or even a single-cylinder engine.

The 17.5-foot *Soot* is one of the best, at once graceful and powerful. Displayed at the Boston Boat Show in 1975, the boat introduced many new enthusiasts to the steam launch kind of boating. At that time, the *Soot* was owned jointly by Richard E. Dickey and Frederick E. Sweetsir. Now the sole owner, Sweetsir has replaced the Rochester-model Shipman engine (see page 297) with another handsome, simple-expansion engine (see page 263).

James P. Grayson's *Whippoorwill Queen* cruises South Holston Lake, near Bristol, Tennessee. A 2.5'' x 3'' engine provides ample power for the 18.5-foot boat. The boiler is an Ofeldt-type water-tube, and both engine and boiler were built by Walt Humphries. The boat was formerly owned by Bill Shaw of Thornbury, Ontario, who installed her power plant and called her *Round the Bend.* (Photos by John Beach, Bristol Newspaper Corp.; courtesy James P. Grayson)

David Conroy's *Eagle* is a pleasing example of old, basic components put together in a balanced ensemble. Her 19.7' x 4' hull was built in 1915. Her vertical fire-tube ''tailor's'' boiler is only half as old. There are 19 one-inch tubes, and firing is by a propane burner that came from a hot-water heater. The engine, 2.25'' x 3'', has good proportions and the special virtue of being all bronze. Some small auxiliary engines, especially on yachts, used to be all bronze to eliminate problems with rust while out of service. The origin of *Eagle*'s engine is not known. The boat won the Lake Ontario Great Steamboat Race in 1973. (Courtesy David Conroy)

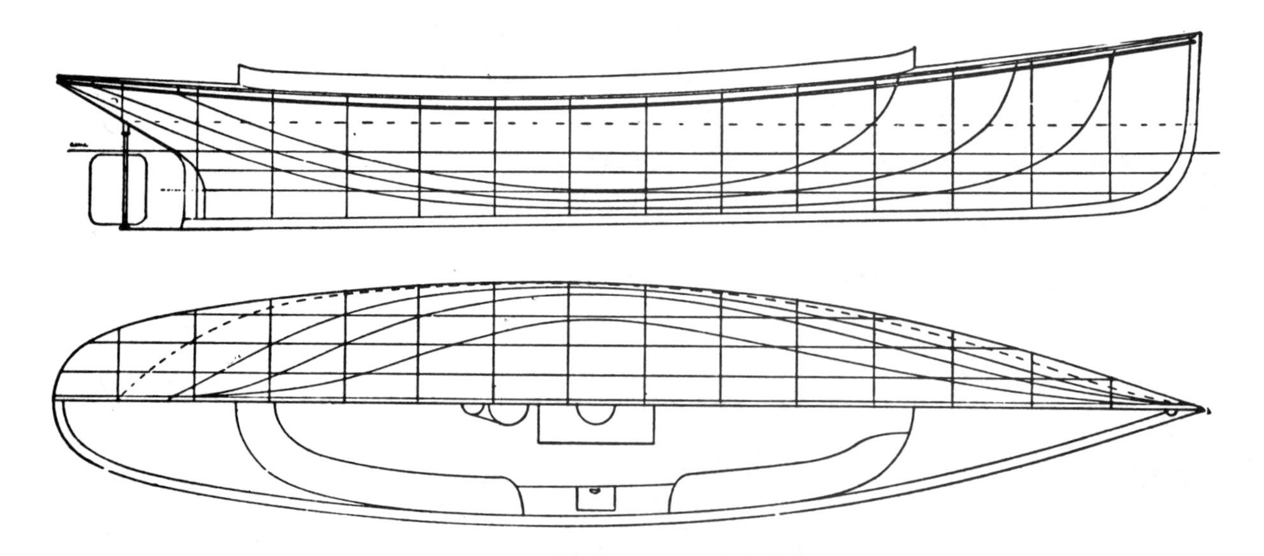

This ''Thirty Foot Open Launch'' design, published by *The Rudder* in 1901, represents for many Americans the archetypal fantail steam launch. She measures 30' x 6' x 2'. (From C.D. Mower, *How to Build a Motor Launch,* 1901)

Are the ladies about to take Linn Taylor's Murray & Tregurtha launch out on Spofford Lake (New Hampshire), or are they just posing? That summer day was a long lifetime ago, and there's no one to ask now.

At one time, builders tried to drive boats very fast *through* the water, not over it. This was a technical dead-end, a lost cause, but for a few years the boats made an exciting splash on the boating scene. Here, a Simpson, Strickland launch cuts smooth water with surgical precision. (From the facsimile edition of Simpson, Strickland and Co.'s Catalogue No. 5)

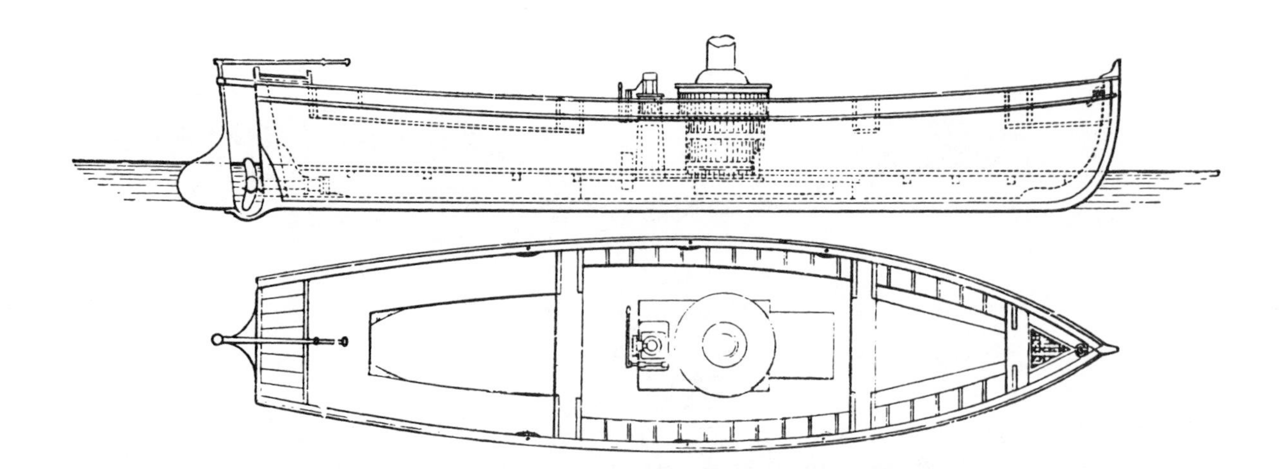

Steam dinghies were especially useful to sailing yachts lacking auxiliary power, since one of these little boats could tow " . . . with at least as much power as two four-oared gigs." This 16' x 4' dinghy (from Simpson, Strickland and Company) was suitable for very small yachts; the machinery and the boat could be hoisted on board separately. If supplied with a water-tube boiler, the 3-h.p. compound machinery weighed under three hundredweight. (From the facsimile edition of Simpson, Strickland and Co.'s Catalogue No. 5)

Cliff Blackstaffe has given pleasure to two generations of steam launch enthusiasts with his generous help and advice and, after 1962, a line of beautifully detailed steam launch power plants. Now retired, Cliff has a very complete triple-expansion plant in a modified fiberglass canoe, and he is doing a few things he's always wanted to do for himself — such as assembling complete trials data on his 1-, 2-, and 3-horsepower marine steam power plants.

Sparrowhawk makes 8 m.p.h. with a 15" x 21" propeller turning 450 r.p.m. (11 percent slip). The triple engine (see page 291) gives a 12 percent increase in fuel efficiency over a standard Blackstaffe-Wood compound engine and a 52 percent gain over a simple-expansion engine. "It's a lot of engine to oil up before a run and a lot more to wipe off after a run, but it's a lot of fun."

Hal and Shirley Will's *Charlie* grew out of lifelong interests in maritime history and working machines. Hal recently served a term as president of the Puget Sound Maritime Historical Society, and he was editor of the Society's journal, *The Sea Chest,* for many years.

Charlie's 15' x 5' hull is a fiberglass copy of a Puget Sound sportfisherman designed for moderate inboard power (3-h.p. air-cooled gasoline engine with reversing propeller). The owner began with a bare fiberglass shell supplied by Skookum Marine, Port Townsend, Washington, and installed a power plant that he obtained on indefinite loan from friends (good friends!). In 1976 he built and installed a 5-h.p. compound engine with Reliable castings. (Courtesy Hal H. Will)

Will Hardy built a miniature version of his *Panatela* design of 1962, and the hull eventually came into the hands of Gordon Sullivan, who likes flashy, out-of-the-ordinary steamers. *Feeble II* is overpowered, and when seen at the steam meets, she is usually characterized by an enormous stern wave and a passenger's white knuckles gripping the coaming. She has been displayed in the Museum of History and Industry in Seattle. *Feeble II*'s engine is pictured on page 262. (Museum photo courtesy Gordon Sullivan)

William Kimberly's third steamer, *Polly*, of 1883, is an all-original (except the 1958 boiler) 22-footer he found in St. Albans, Vermont. Her 2" & 3.5" x 3" engine (see page 284) has a bronze frame, and it's just the right size to drive *Polly* to her natural speed of 8 m.p.h.

Panatela is a plain, sturdy, open launch, full of useful ideas and first-rate workmanship. The hull is a simple structure in marine plywood, with length and easy lines to make it as efficient with low power as the rounded hulls of an earlier era.

The owner, Peter Moale, chose to build his modern water-tube boiler (see page 312) all of copper. This choice is becoming increasingly popular as the cost of copper boilers declines relative to welded steel, and as accumulating experience indicates that copper is reliable and exceptionally efficient at working pressures as high as *Panatela*'s 150 p.s.i. Twenty-seven square feet of heating surface supplies abundant steam for a Moale-built Reliable engine (see page 275), 2.5" & 3.75" x 2", turning a 14" x 13" wheel 900 r.p.m., for 9 m.p.h. top speed. (Courtesy Peter Moale)

Percy Stewart assembled two fine fantail steam launches, using antique hulls that he found in New England. *Gemini (above)* of 1961, had a 17.5-foot hull of traditional wood structure and a homemade 2.5'' x 3'' engine. *Gemini II (below)* has a 23' x 6' sheet-steel hull, a Stickney compound engine (see page 285), and a Levitre combination water-tube/fire-tube boiler. The latter boat was sold to Frederic C. Merriam of Danvers, Massachusetts, and Robert Dickey of Haverhill, Massachusetts.

Above: A native of St. Louis, Fred Semple has summered in Maine since childhood. By 1937 he had his first steam launch in operation at Lovell, Maine. (Courtesy Fred Semple) *Below:* During the early 1950s, Fred Semple employed a rakish, 1915 Fay & Bowen launch, 24 feet long, as test bed for numerous experiments leading up to the founding of the Semple Engine Company. Fred still enjoys steaming his *City of Lovell.* (Courtesy Fred Semple)

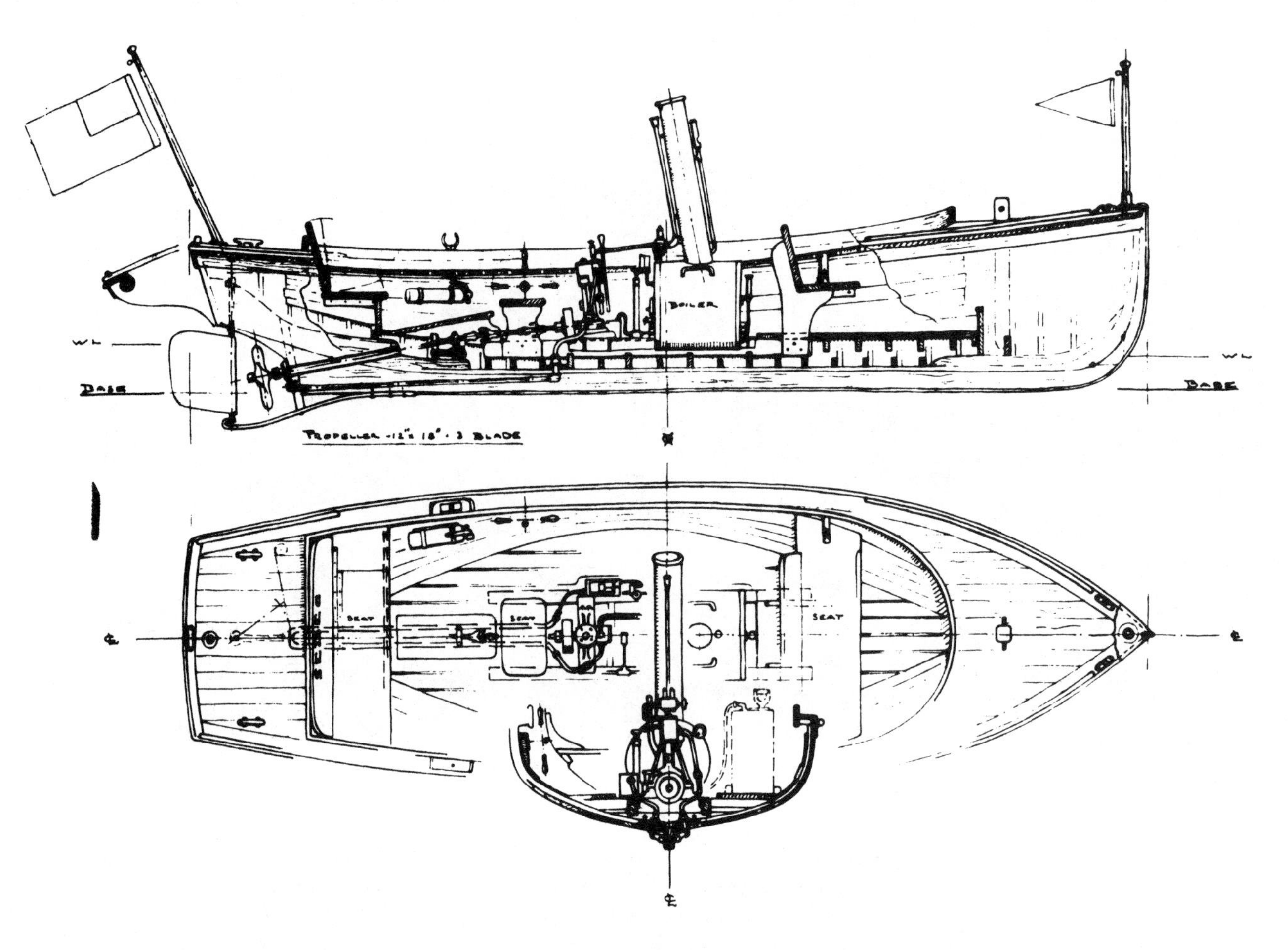

Steve Pope's *Sebec* is the result of several years of careful planning and construction. The preparation of detailed drawings no doubt contributed to her neat finish and practical layout. Some of the men who put the most dedicated effort into their steam launches prefer to work on the personal, portable scale of *Sebec.* The power plant is a Blackstaffe-Wood, including a rare simple-expansion Blackstaffe engine, 2'' x 2.5''. (Courtesy Steven Pope)

Most launches in the Golden Age were either wood or coal fired, but this one was fueled with kerosene. This good-looking boat was owned by Charles Harris, a Brattleboro (Vermont) banker, and used at his cottage on Spofford Lake about 1900. At one time there were seven little steamers on the small New Hampshire lake.

Evers Burtner graduated from M.I.T. as a naval architect and marine engineer in 1915, then taught these subjects at the school until 1963. His deep professional understanding of the field and his lifelong interest in steam launches led him to create plain, unassuming little steamers for himself — a *thoroughly* perfected mechanism is more likely to be simple and minimal than showy or elaborate.

The plainness of *Ala I* is perhaps reminiscent of Professor Burtner's experience as Marine Superintendent and Port Engineer for the John S. Emory Company, Boston, from 1917 through 1919. He supervised the maintenance of 12 "Lakers" and four 400-foot freighters of U.S. Shipping Board design — ships remembered for their plain, commonsense qualities. The lengthening of a conventional aluminum motorboat hull for *Ala II* is no more than naval architectural common sense.

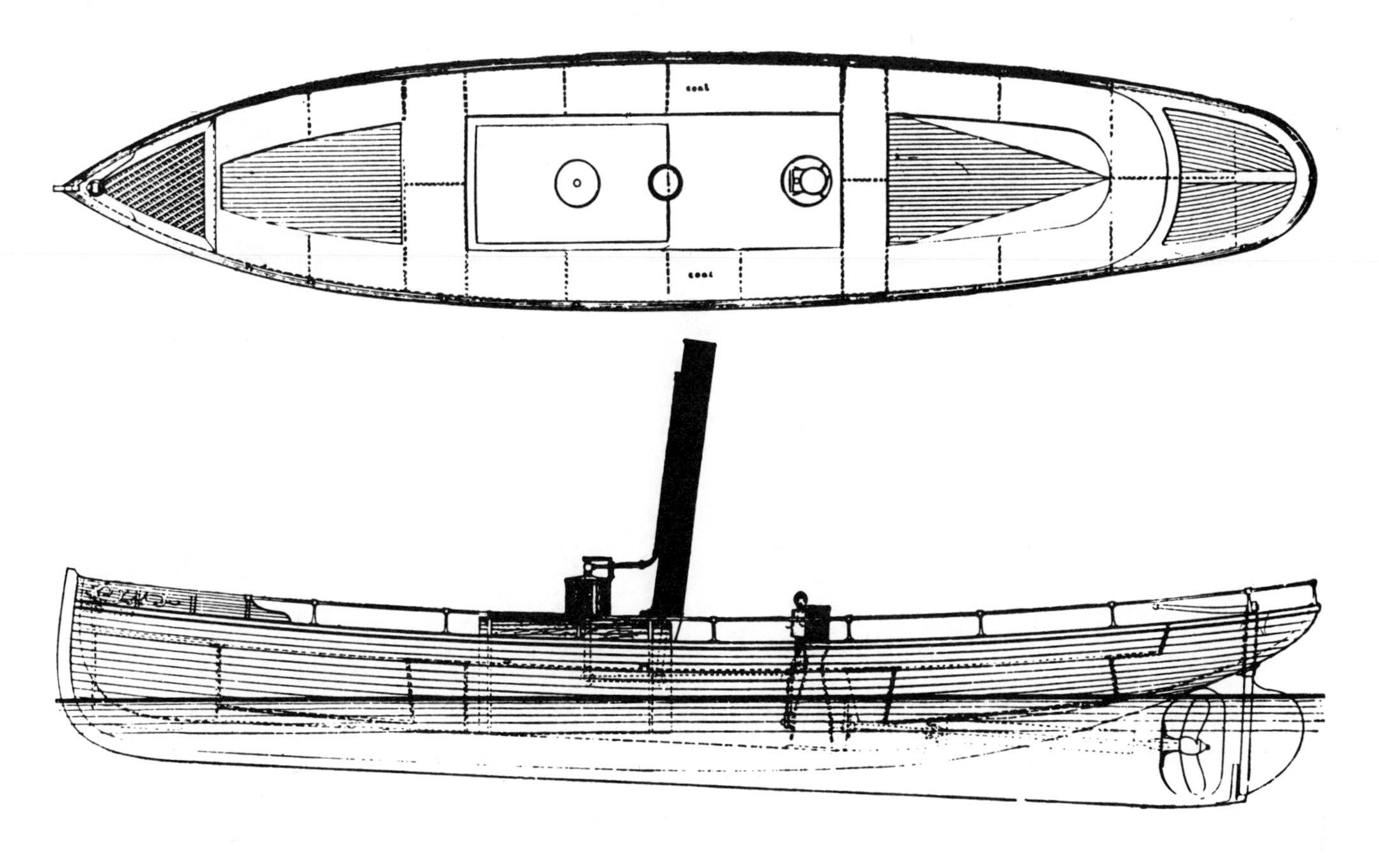

Charles P. Kunhardt thought these open-launch designs worth including in his book, *Steam Yachts and Launches. Mohawk (above)* proves that a dead-amidship funnel — as required here by machinery design and weight — can be pleasing. Some men admire a reverse rake to a steamer's stem, and this one is about right. *Mohawk*'s counter appears high, but wait 'til the full four horsepower on that 24-inch screw piles a big stern wave under it!

The shallow-draft boat *(below)* with geared triple screws is an oddity — a reminder that the first solutions to new design requirements (shallow draft, in this instance) are rarely the best solutions.

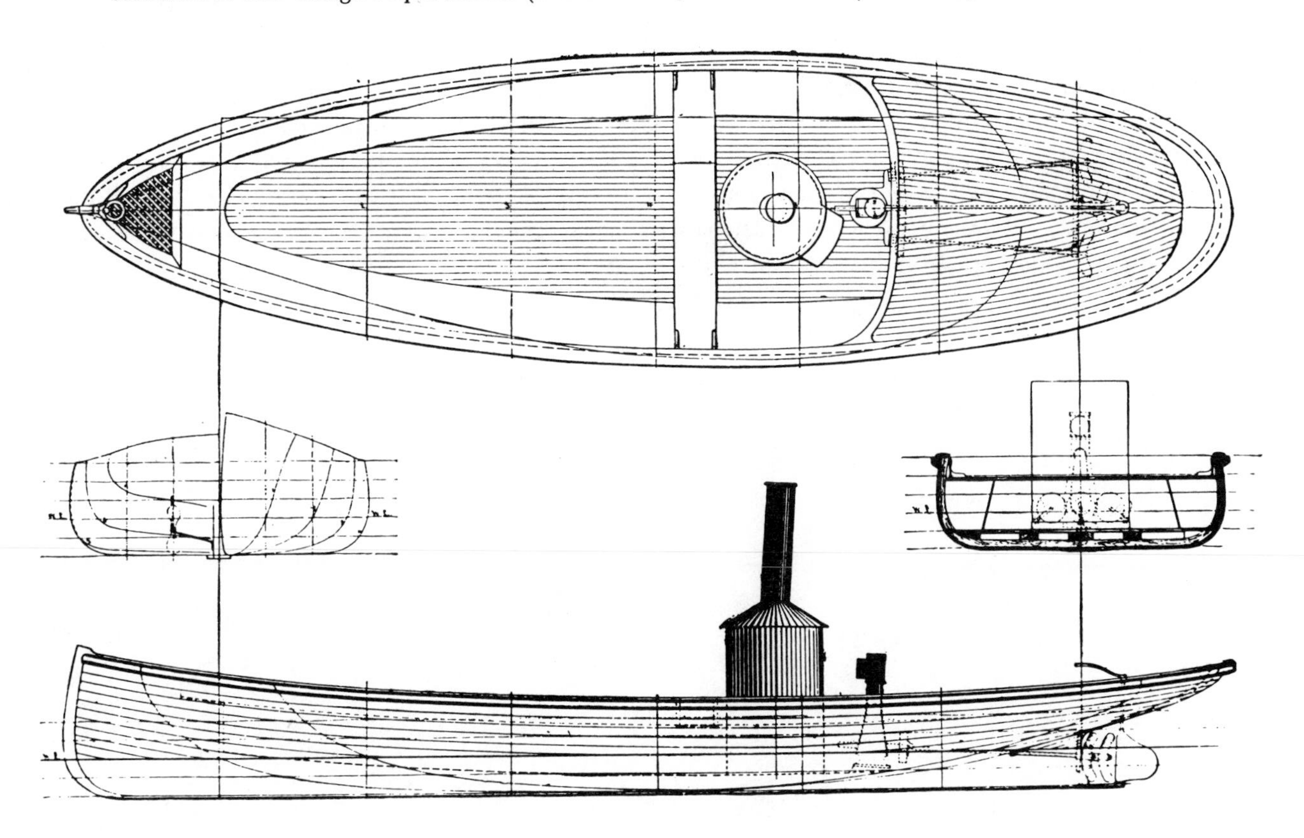

Canopy Launches

For most American admirers of steam launches, the "classic" boat is a canopy launch. There is an uncluttered simplicity and spaciousness to the boat, yet passengers are sheltered from rain and soot.

Canopy launches became favorites in America because of the hot summer sun in much of the country and because, for many years, the boats were much used for running guests to summer resorts. These boats — simpler and friendlier than decked yachts or modern cabin cruisers — suggest halcyon summer days and an era when powerboating was a quiet and gentle activity.

Probably the oldest steam launch in America still in her original state, *Nellie* was built by the Atlantic Works, East Boston, in 1872. The 30' x 5' beauty steamed on Lake Winnipesaukee, at Portsmouth, and on other coastal waters of New England under the names *Old River, Clermont,* and *Glory B,* as well as *Nellie.* She was in use right up to 1957, when George B. Lauder presented her to Mystic Seaport. *Nellie*'s engine is shown on page 265. (Courtesy Mystic Seaport)

This 19th-century watercolor of the *Frank A. McKean* (apparently done from a photograph) fixes forever one artist's view of a boat, a time, and a place. (Courtesy John S. Clement)

Ah, summer! The DeWitt family's *Pastime* nosed ashore in the Owasco River, New York State. (Courtesy Frank P. DeWitt)

The Queen's hull is more than 80 years old, but her lines are fine enough to make her faster than most of the youthful competition on Lake Winnipesaukee. Donald O. Beckner adapted a 2.5'' x 3.5'' double-simple Mason steam-car engine to drive her 18'' x 24'' propeller. The boiler has firing doors fore and aft, permitting passengers to help with the stoking.

Dark topsides or pale, *Zephyr* has made open-sea passages along the New England coast and has been a much-used boat throughout her 30-year career. The 26-foot ex-navy whaler is powered by a Navy B3 compound engine (see page 274) and a Roberts-type boiler built for the boat by George E. Whitney. Dr. Henry Stebbins, a heart specialist, has operated the boat since 1957, and he's had time to get everything just the way he wants it — plain, rugged, and homey. (Photo above by Judy Buck, *Boston Record-American.* Photo at left by Samuel Chamberlain; courtesy Dr. Henry Stebbins)

Above: The *San Juan Queen,* formerly owned by Russ Hibler, shows what a skilled craftsman with a good eye can do with a ''tin lifeboat.'' The boat boasts one of those rugged, foolproof power plants that every steam launch should have. Russ built a 3'' & 6'' x 4'' compound engine to replace the Upton 3'' x 4'' simple that gave him years of trouble-free service. The cross-drum Roberts-type boiler can accept sizable chunks of driftwood, and the boat has proved seaworthy in the broad waters of the Straits of Juan de Fuca. She is now owned by Jerry Batchelor of Langley, Washington. *Below: Pelican*'s monkey-rails and canopy are identical with those fashionable in America in 1895. In addition to creating this 1973 replica from the keel up, the Bill Richardson family of Sheboygan Falls, Wisconsin, is well known for restoration work. Among other boats, they have restored (and now operate) a naphtha launch (see page 210). (Jean Nelson photo; courtesy *Antique Boating*)

Above: Lillian, a 22.5-foot stock steam launch of 1894, was near the pinnacle of low-power pleasure-boat design. Soon thereafter, small-craft design came under the influence of much lighter, cheaper gasoline power. When a boat with an ''explosive motor'' first crossed his bow, did *Lillian*'s owner sense the epic changes that internal combustion would bring to pleasure boating? *Below:* Among the most enthusiastic of steam launch men is Dr. Mason Saunders of Albany, New York, here running his 28-foot *S.S. Pamelaine* on Lake George. The Fay & Bowen gas-boat hull, of 1918, now has a wood-fired Semple F-40 boiler and a 3'' & 5'' x 4'' Semple V-compound engine, installed by Dr. Saunders. The speed made possible by *Pamelaine*'s long, slender hull brought him the blue ribbon in the 1973, 1974, and 1975 Lake George Steamboat Races. Fred Semple built the engines for the new stern-wheel excursion steamer, *Minne-Ha-Ha,* seen in the background. (Photo by Shangraw)

Lou De Young's 27' x 6' *SS Ramona* was built in 1911 by the Racine (Wisconsin) Boat Company (as the *Wahoo*) for a member of the Ball (glass jar) family. The boat has been meticulously restored by the present owner, who acquired her in 1974 for use on Gun Lake, Michigan. The old hull was rebuilt and fiberglassed, and professional rebuilding of the power plant (see page 308) was carried out by a Grand Rapids boiler works. *Ramona* also appears on the dust jacket of this book. (Photos by Bernie Photographs; courtesy Lou De Young)

In this scene reproduced from a stereopticon card, the *Willie Brace* noses into a tranquil backwater.

George Whitney built steam launches to order, to suit the needs of the purchaser and the intended service. *Anemone,* 1898, 36' x 9', was given a plain 6'' x 8'' engine and a tall industrial boiler, to fit the requirements of service on the backwaters of the St. Johns River, Florida. (George Whitney collection)

Di Vernon's power plant was located aft to get Dad out from behind the boiler and together with the family. (Nobody asked Dad where he'd rather be.) The passengers' garb indicates that Merwin, Hulbert and Company hoped to sell their boats to a broad segment of a democratic populace. Contemporary English steam launches were available only to the rich.

Jack Fowler's *African Queen II,* 34' x 10', received a 4'' x 6'' simple engine and a new Hodge Company vertical boiler in the early 1950s. She ran on Cobbosseecontee Lake, Maine, until her owner gave her to the Maine Maritime Museum. (Courtesy Jack Fowler)

When David Conroy of Lyons, New York, steams his *Charmaine* on the Erie Division of the New York State Barge Canal, he does not have to wait for other boats to gather before the lockmaster will open the gate. It is opened specially for *Charmaine,* because Dave is the lockmaster's boss. This very beautiful old hull is 26 feet long and is powered by a Semple F-40 boiler and a single-cylinder engine that Dave converted from a stationary to a marine version by adding a Stephenson link. *Charmaine* is shown here in Kingston, Ontario, with Bob Day at the wheel. (Bob is the Chief Interpretive Officer for the Rideau Canal.) Dave, who has a long background in steam, is a member of the Frontenac Society of Model Engineers and served as chief engineer aboard *Phoebe* when she steamed all the way to Ottawa and back, on the Rideau Canal, in the spring of 1980.

Hugh Cawdron's *Bubbly Jane* is representative of the new generation of steam launches in England. Cawdron, an architect, sees steam launches as a charming facet of "industrial archaeology," useful in illustrating the age of steam for space-age youngsters. "It seems very sad that what was once such a great and vital form of power should have died so quickly."

The 40-year-old mahogany river-launch hull is 26' x 4', equipped with a 2.5" x 3.75" double-simple engine (see page 269) and a Merryweather "B" boiler. *(London Evening Standard* photo)

UNITED STATES

SERIAL NUMBER 3714

ISSUE NUMBER

INSPECTORS' LICENSE TO ENGINEERS

OF VESSELS PROPELLED BY GAS, FLUID, NAPTHA OR ELECTRIC MOTORS.

This is to Certify that John M. Thompson has been duly examined by the undersigned Local Inspectors of Steamboats for the District of St. Michael, at Seattle, Wn. as to his qualifications as ENGINEER of Gasoline Motor Vessels, and found to be a competent and reliable person to be entrusted with the powers and duties of Engineer of such vessels of 25 gross tons, and he is therefore hereby Licensed to act as such for the term of five years from this date.

Given under our hands this 10th day of October 1907

INSPECTOR OF HULLS.

INSPECTOR OF BOILERS.

In 1906, the Steamboat Inspection Service was required to begin licensing the engineers of any boats that employed one of the newfangled motors. It was not clear then that gas boats would be more than a passing fad (petroleum reserves known at that time could not last many years), so the Service thriftily used the engraving from a steamboat-engineer license to ornament the " . . . gas, fluid, naphtha, or electric motors" license. (Puget Sound Maritime Historical Society)

Cedric Witham was 70 years young when he decided to assemble a steam launch for use at his Cow Island, Lake Winnipesaukee, summer home. He rebuilt a 24-foot hull, built a Roberts-type WT boiler, and adapted a stationary engine for marine use by adding a Stephenson link. *Cow Island Belle,* shown here in 1980, is the lively result.

The author brought intense enthusiasm and quite a bit of accumulated experience and knowledge to the designing of his ''lifetime steamer,'' *River Queen*. Built of New England oak and pine, she incorporates suggestions and help from dozens of friends. Lowell Patch (in suspenders, with Dick Mitchell and George Whitney) made the drawings for the 23-foot boat from Whitney's design for the 38-foot *Siesta* (see page 73), and when Mitchell was stricken with rheumatic fever, in 1956, Patch moved the *Queen*'s bare hull to his shop in Bernardstown, Massachusetts, and finished the boat in an act of spontaneous generosity. (Photo at left by R. Loren Graham; photo below by Harold A. Barry)

Although they were not always in keeping with "tradition," the strongest boats that the author has seen were built by Pete Levitre of Hatfield, Massachusetts. Pete churned the waters of the Connecticut River with four different steamers between 1912 and the 1950s, when his last boat was sold.

His first launch, built when he was a boy, was only 12 feet long, and both the engine and boiler were homemade. The engine's crank had once operated a cutter-bar on a horse-drawn mowing machine; the single-acting piston was from a Stephens-Duryea car; and the propeller was carved from beechwood. The porcupine boiler lacked a feed system. Pete's first few trips were not very successful, since he regularly had to dump the fire, release the steam, remove a plug in the top of the boiler, pour in water, and relight the fire. Young Levitre knew about pumps and injectors, but lack of money was a handicap — so he found a strong little tank in a junkyard and mounted it vertically, alongside the boiler. Cross-over pipes equipped with shut-off valves connected the tank to the boiler's top and bottom. With this rig, Pete could shut the valves, open the plug in the top of the tank, and fill it with buckets of water from the river. When the boiler water got low in the glass, he opened both valves and equalized the water levels in boiler and tank. By repeating this cycle over and over, Pete was able to keep steaming.

His second steamer (shown here) was built in 1917. A "yacht" by comparison with Pete's first boat, she was 16 feet long and had a boiler and engine from a Stanley car.

What his third and fourth boats lacked in style, they made up for in efficiency and performance. By that time in his life Pete could afford all the components he needed. His boats were floating laboratories for many types of engines and other equipment. (Courtesy Steamship Historical Society)

When steam launches first returned in numbers, in the 1960s, they were often featured in nostalgic tableaux. In 1970, Milton Gallup's *Fanny Dunker* was a centerpiece at the 97th anniversary party of the Cleveland Yachting Club. The VFT boiler supplied a Mumford compound engine of 1885, evidently a powerful one. The launch has since changed hands several times, and the engine is now on display at the Whistle-in-the-Woods Steam Museum, Rossville, Georgia. (Frank Reed photo; courtesy Milton Gallup)

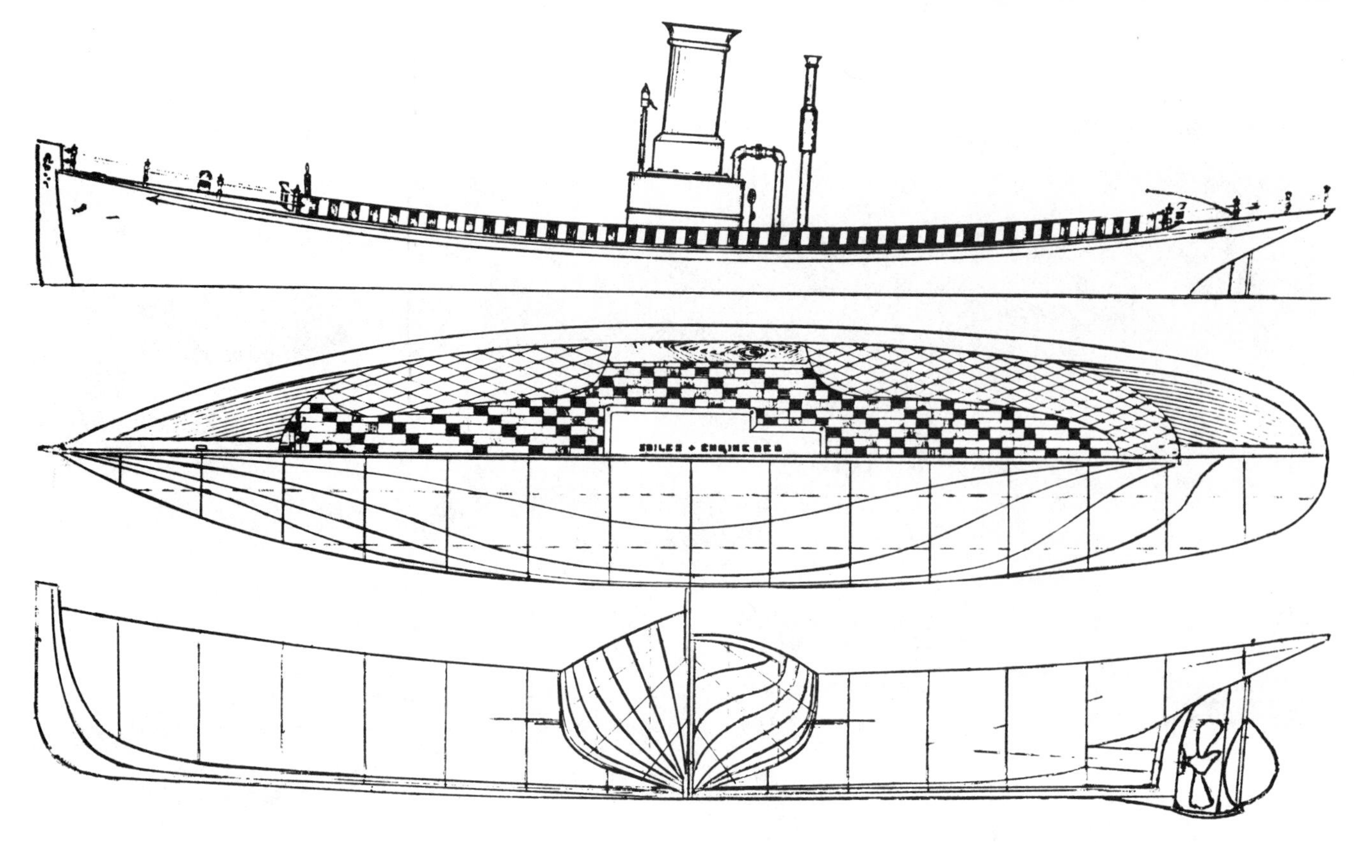

The use of contrasting woods for striking effects was a fad in the Gay Nineties. The coaming of the 25-footer *(above)* was staved in mahogany and maple, or perhaps cherry and basswood. Most cabin floors were linoleum-covered, with regular rectangles being the favorite pattern. Tufted horsehair (black) and leather were popular seat coverings in high-class boats; corduroy and duck were common in plainer service.

The botanical floor decorations in *Madge (below)* are a nice touch, but it is not clear whether these were carpets or inlaid linoleum. *Madge* was recommended " . . . for the use of a family of means . . . affords all the comforts of a modern home and may be furnished as elaborately or as cheaply as the wishes of our customers may desire." (Courtesy Allie Ryan)

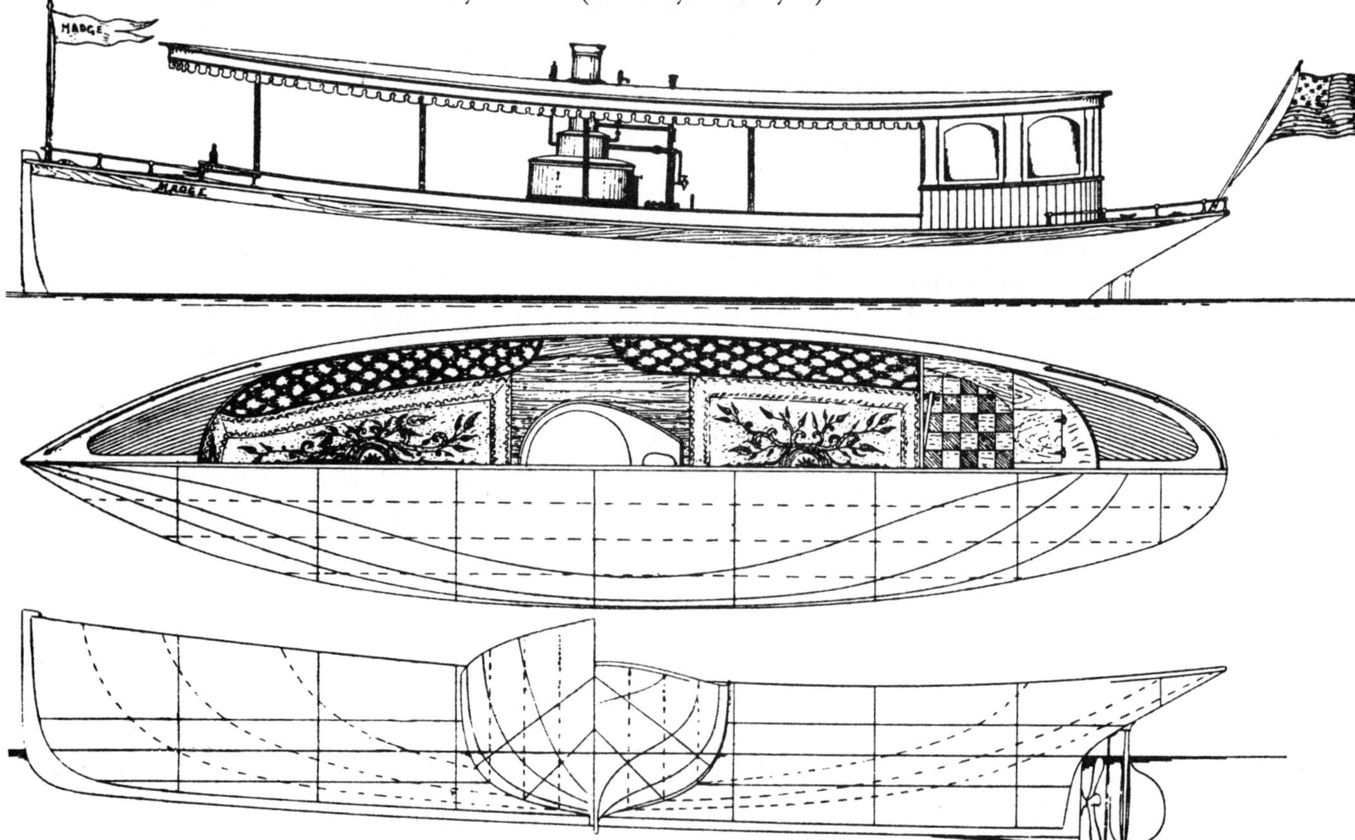

The *Ida F,* early and late. *Above: Ida F* in a workaday role, maintaining the buoys on Big Island Pond for the state of New Hampshire. *Below:* Whitney is the man with the cap, his hand on the aft stanchion. *Ida F* was still in her youth on this summer afternoon. (There's more on the *Ida F* on pages 44-46, and her engine is shown on page 277.)

STEAM LAUNCH "LEONE."

This engraving represents the Steam Launch "Leone," built by us in 1885, for Mr. C. H. Prior, Gen. Supt. Chicago, Milwaukee & St. Paul Railroad, Minneapolis, for use on Lake Minnetonka.

She is 34 ft. long, 8 ft. beam, by 3 ft. 4 in. deep. As will be noted, her machinery consists of a Clyde Boiler, 40 in. in diameter by 50 in. long, supplying steam to a Willard Special High-speed Engine, 6 in. in diameter by 8 in. stroke. Her outfit of machinery includes every device that can add to its successful operation. She is elegantly fitted up inside, her wood-work being of Birds-eye Maple with Spanish Cedar Trimmings, all finished in oil. Her steering apparatus is located as shown, just aft of the engine, enabling the boat to be steered by the engineer or by any of the passengers who may enjoy such sport.

The *Leone,* proudly featured in an 1880s Charles P. Willard and Company catalog, was fitted with a ''marine'' boiler, but she steamed on the lakes of the Midwest.

Bill Downey of Kittery, Maine, owner of this little (17.3-foot) launch, trails her to where the action is. Her Stuart 5A engine does a good job, and the owner and friends have a lot of fun. Among her past journeys is a 76-mile cruise on the Hudson River and Lake Champlain. The canopy stanchions are an occasional addition. (William Downey photo)

Some connoisseurs regard the 25' x 6' canopy launch *Nipissing III,* formerly owned by Keith Holmes, as having the best proportioned small fantail hull now in service. It is surprising to learn that the boat was built for gasoline power in 1910 — perhaps by a boatbuilder who had loved steam launches in his youth. The tasteful sumptuousness of the low canopy and tieback curtains and the uncluttered lines of the stack and sheer reveal Holmes' sensitivity to the requirements of the hull. J.W.B. Coulter, of Toronto, Ontario, is the present owner. (Courtesy Keith Holmes)

This 26-foot double-ended steel lifeboat was converted to a steamer by Joseph Pike of Bridgton, Maine, with the assistance of P.L. (Pete) Levitre. The fire-tube boiler burned wood, and the engine was a 3'' x 4'' Shipman stationary that Joe converted to a marine engine with the addition of a Stephenson link. The steamer ran on Long Lake, in Bridgton, during the 1960s, and was eventually sold and taken to the Southwest.

This snapshot (original in sepia) of a nameless boat in Maine (probably Shipman-engined) shows what the mailorder steam launches looked like when they arrived at their destinations. (Courtesy Steven Lang)

The *Polly,* seen here on the Siuslaw River, Oregon, is entirely the fruit of the steam launch revival of recent years. The hull is a beamy plywood structure that was offered as a modern steam launch design in *Sports Afield*'s *1962 Boatbuilding Annual.* The engine is a Semple V-compound and the boiler a new industrial VFT ordered from Portland. The John Tiffany family, of Eugene, Oregon, was well satisfied with the ruggedness of the power plant and the stability and capacity (12 passengers) of the 20' x 7' boat. Edwin Todd, of South Colby, WA, is now the launch's owner. (Courtesy John Tiffany)

Some of the people who, like John Winters, have the most fun with steam launches cruise in boats like *Soar Point,* in places like this (the River Soar, at Hainley, England). The true Scotch marine boiler (see page 305) and the workmanlike piping indicate careful planning by a knowing owner. (J. Harvey photo)

This photo barely made the ''Canopy Launch'' section. In those days they figured that a little sun and rain wouldn't hurt passengers, but the boiler cost *money!* (Courtesy Charles Hill)

Above: This clinker-built launch, seen on the Ottawa River, Ontario, about 1895, deserves close study by anyone who is seeking that elusive quality — style — in a small boat. The curve of the sheer is perfectly matched by the full-length canopy. The stack is the right size and rake; the 2.5'' x 3'' Shipman power plant is just right for the job. Barry Price reports that the shirtsleeved helmsman was his grandfather, chief mechanic for a chain of lumber mills. The bearded engineer was a pill manufacturer who died a millionaire. (Courtesy Barry Price) *Below:* Pat Spurlock's *Willow* has a 2-h.p. Blackstaffe-Wood compound engine in a 20-foot Bill Durham fiberglass hull. Pat has trailered the boat thousands of miles and cruised extensively on the lower Columbia River and on lakes in Washington, Idaho, Montana, and Wyoming.

Much of the history of this long-ago steam launch is lost, but this much remains: the *Cygnet,* once a passenger boat, had suffered the indignity of towing logs for the Inman-Polson Lumber Company in Portland, Oregon, when this picture was taken. (Courtesy Thomas C. Graves)

Cabin Launches

About 100 years ago, it became possible for a middle-class citizen to consider buying a powered vehicle with an enclosed cabin and even, perhaps, a toilet — the "Winnebago" idea — but there were no cars and few roads then, so the dream had to be realized on the water. Cabin steam launches were the first precursors of the half-billion automobiles, motor cruisers, motor homes, and aircraft that have since burned fuel to provide their owners with mobile shelter.

It is impossible for us to know what these first privately owned, mechanically powered vehicles meant to their owners. Did our great-grandfathers sense the epochal importance of putting heat to work for private pleasure?

The accommodations of early cabin steam launches were most inadequate. Narrow settees along the sides dominated the furnishings; galley and toilet facilities were primitive and awkward. Most present designers of cabin steamers choose convenient and comfortable shelters that look much like those of motor cruisers. There is a large, unexploited potential in adapting the 20th century's revolutionary improvements in creature comforts to the graceful and airy steam launch enclosures of 100 years ago.

Scudder, of 1903, is a fine 36' x 7' specimen from a major North American steam launch builder, Davis & Son, of Kingston, Ontario. Until 1967, she had a Davis water-tube boiler, a compound engine, and a 36'' x 34'' propeller. *Scudder* was part of the collection of D. Cameron Peck, a savior of many steam launches during the 1930s and 1940s, when other people were breaking them up. George Lauder bought the boat for use at Watch Hill, Rhode Island, after he gave his 100-year-old *Nellie* to Mystic Seaport. The old boiler was replaced by a new Funk boiler.

Many a rich man's private launch spent her declining days hauling passengers for hire. *Rowena,* a Lake Winnipesaukee party boat in the 1920s, was bright mahogany all over — hull, cabin, and decks. Captain L.C. Brock gave this picture to the author in 1943. (Courtesy Capt. L.C. Brock)

Of all the small steamers built in recent years, the author knows of none that have been more thoroughly planned, executed and documented than Ike Harter's *Susan Gail.* Educated in naval architecture and marine engineering and employed at the Newport News Shipbuilding and Dry Dock Company, Ike had the background to do the job. This powerfully built 28-foot cruising launch, the product of his own hands, is powered by a coal-fired Semple fire-tube boiler and a Semple V-compound engine. The *Susan Gail* is a wonderful example of what can be done with a lot of know-how and hard work.

On this occasion in 1974, the *Susan Gail* was flying the flag of the Steamship Historical Society of America. She has since been sold to Hershey Lerner, of Hudson, Ohio. Lerner steamed her from Newport News to Hudson, a 1,200-mile voyage (including a part of the Intracoastal Waterway, the Hudson River, and the New York State Barge Canal) that consumed less than $150 worth of coal. (Photo by Alexander C. Brown, Cdr. USN Ret.; courtesy Isaac Harter III)

Chester Belle's "torpedo stern" was a passing fancy of 1910, but her glass cabin was a good idea, providing for airy but comfortable boating. Originally a steamer, the 40-footer ran many years on Lake Winnipesaukee as the gas-powered passenger boat *Foxy.* She became *Slo-mo-shun* in 1954, when Dick Mason hired George Whitney to install the steam plant that Whitney had designed and built for the never-built *Siesta.* Lee Graves took the boat to Maryland and renamed her *Chester Belle.* (W. Lee Graves photo)

Artú Chiggiato, of Venice, has only sketchy knowledge of these cabin launches in old family snapshots. The 45-footer, *Tuna-Puna (top),* was English-built for a Parisian nobleman, Le Marquis de Polignac, who summered in a Venetian palazzo. The boat was sold in 1895 to G. Stucchi, a relative of Chiggiato.

The 24-footer *(bottom)* was American-built, about 1885, and served a Venetian owner. The low funnel height on these boats was necessary because of the canals' low bridges. (Courtesy Artú Chiggiato)

The 38-foot *Cyrene* is all business — a stout workboat hull, a navy boiler, and an imposing Canadian triple-expansion engine. Allen Rustad's big, heavy boat lolls around quietly in the background at Puget Sound meets, but still she dominates the scene — especially when her steam siren whoops!

The 5'' & 8'' & 12'' x 8'' Davis & Son engine is lugged down to 120 r.p.m. by step-up chain drive (60:22 gearing) to the tailshaft and a 28'' x 36'' propeller. Rustad is the builder and former owner of the surfboat steamer *Lucifer* and also serves as engineer for the Sea Scout steamer *Oceanid.* (David Rucker photo)

Al Giles's navy whaler, *Crest,* is the champion cruising steam launch on the Pacific Coast. Since installing the steam plant in 1962, the owner has cruised nearly 20,000 miles in Puget Sound and in British Columbia waters. The boat routinely makes long passages and plows through rough weather.

Crest has a 3.5'' & 6'' x 4.5'' engine (see page 282), a Merryweather boiler (see page 315), and a 25'' x 34'' propeller. Feedwater and fire controls are automatic, and the boat is equipped for comfortable cruising, with steam heat, electric lights, and modern cooking facilities. (John Bailey photo)

Tom Rhodes, of Saratoga Springs, New York, appreciates old steamers of every type, whether they be large steamboats, tugs, or launches, and when he built his 29-foot *Tuscarora* he incorporated the best of them all. Building a hull of this size is a big project when you already have a full-time job — even more of a project when it includes scrollwork, paneling, ornate deck timbers, and just about everything you would find on a 200-footer. *Tuscarora* was launched on Lake George, New York, in 1979 and is powered by a Semple woodburning boiler and a V-compound engine. Every convenience for long cruises has been incorporated into this vessel, and Tom looks forward to a lifetime of fun.

The author's little cabin steamer, *Lourick,* is shown steaming up the West River in Brattleboro, Vermont, in 1946. She wasn't fast (her 2.5'' x 3'' engine is shown on page 257), but she was satisfying.

THE WINDERMERE STEAMBOATS

A deep debt of gratitude is due to George Pattinson, the moving force behind the Windermere Steamboat Museum. Located in the Lake District of England, the Windermere museum officially opened in 1977. Pattinson was largely responsible for the acquisition of the boats on display there. The four boats pictured on the following two pages show the development of Lake Windermere steam launches. Pattinson writes in a museum booklet that "the oldest, S.L. *Dolly,* shows the primitive application of steam power . . . . T.S.S.Y. *Esperance* and S.L. *Branksome* exemplify the development and ultimate refinement of steamboats."

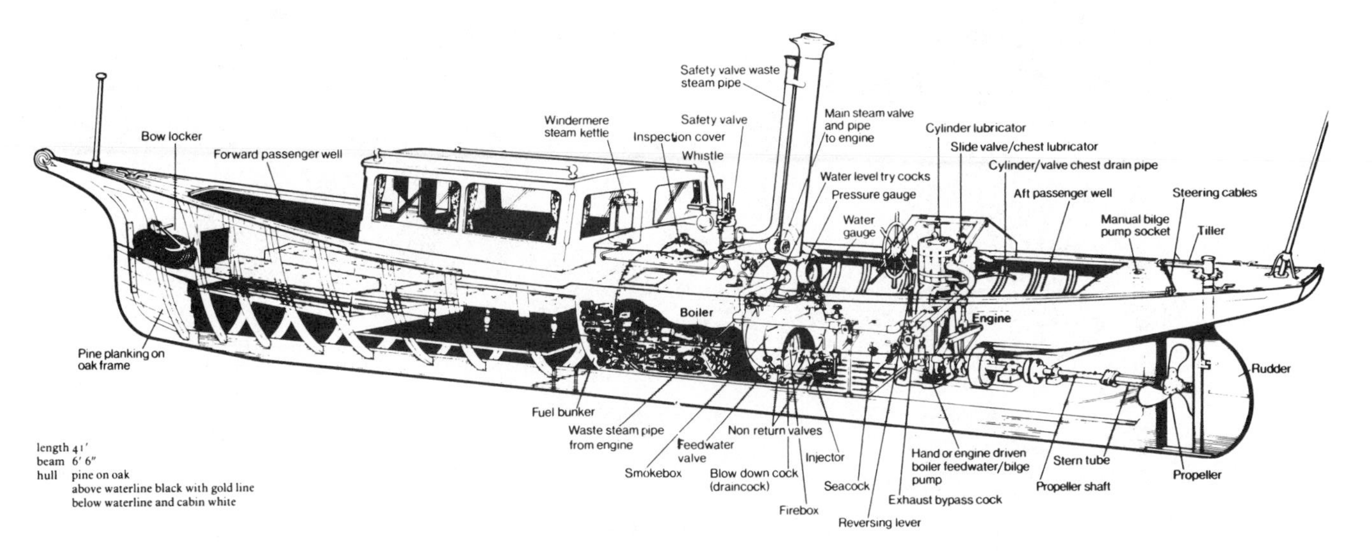

Based on the measurements and drawings of David Collins, a marine surveyor, this ink drawing of *Dolly* is given prominent display at the Windermere Steamboat Museum. According to a museum booklet, *Dolly* was built around 1850 (nine years before Thornycroft's *Nautilus!)* for a gentleman named Alfred Fildes, who lived on the west shore of Lake Windermere. A few years before, a railhead had been established at Windermere; tourists were attracted to the region, and some of the most privileged built summer retreats and introduced steam launches and yachts to the lake.

In 1894, in her "old age," *Dolly* was sold, and she was taken to Ullswater. There, in February 1895, during "the great frost," she was damaged by ice and sank at her mooring.

Covered by 45 feet of water, *Dolly* lay forgotten until 1960, when she was discovered by two members of the Furness Branch of the British Sub-Aqua Club. Her historic value was immediately apparent, and the Furness Branch salvaged her after much time and effort. George Pattinson undertook her remarkably successful restoration, and today she "remains an astonishing glimpse into Britain's industrial past, all the more so for being restored to perfect condition and kept in working order." (Courtesy Windermere Steamboat Museum)

Dolly, the oldest existing steam launch, displayed her original power plant when these photos were taken in 1966. That engine was *meant* to last 120 years! The boiler has since been replaced.

OPPOSITE PAGE

Top: The iron-hulled *Esperance,* 75' x 10', was built on the Clyde, in 1869, by T.B. Seath & Co. Her first owner was H.W. Schneider, a wealthy industrialist, who used her for one leg of the journey from his Lakeside, Lake Windermere, home to his offices in Barrow. *Esperance* steamed in all weather, so her bow was sharply raked for ice-breaking. She sank in 1941 but was later salvaged by George Pattinson's father.

Center: George Pattinson was not exaggerating when he wrote that "*Branksome* represents the ultimate in Victorian elegance, and there is probably nothing to equal her in Europe or America." Her 50' x 9' hull is teak, and some of the paneling is walnut. The wash basin is solid white marble. With her side-fired locomotive boiler and her 7.5" & 11" x 7" engine, *Branksome* can steam at 14 m.p.h.

Bottom: Swallow, of 1911, was the last of a long line of Windermere steam launches. The 45.5' x 8' teak hull was built by Shepherd's, of Bowness; Messrs. W. Sisson and Company, Ltd., built the 4.5" & 6" & 8" x 5.5" triple-expansion engine.

Below: Harold Lanning designed his 37-foot *Dodo* in 1915 to be a plain, functional, working steam launch, with no concessions to nostalgia or superficial prettiness. The deckhouse is characteristic of Puget Sound gasoline workboats of 65 years ago, the small openings revealing more interest in keeping cozy in a gray climate than in viewing the scenery. Under her original owner, *Dodo* had a Navy K engine (bought as unused surplus after World War I) and a wood-fired U.S. Navy Ward boiler. An O'Connor WT boiler now supplies steam to the same engine. Mr. and Mrs. Bill Dessert of Marysville, WA, acquired *Dodo* in 1977. (Everett Arnes photo)

Right: The Racine Boat Co.'s model #4 was a 25-foot "family" boat. A 3-h.p. plant gave her 8 m.p.h. *Below:* Bob Thompson's *Rum Hound* glows in the afternoon sun on Squam Lake, New Hampshire. This is pretty close to the vision many men have of steamboating — a personal boat, the way they want it, homey and containing very little that could be bought at a gas-boat marina. An old kitchen-sink pump does bilge duty on *Rum Hound.*

The *Susquehanna* relaxes at her wharf in Grapeview, Washington, about 1950. Built for W.G. Clayton, the 23-foot clinker-built hull contained a Seabury compound engine, 2.75'' & 5.75'' x 4.5'', and a VFT boiler built by Clayton. (Courtesy E.A. Middleton, Jr.)

Richard Hovey, of Pride's Crossing, Massachusetts, employed an experienced boatbuilder and designer, Pete Culler, to design his 34' x 8' *Skookum Jack.* The hull lines were adapted from those of the U.S. Navy steam cutters of 1900. The boat was built by Robert Rich at the Bass Harbor Boat Shop in Maine.

When launched, in 1974, *Skookum Jack* was equipped with a new Amoskeag Iron Works water-tube boiler and a 4'' & 8'' x 4.5'' engine (see page 279) turning a 28'' x 30'' wheel. Allen Bigelow of Beverly Farms, Massachusetts, installed the steam plant. Later, a fully automatic, enclosed high-pressure power plant was built for the boat by O'Connor Engineering Laboratories of Costa Mesa, California.

Dr. Magnor Vatne, of Springfield, Vermont, designed his 35-foot *Vatnefjord* and handbuilt the hull of solid fiberglass. The deckhouse is of fiberglass-covered plywood. A woodburning Clyde-type boiler supplies steam for a 3.25'' & 6'' x 4.25'' antique engine, turning a 24'' x 36'' propeller. *Vatnefjord* cruises the Connecticut River at speeds up to 9 m.p.h. (Everett Mitchell photo)

The steam tugboat format appealed to many during the 1960s, and several good specimens were built. *Benj. F. Jones,* a 26-footer designed and built by Bobby Rich, of Maine, in 1965, is now owned by William W. Willock, Jr., of Maryland. The color scheme and detailing are true to harbor tugs of a generation or two ago. Miniature tugs provide limited accommodations for the investment and may be useful only as day boats. The *Benj. F. Jones'* engine is shown on page 256. (Courtesy W.W. Willock, Jr.)

Phoebe comes home. The Frontenac Society of Model Engineers, with headquarters at the Pump House Museum in Kingston, Ontario, acquired the much-publicized *Phoebe* and returned her to Kingston, her 1914 birthplace. The 1952 photo above (by Milton Gallup) shows her at Put-in-Bay, Ohio, when she was owned by Frank Miller and Warren Weiant, Jr. The condition of *Phoebe* today attests to the loving care lavished on her for nearly 70 years. Her engine is shown on page 279, her boiler on page 310. (Photo at left courtesy Frederick G. Beach)

Historical Note: In 1901 Mat Davis, of the Davis Dry Dock Co. in Kingston, Ontario, designed and built the steam launch *Alleghenia* for Dr. John Brashear, president of the University of Pittsburgh. The *Alleghenia* burned at her dock in 1903. In 1904 Andrew Carnegie presented a new boat to Dr. Brashear, named *Phoebe* for Brashear's wife, but this boat also was lost to fire. In 1914, upon his retirement, friends, students, and colleagues secretly collected funds and had Davis Dry Dock Co. build still another boat, also named *Phoebe,* as a gift to their beloved president and friend.

Above: Henry Hoffar, of British Columbia, had a ride in a steam launch named *Antic* in 1898, when he was 10. This event fixed his interest on boats and boatbuilding, and he went on to build 2,206 yachts, fishing vessels, tugs, and other boats. Hoffar Hull #2,207, 1963, is his own *Antic.* The 28' x 7' boat, built to the highest standards of wooden boatbuilding, is equipped with a Semple F-40 VFT boiler and a 1911 Lune Valley engine, 3'' & 6'' x 3.5'', restored to original condition. Hoffar donated *Antic* to the Vancouver, British Columbia, Maritime Museum in the mid-1970s. (Everett Arnes photo)

Opposite, top: Andrew Culver's *Maggie* unites several Pacific Northwest elements in a pleasing boat suited to the region — a part of the world where even a wee steamer wants a cabin most of the year. The hull is a type designed for sportfishing on the choppy waters of Puget Sound; it is jaunty, spacious, and buoyant for its 15½-foot length. *Maggie* is powered by an early Blackstaffe-Wood 2-h.p. steeple-compound power plant, which has given good service since 1962. (Everett Arnes photo) *Bottom:* Once the property of Horace Nelson, who called her the *Meadowdale Queen,* the *Island Queen* has the most refined, most thoughtfully designed and built steam launch power plant in North America. Nelson's experience in rebuilding steam cars may have some bearing on the unusually high standards he set for his first steam launch. For example, in rebuilding the *Queen*'s Mare Island Navy K engine, he had all journals hard-chromed and precision ground to original specs. Valve rods and the high-pressure valve are chromed. The Navy Type A express boiler (see page 312) with 200 square feet of heating surface is as fully engineered and detailed as a destroyer's. There is a one-kilowatt, 32-volt dynamo, and converted 110-volt AC power is provided for the controls of the boiler's two burners. The quarters will sleep five in comfort, and the boat's accommodations and equipment complement the excellence of her power plant. Richard Robbins, of Friday Harbor, Washington, is the present owner.

EAGLE
MAGGIE
SHAW ISLAND

WHITNEY CABIN LAUNCHES

When the author visited George Whitney's apartment in Bridgeport, Connecticut, he saw that the walls were covered with photographs of boats and engines Whit had built. The author then enlisted the help of a number of friends to have good copies made of the photos, in case of future loss (which later occurred). These pictures of Whitney cabin launches were copied from his wall at that time.

Zuellr (or is this Whit's rendering of *Zuella*?) on the ''Missippi'' River in 1888. There are some nice whimsical touches in the Turkish cupola with American eagle on top of the ornamental studs, and the pilasters on the cabin.

Minnie May, 1894, was the third steamer built for George F. Sweet, of Middletown, Connecticut. A toilet cubicle and an enclosed space to starboard — perhaps a galley — are clustered near the machinery. Except for accommodations, *Minnie May* was identical to *Mohawk,* even to the swivel fishing chairs bolted to the afterdeck.

Mohawk, 1889, 40' x 9', had a Roberts boiler and 5'' & 9'' x 7'' engine. Built for Dr. H.F. Libby, of Wolfeboro, New Hampshire, her accommodations were planned to be just right for family use on Lake Winnipesaukee. The clerestory windows appear to be a conceit here; in some excursion steamers they were useful for regulating cabin temperature.

West Wind, 50' x 10', was built in 1891 for John H. Libby, for use on Lake Winnipesaukee. She made 11 m.p.h. with a 6'' & 10.5'' x 8'' engine. Whitney made a sliding-piston whistle for the boat, and the owner had a lot of fun playing ''How Dry I Am'' on the whistle whenever he steamed past a large camp of religious teetotalers.

Karl Carlson's *Cannonball Special* has a nicely balanced layout and a character all her own. The 26-foot ex-navy motor whaler, of Palo Alto, California, is powered by a simple engine and VFT boiler.

David Kyle's *Souvenir d'Antan* is a sleek, gracious river launch in which the owner has taken innumerable cruises up and down the Thames. The original engine is in the 45' x 7' x 2.5' hull, launched at Kingston-on-Thames in 1900. The present crisp, authentic state of the launch is the fruit of 40 years of planning, hull restoration, boiler building, and engine rebuilding by a man of energy and vision. When Kyle bought the boat in 1935, she had lain neglected and unused for 20 years. If he gets the 12 knots he expects out of a new 250-p.s.i. water-tube boiler he is building and the complete overhaul he is giving the engine, he will rename the boat *Scalded Cat*.

Aries, originally from Lake Winnipesaukee but now owned by Bill Harrah's museum in Reno, Nevada, is a fine example of a small glass-cabin launch. Built in 1905, the boat is 35 feet long, with a Murray & Tregurtha water-tube boiler and a Stickney compound engine. She was refurbished and the power plant installed by John S. Clement, a steam launch expert.

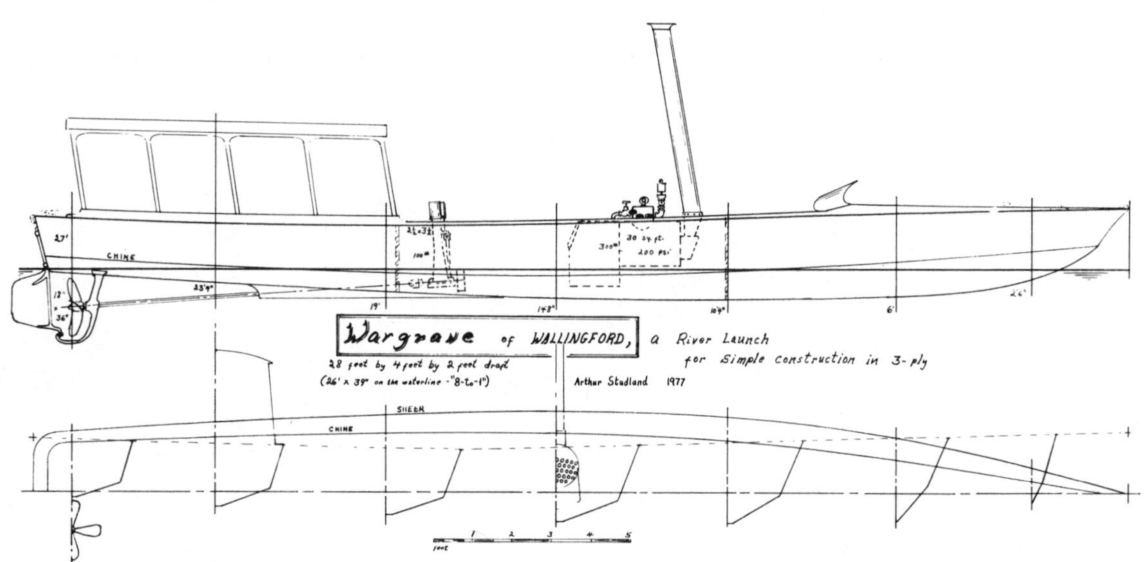

Arthur Studland, of England, is planning a river launch in the aristocratic tradition of long, lithe grace and separated cabin and engine spaces. He expects *Wargrave* to run 10 m.p.h. with three horsepower on 1,500-pound displacement. Studland also intends the launch to demonstrate that a displacement boat capable of high speed will cruise at moderate speed with very little power and fuss. The hull is a simple fabric of softwood longitudinals and 3/8-inch plywood. The boiler is a composite version of the pinnace type, with a dry firebox joined to a horizontal fire-tube pressure drum, and a dry-steam dome for the boiler fittings.

Throughout the 1950s, *Staurus* was the happy ship of Lake Winnipesaukee. Widely known among New England steam fans and always friendly to visitors, the old launch fostered warm feelings about steamboating and countless dreams about "building my own steamer."

The 34-foot boat was launched as a passenger steamer on Squam Lake, New Hampshire, in 1908. Thirty years later two good friends, Bill Viden and Bob Bracchi, restored steam power to the old hull with ex-navy components, a Ward water-tube boiler (see page 314) and a Type B compound engine. The hull and boiler eventually wore out, but the engine went into David Thompson's *O.D. York.* (Loren Graham photo)

David Thompson's 46-foot *O.D. York* was originally the Lake Winnipesaukee camp boat *Recruit,* with a gasoline engine. He installed a homemade Roberts-type boiler and the Navy Type B cutter engine that was formerly in *Staurus.* A roomy firebox permits easy, silent steaming on wood fuel. David grew up in a steam launch family and has floated several boats, and probably a dozen different steam power plants, on Lake Winnipesaukee. For many years, he has been the New England host for steam launch meets in Moultonboro, New Hampshire.

Elmer Brooks was approaching retirement age as a machinist for a heavy-equipment builder when he built a 10-h.p. compound engine (see page 274) and boiler (see page 309) and installed them in a Navy whaleboat, which he named *Artemis.* The boat is still steaming — several owners and name changes (*Success, Steven Douglas,* then back to *Artemis*) later. (Courtesy Ric Shrewsbury)

The 32-foot *David T. Denny* flies her flags proudly in Seattle harbor. The hull, built many years ago in the Bremerton Navy Yard for the U.S. Bureau of Indian Affairs, took part in the Byrd Antarctic expeditions. *David T. Denny* is powered by a Scotch marine boiler and a U.S. Navy Type B3 compound engine, 3.5'' & 7'' x 6''. Richard Robbins rebuilt the hull and machinery in 1971 through 1972. (Melvin Fredeen photo)

Naphtha Launches

Once in a while, a fully realized and wonderfully apt product design or merchandising idea suddenly appears, and it is cited forever after as a model for inventors and businessmen to reflect upon. During the past 20 years, the naphtha launch has received a good deal of attention as just such a model. The best justification for writing about "those wonderful naphtha launches" one more time is to reassert their identity with steam launches and to assess the new power sources that made boating full of variety and promise at the turn of the last century.

The 20-year career of the naphtha launch coincided with the crude beginnings, uncertain rise, and final triumph of gasoline power in small boats. Naphtha launches burned the same fuel as early internal-combustion boats, so they are sometimes erroneously placed on the same family tree, but they were not gas-boat ancestors. They were, instead, refined and prettified steam launches, an "improved product" that brought deserved rewards to its inventor.

There were gasoline motorboats before there were naphtha launches, but they evolved slowly and were clumsily promoted, at first scarcely reaching the consumer market. Their superiority was not widely perceived for 20 years, while men with flair and imagination could find a ready market for a concept as far-fetched as boiling gasoline in a steam launch's boiler.

By the late 1870s, railroads, steamboats, and transatlantic liners had profoundly affected the lives of most people, but these steam-powered conveyances were all coalburning. Petroleum was commonplace in the form of kerosene, marketed worldwide for illumination, but it was not used to generate power. Gasoline, or naphtha, was a cheap and dangerous surplus commodity, in some demand as an industrial solvent or an odorless fuel for cooking stoves, but mostly burned off or dumped in creeks near the refineries.

In 1878 the U.S. Treasury Department issued a circular recommending that "special engineer" licenses be required for the engineer and pilot of every steam launch in the country. Given the tens of thousands of launches, large and small, that in those days steamed the coasts, rivers, lakes, and ponds, there is no doubt that the licensing effort was honored mostly in the breach. However, the possibility that operation of steam launches would be subject to licensing and other restrictions intensified the search for alternatives to steam power.

In 1883, Frank Ofeldt patented an ingenious power plant that did not use steam — he specified "naphtha or gasoline vapor." Today one's imagination recoils at the thought of lighting a fire under a vessel filled with gasoline, but the response was different then. According to Malcolm MacDuffie, the U.S. government, "which deemed the confinement of water vapor in a pressure vessel so dangerous as to require a licensed attendant, never turned a hair at the prospect of white-flanneled, starched-collared amateurs regulating high-pressure gasoline vapor in contact with sizzling hot copper tubing." MacDuffie speculates that it was "not then thought necessary to render all of life foolproof."

Ofeldt's fuel-burning system, his single-acting enclosed-crankcase engine, and the boiler arrangements and controls were of original, patent-protected design. The heat cycle was not patentable, since this steam engine principle was already two centuries old. The inventor found the most direct and logical way of achieving his ends. All systems were in balance and harmony, and each potential weakness was countered in an effective way.

Boiling gasoline (or alcohol, or freon, or . . .) instead of water to drive a "steam" engine has real thermodynamic advantages, and enthusiasts still experiment with this idea. The "freon cars" mentioned in the newspapers from time to time are nothing more than replays of Ofeldt's 1883 concept, although gullible journalists always hail them as "new breakthroughs."

Ofeldt's original patent drawing was of a 2-cylinder single-acting engine of 3-inch bore and stroke. This was soon revised to three cylinders, with cranks spaced at 120 degrees to ensure self-starting. Slide valves moved transversely across the cylinder heads, and reversing was by slip eccentric. The engine exhausted naphtha vapor into the crankcase, then through a long keel condenser to the storage tank in the bow.

The inventor joined forces with a wealthy oil man, Jabez A. Bostwick, and the two founded The Gas Engine and Power Company in New York City. This establishment built 2,000 naphtha launches between 1885 and 1905, launches that were as much a marketing achievement as they were a technical innovation. A beautifully unified combination of boat and power plant was designed to suit a specific, prosperous market and promoted with skill and determination. They were Buick and Chris-Craft precursors as no American steam launches ever were.

Before the naphtha launch, heat engines were operated by men from the lower classes. Upper-class men might command machines, but they did not touch them. The naphtha launch builders saw that this antique social narrowness went against human nature, and they at last made it respectable for the well-to-do to operate mechanically powered machines (for amusement only, of course). The promotional literature claimed that "any man of leisure and means who wishes for a chance to run a neat little bit of machinery himself — and it is a passion strong in most of us — will find the Naphtha Launch the very thing his fancy has painted."

The boats had everything, including grace and comfort, cleanliness (in an era of abounding coal soot and horse manure), mechanical ingenuity, and ease of operation. Even humor sparkled in the instruction book that went out with each boat from the factory on the Harlem River: "Don't shoot through your tank. Don't light a match to look into your naphtha tank."

The fuel consumption per horsepower-hour of small boat engines in 1885 was about as follows: A simple steam engine burning wood or coal required 60,000 B.t.u.'s, or a penny's worth of fuel; a naphtha launch needed 40,000 B.t.u.'s, or 2 cents worth of 6-cent-per-gallon gasoline; and a gasoline ("explosive") motor burned 12,000 B.t.u.'s, or about a half-cent worth of fuel.

Ofeldt left the Gas Engine and Power Company in 1887 and set up a factory in New Jersey to manufacture a new engine of his design. This engine used alcohol instead of naphtha. The "Alco-Vapor system" at first employed a 3-cylinder radial engine with a square-sectioned pipe boiler on top of it. This was succeeded by a V-shaped engine employing an alcohol-water solution as the working fluid. Firing was by kerosene. President Grover Cleveland was an Alco-launch owner.

For 20 years after Ofeldt introduced them in 1883, naphtha (and other "vapor") launches were very popular. In England, the Yarrow Company brought out a "petroleum-spirit vapor launch" in 1888. Like the Alco-Vapor launch, this used kerosene fuel, separate from the small quantity of gasoline in the system. The Escher Wyss firm, with factories in Switzerland and Germany, obtained a license and built naphtha launches for the European market with good success.

Naphtha launches were nearly all of a single hull type — the classic fantail launch with a standing canopy or, on the larger boats, a glass cabin. They are remembered as a sort of ideal, an epitome of the best powerboat design for recreational uses when one to four horsepower per ton was the norm. They were graceful, well proportioned, highly finished, and always recognizable. In Malcolm MacDuffie's eulogy, the naphtha launch was " . . . that sweet little, plumb-bowed, fantail-stern launch with her warm brightwork, her shining brass funnel, her scalloped canopy and her lazy flags. There [was] a kind of charm about her like that of a pretty girl with . . . her head held high, dainty, refined, . . . adorable."

Two hand pumps were used in starting up the naphtha engine, one to pump liquid fuel into the coil boiler and another to pump air to the storage tank in the bow and force naphtha vapor to the burner for initial heating. Two or three minutes after "lighting off," the small quantity of gasoline in the boiler generated enough vapor so that the engine could be rolled over.

Once underway, a pump driven off the crankshaft forced additional naphtha into the boiler. A small quantity of vapor was bled off the top of the boiler and blown down an air-mixing tube at the front of the casing, to reach the burner. A single control wheel was rotated a quarter-turn to the left to go ahead, or to the right to go astern. The three pistons at 120-degree intervals on the crank ensured positive starting and reversing.

The naphtha tank in a 2-h.p. boat held a barrel of fuel. In many boats the hull was perforated forward of a watertight bulkhead, so that the fuel tank was awash in cooling water. This weight in the bow was not much less than that of the engine at the stern (200 pounds for a 2-h.p. engine, 600 pounds for the 8-h.p. size), and the live load of passengers was carried amidships, making for a roomy, well-balanced boat.

The Gas Engine and Power Company advertised launches from 18 to 76 feet long and power plants of 2 to 16 horsepower. (There were also 1-h.p. 16-footers, apparently rare.) The 64-, 67-, and 76-foot boats were twin-screw. The working pressures and propeller speeds of the naphtha launches were those of conservative 1885 steam launches — 65 p.s.i. and 300 r.p.m. for a 4-h.p. unit. An 18-foot launch could make 7 m.p.h., and the largest boats, 12 m.p.h.

Naphtha launches sold on their own merits at prices comparable to the highest-quality factory-built steam launches. The 1-h.p. 16-footer cost around $500; a 4-h.p. 25-footer, $1,000; and a 10-h.p. 36-footer, $2,400. Cabin launches were more expensive, with the 36-footer costing $3,600; a 12-h.p. 42-footer, $5,200; and the twin-screw 76-footer, $12,500.

By 1895, it was becoming clear that ordinary pleasure steam launches were on the wane, and new motive powers would serve the new century. Naphtha launches were near their peak, and they were universally admired (except, as Kenneth Durant says, by a few diehards off in the woods who objected to " . . . the sullen roar of those terrible Naphtha Launches").

Battery-electric boats were briefly ascendant — Edison was an American, and the gas-engine pioneers were foreigners. Gasoline marine engines were well established in California before 1890, proving their superiority in working boats, but they were not yet widely accepted in the East. "Explosive motors" were considered a coarse and noisy breed, sometimes absolutely unstartable, and their claims of remarkable fuel economy were still met with skepticism.

When competition finally forced the Gas Engine and Power Company to offer gasoline engines, they could not summon up much enthusiasm in their announcement:

> The demand for boats fitted with motors of the explosive type, in which the impulse is obtained by a series of explosions of a combination of air and gasoline vapor, has induced us to build a motor of this description. In it we have embodied all of the

latest improvements, which, together with our exceptional facilities, has enabled us to produce a nicely fitted, well-balanced motor, which we can guarantee to give satisfaction.

Given their popularity in their day, it is surprising that so few naphtha launches have survived. As far as I know, the only presently operational naphtha launch in the world is *Anita,* of 1902, restored and owned by William H. Richardson, Sr., and his son, Bill Richardson, Jr. The 6-h.p. *Anita* plies the waters around Sheboygan Falls, Wisconsin. Six other naphtha launches are on display in public places. The *Lillian Russell,* of 1904, was among the last naphtha launches built; she has a transom stern and can be seen at Mystic Seaport. *Chiripa,* of 1899, is displayed at the Henry Ford Museum, in Dearborn, Michigan. *Frieda* can be viewed at Yellowstone National Park. The Gas Engine and Power Company's Hull #1806 is at the Adirondack Museum, Blue Mountain Lake, New York; Hull #1744 is at the Philadelphia Maritime Museum; and Hull #1168 is being restored by Richard Morrison and others in Old Forge, New York.

Forty years before Americans acquired the habit of washing the car on Sunday, the naphtha launches' stacks and engine casings were always polished and their paint and canopies fresh. It was part of the naphtha launch image. The great locomotive headlight on this Fulton Lakes, New York, launch (in 1896) indicates its usefulness in taking late guests home across the lake or making the rounds with a party in the still of the night. (Courtesy Richard Morrison)

A naphtha launch in the Thousand Islands. (Courtesy Frederick G. Beach)

With its naphtha tank forward and seating amidships, this 30-footer made a roomy day boat. (From C.P. Kunhardt, *Steam Yachts and Launches,* 1891 edition)

Below: This "warning" appeared in a Gas Engine and Power Company catalog. The company's good idea had spawned imitators, and the board of directors was getting testy.

Bottom: Out of several thousand naphtha launches built between 1885 and 1905, only a handful survive today. *Chiripa,* of 1899, is in near-original condition. Note the large condensing pipe extending forward near the keel. (Reproduced with permission of the collections of Greenfield Village and the Henry Ford Museum, Dearborn, Michigan)

A WARNING.

There is but one Naphtha Launch built; the distinctive feature of which is using Naphtha for both **Fuel and Power.** The system has been amply demonstrated absolutely safe, and to this owes its world-wide popularity.

Do not be misled by advertisments of other parties pretending to manufacture Naphtha Launches, and do not confound our boats with the so-called improved (?) Naphtha Launches, making steam by burning wood or coal or petroleum or naphtha. Do not be deceived by stories circulated by competitors concerning the safety of our Launches. Rely rather on Testimonials of those who have by experience learned to speak advisedly on the subject.

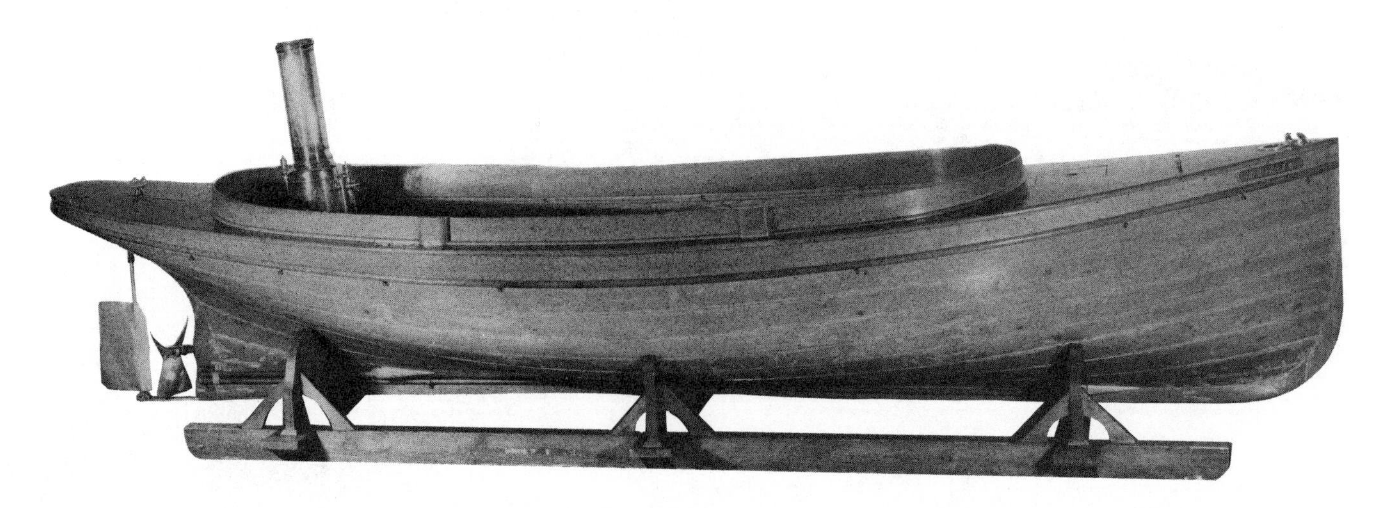

THE ONLY NAPHTHA LAUNCH.

OVER 700 IN SUCCESSFUL USE.

The Only Safe, Comfortable and Reliable Pleasure Boat.

No inspection or engineer required. Can operate it yourself. This system is endorsed by leaders in social, financial, literary and commercial circles.
Send 5c. stamp for catalogue containing all particulars, including some hundreds of testimonials.

GAS ENGINE AND POWER CO., Morris Dock Station, New York.
PACIFIC NAPHTHA LAUNCH Co., Tacoma, Washington, Agents for Pacific Coast, China and Japan.

(From C.P. Kunhardt, *Steam Yachts and Launches,* 1891 edition)

Below:
Naphtha launch advertisements offered a vision of happy, prosperous people enjoying the day in a beautiful boat. There was rarely any suggestion that the passengers were in any hurry to arrive someplace, or that the boat might be used for any purpose but recreation. This chosen image was a merchandising gambit; the boats were just as reliable and capable of useful work as other steam launches. (Courtesy New England Wireless and Steam Museum, East Greenwich, Rhode Island)

Dorcas, circa 1912. When these pictures were taken F.L. Brigham owned *Dorcas,* and he operated her in the Thousand Islands region of the St. Lawrence River. She still cruises the same waters today, but her naphtha plant was replaced by a gasoline engine years ago. (Courtesy Frederick G. Beach)

The only naphtha launch in operation at this writing is *Anita,* Gas Engine and Power Company Hull #1754, owned by the William H. Richardsons (father and son) of Sheboygan Falls, Wisconsin. The 25' x 6', 6-h.p. boat has been extensively rebuilt and beautifully maintained. Here she is shown participating in the 13th Annual Antique Boat Show, in Clayton, New York. (Courtesy Richard Morrison)

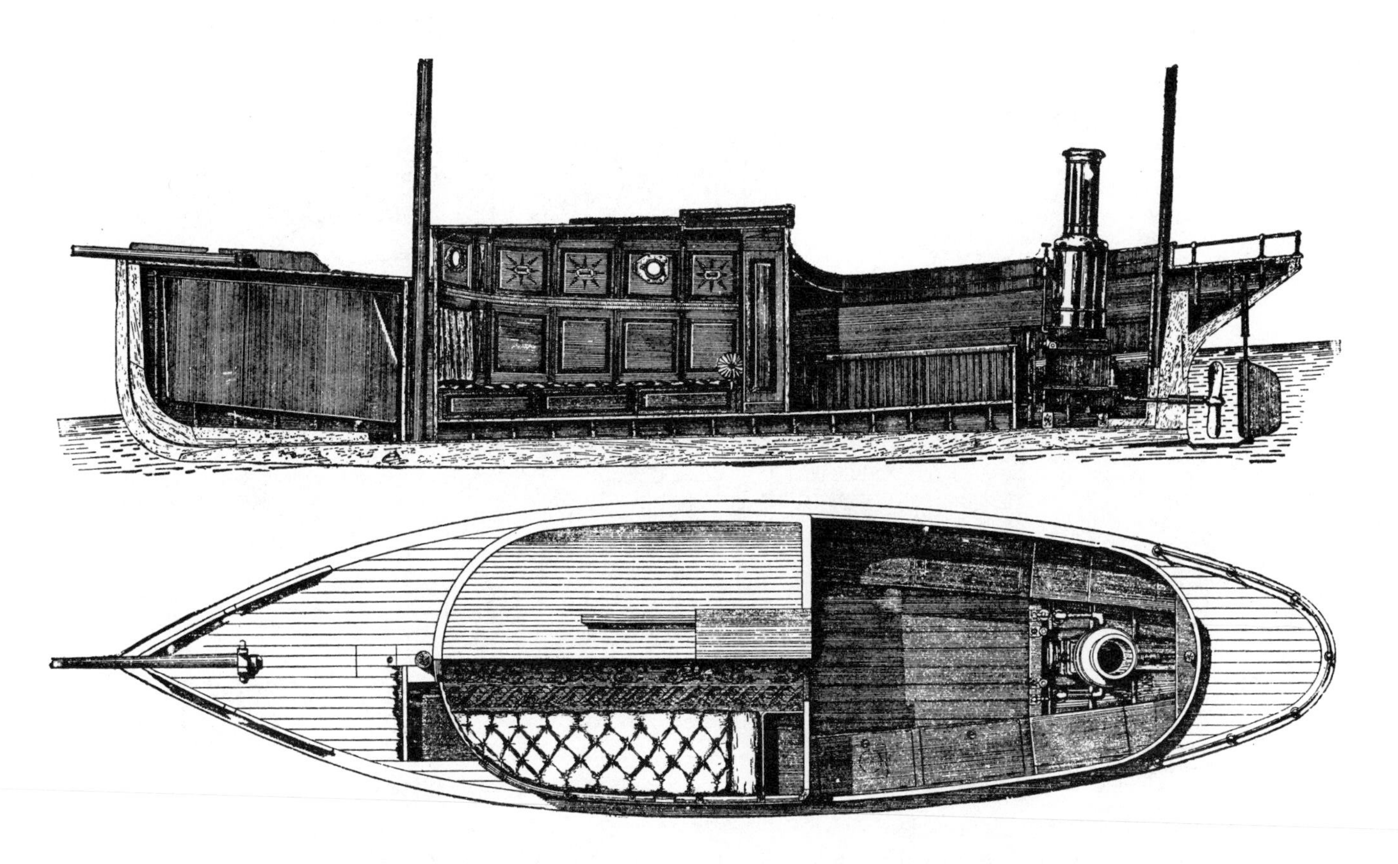

Naphtha power plants were compact enough to suggest, for the first time, the practicability of auxiliary power in small sailboats. The tasteful barrenness of *Etcetera*'s settee cabin in 1886 reveals how many necessities for cruising we've discovered since then. The huge naphtha tank in the forepeak was counterbalanced by the engine's weight aft. (Photograph from W.P. Stephens, *Traditions and Memories of American Yachting, Complete Edition;* drawings from C.P. Kunhardt, *Steam Yachts and Launches,* 1891 edition)

Frieda is now on display at a hotel in Yellowstone National Park, but in 1960, when this photograph was taken, she was owned by E.F. Coleman of Fargo, North Dakota. Coleman operated her on Pelican Lake, in western Minnesota. (Courtesy Wilbur Chapman)

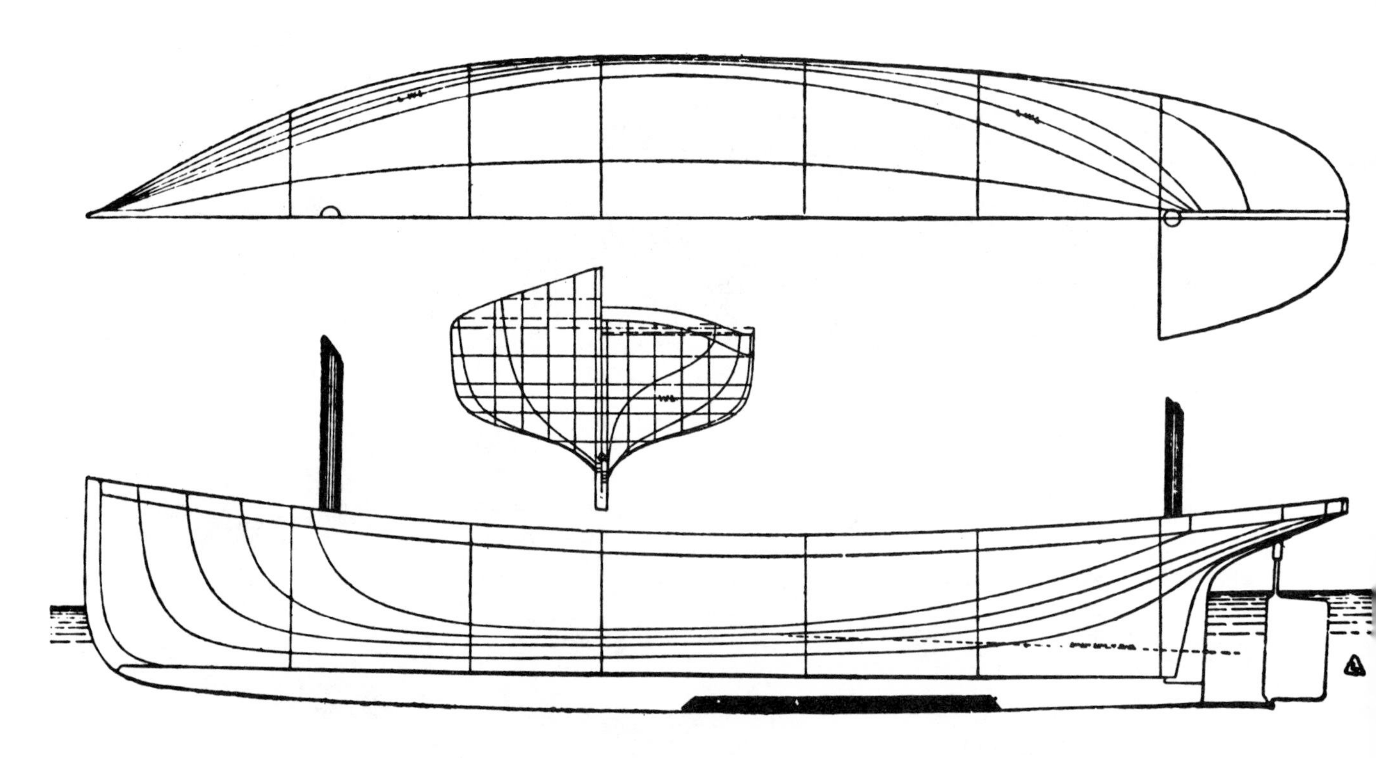

The hull designs of the early naphtha launches were carefully thought out by professional designers. This 35' x 8' x 3' naphtha auxiliary was one of the best. A beautiful boat, she was capable of swinging a 26-inch wheel, but she had the added beam, deeper keel, and hardened bilges to make her a stable sailer. The screw was two-bladed to reduce drag while sailing. (From C.P. Kunhardt, *Steam Yachts and Launches,* 1891 edition)

The 27-foot *Nina* had a glass cabin — with curtains — forward and a fixed canopy aft. (Courtesy Frederick G. Beach)

The flags and greenery that bedeck *The Saunterer* on this occasion nearly obscure the white naphtha stack in her stern. The time is circa 1910, the occasion may be a regatta or boat parade, and the place is Lake George, New York. The owner of *The Saunterer* (white suit and beard) has decked himself out for the occasion, too. (He was Colonel W. Dalton Mann, publisher of a New York fashion rag called *Town Topics.*) The *Saunterer*'s hull was built by Jesse E. Sexton, who had a boat shop near Hague, on Lake George. (Courtesy Dorothy Offensend)

A WEEKLY JOURNAL OF PRACTICAL INFORMATION, ART, SCIENCE, MECHANICS, CHEMISTRY, AND MANUFACTURES.

Vol. LXXII.—No. 11.]
ESTABLISHED 1845.

NEW YORK, MARCH 16, 1895.

[$3.00 A YEAR.
WEEKLY.

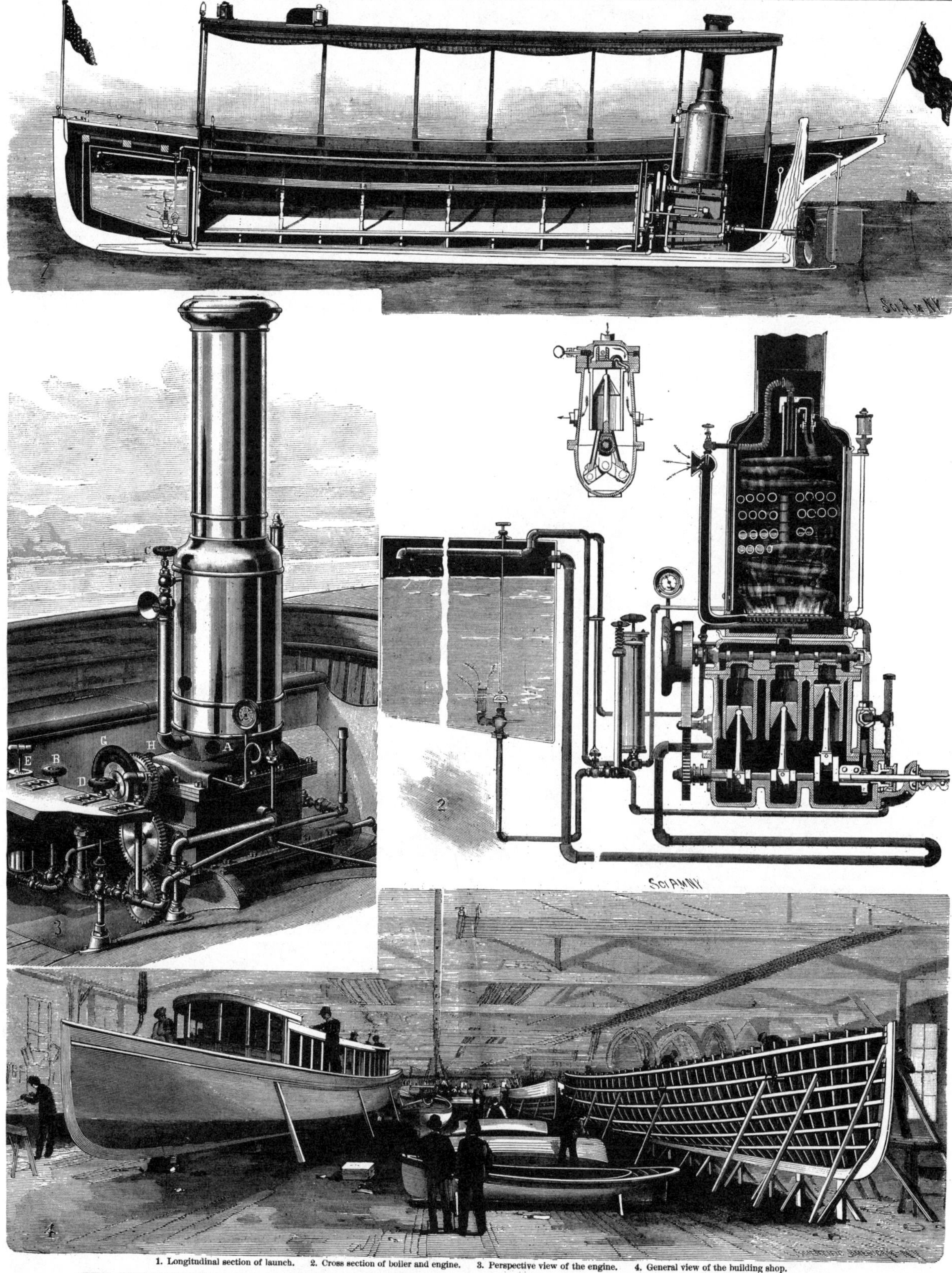

1. Longitudinal section of launch. 2. Cross section of boiler and engine. 3. Perspective view of the engine. 4. General view of the building shop.

THE MANUFACTURE OF NAPHTHA LAUNCHES BY THE GAS ENGINE AND POWER COMPANY

Opposite: By 1895 the wise heads at *Scientific American* should have known that the future belonged to ''explosive motors,'' but they devoted a magazine cover to the naphtha launches that were manufactured a few miles uptown from the editorial office. The accompanying story stated:

> No type of power-propelled boat has acquired such popularity in so short a space of time as has the naphtha launch. Whiie these boats are used extensively in all waters, and have become a most familiar object to all, it is fair to say that very few people know how they are operated . . . . These little boats have won for themselves an astonishing record. They seem to be as absolutely secure from accidents as any kind of power-driven craft can be.

STEAM "YOTS"

George Eli Whitney had a clear notion of where steam launches left off and steam yachts began, and he always described boats a little beyond the steam launch pale as "yots."

Erik Hofman opens his handsome book, *The Steam Yachts,* with the bald assertion that, "The steam yacht was the most striking personal possession ever produced by man . . . ." True, but the great steam yachts were a little monotonous in their uniform designs and uniform pretentiousness — clipper bows, flush decks, marble baths, tall imperial stacks, and so on.

The "yots" excelled in variety and individuality. They were often the creations of people who, eschewing popular fashions in conspicuous consumption, had the courage to order boats to suit their needs or whims.

A 60-foot steam "yot" can be as relaxed and amiable as a home-built 20-footer, if the heart's in the right place. The men are discussing the Boer War or the women of Southampton, and the schoolboy with the helmsman imagines he's conning *Turbinia.* The old boat glides along at her usual 4 or 5 knots, and the peanut whistle abaft the funnel speaks volumes about unpretentiousness. (Beken of Cowes photo)

George Whitney's 61-foot *Montauk,* a conventional "big-family yot" of 1900, had a stateroom, a galley, and 12 berths. Her chief claim to fame lay in the 7" & 14" x 9" triangle compound engine (see page 276), one of only two ever built for marine use. (George Whitney collection)

Lady Hopetoun, of Sydney, Australia, is a plain, unremarkable steamer, but full of interest nonetheless. Built in 1901-02 as a Vice-Regal yacht, she is planked and decked with New Zealand kauri on American elm frames. The machinery carries honored names. The 1901 Yarrow boiler has 383 1⅛" tubes, incorporating 622 square feet of heating surface; 21 square feet of grate area is available for hand-fired coal fuel. The main engine is a 1901 Simpson, Strickland triple, 8" & 13.5" & 20.25" x 10". A Tangye engine drives the dynamo by belt.

The Sydney Cove Maritime Museum owns the vessel and has fully restored it, with a reduction in pressure in the original boiler from 260 to 200 p.s.i. (Courtesy Warwich Turner)

Caprice (above) and *Sally (below)* were built just before the turn of the century by the Fore River Engine Co. of Weymouth, Massachusetts. *Sally* went to a Mr. Emmons of Marblehead. (Photos courtesy H. Hobart Holly, Quincy Historical Society, Quincy, Massachusetts)

John I. Thornycroft's *Miranda* of 1871, 49.75'x 6.5' x 2.5', was the first small craft to greatly exceed the speed of the fastest Atlantic liners. (In America, some big river boats were faster than any steamships afloat, being capable of 22 m.p.h. or a little more.) *Miranda* was put into service hastily in June 1871 to serve as umpire boat at the Henley regatta. Later, during timed runs between Vauxhall and Westminster Bridge, and between Barnes Bridge and Chiswick, she was credited with a corrected top speed of 18.4 m.p.h. The hull and nearly everything in it was made of Bessemer steel. The 6'' & 6'' x 8'' engine received 120-pound steam from a side-fired locomotive-type boiler with 116 square feet of heating surface. *Miranda's* surprising performance influenced naval architecture and naval tactics for years thereafter. (From William H. Maw, *Recent Practice in Marine Engineering,* 1883)

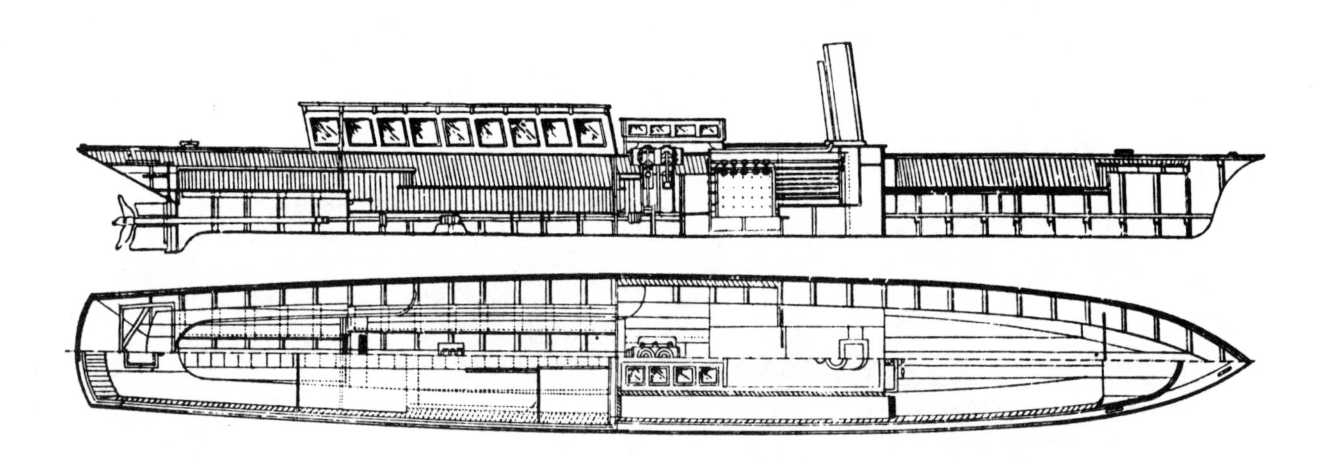

This 60' x 7.5' boat, listed in Simpson, Strickland & Co.'s catalog of 1902, shows a striking resemblance to Thornycroft's *Miranda* of 1871. She had a double-simple engine and a speed of about 20 m.p.h. (From the facsimile edition of Simpson, Strickland & Co.'s Catalogue No. 5)

The Racine Boat Manufacturing Co. offered *Cleopatra* in their 1894 brochure.

Robert Ellis' 60-foot yacht *Oceanid* steams past South Pender Island, British Columbia, in 1967. Ellis later donated the steamer to the Seattle Sea Scouts. The boat is a Royal Navy Harbour Service Launch of 1944, extensively modified and rebuilt in 1960-1961. *Oceanid*'s career and machinery were described in some detail in *Steamboats and Modern Steam Launches,* January-February 1962. (Everett Arnes photo)

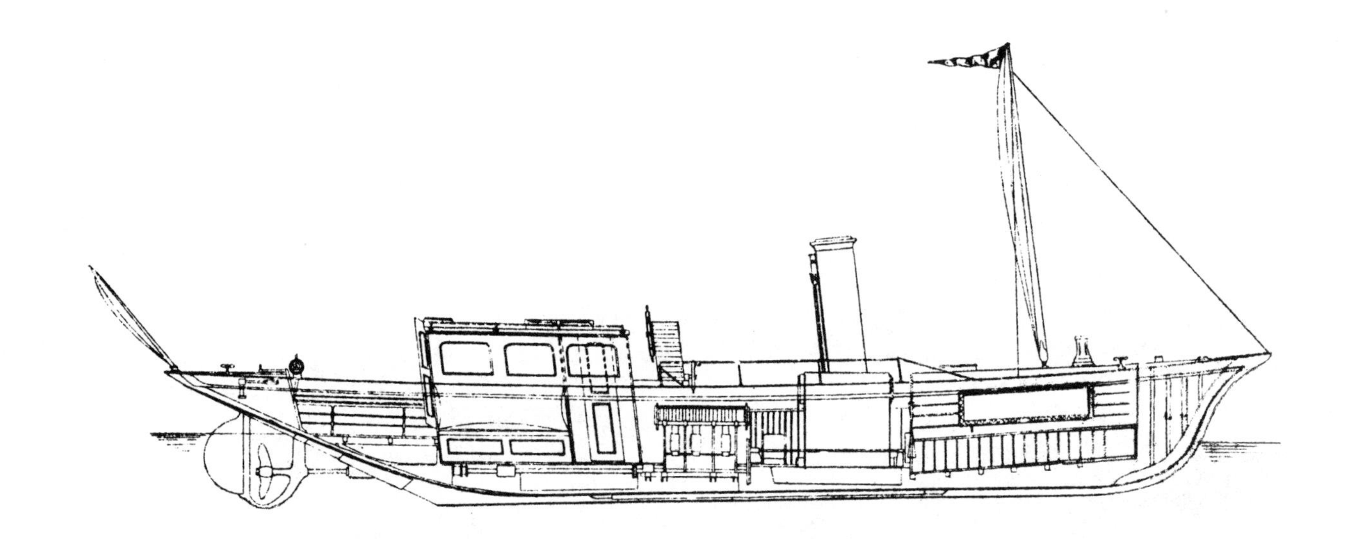

L'Annett III (formerly *Snowdrop*) was a Simpson, Strickland product launched at Dartmouth, England, in 1901. All teak, bronze, steel and copper, " . . . construction très robuste et très soignée . . . d'une grande élégance," with hot and cold running water and flush toilet. The owner's cabin in this 62-foot, 3-man-crew launch was seven feet square. Now, *that's* conspicuous consumption!

The triple engine generated 90 horsepower using steam from a Thornycroft-type WT boiler. The boat could attain 14 knots at 500 r.p.m. Locating the steering station on an athwartships runway above the main engine with fire and gauge-glass in view is a detail worth considering for a modern owner-operated boat. (Courtesy Artú Chiggiato)

The section of Connecticut's Thames River between Norwich and New London provides good boating for small vessels, and some not so small. Here we see the "yot" *Katrina* in Norwich, ready for a cruise on Long Island Sound. (Courtesy *Norwich Bulletin*)

Nat Goodhue, one of the old-time steamboat men on Lake Winnipesaukee, has passed from the scene. His 60-foot steam yacht *Swallow,* originally the flagship of the Eastern Yacht Club in Marblehead, Massachusetts, was converted to diesel power many years ago. The old boiler with its port and starboard firedoors was sold for other use, and the lovely Fore River compound engine with its Joy valve gear (see page 299) is in storage in Maine. *Swallow* was later left to rot on the edge of the lake. David Thompson of Moultonboro, New Hampshire, bought the hull and towed her to his home town, but his dream of restoring the yacht to her former glory has gone unrealized.

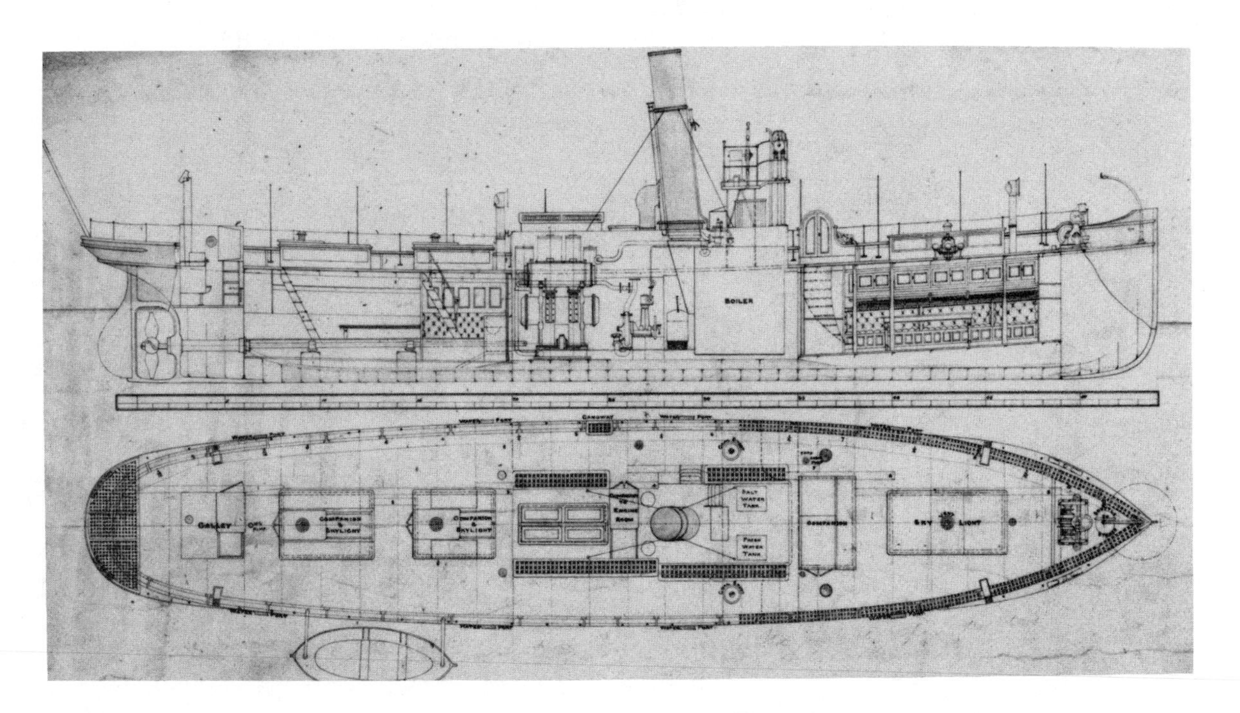

Most British steam launches, except those designed for inland waters, had a seamanlike, battened-down look. The *Snark,* of 1882, was built by the Denny Co., of Dumbarton, for use as a company yacht, tender, and experimental vessel. The design is in the spirit of the oceangoing iron steamships that were the yard's principal product. The compound engine operated on 90-pound steam. A large centrifugal pump was installed for fighting shipyard fires. As an experiment, the iron hull was sheathed with zinc. (National Maritime Museum, London)

Kestrel was built in 1892 by George Lawley & Son of South Boston and engined by the Fore River Engine Co. of Weymouth, Massachusetts. In her long life, she has steamed Lake Champlain and the coastal waters between Florida and New England. *Kestrel* has always been a coalburner, and she is now on her third boiler, installed in 1967. The author had a memorable cruise in the steamer in 1971, when he served as appraiser and agent for the sale of the yacht to Robert Scripps, of Balmorhea, Texas. Scripps donated *Kestrel* to the American Maritime Academy in 1975.

The boat was used only as a private yacht during her first 83 years, and rebuilt or reboilered from time to time to preserve her original qualities. The 48-inch-diameter vertical boiler holds 300 1½'' tubes. The handsome 6'' & 12'' x 10'' engine (see page 298) generates 50 horsepower. (Philip Teuscher photo; courtesy Steamship Historical Society of America)

D. Cameron Peck of Evanston, Illinois, bought up dozens of classic steam, naphtha, and gasoline boats between 1920 and 1950, when modern high-speed engines were making the beautiful old boats seem antiquated and feeble. Peck sold his important collection in the 1950s and moved to the Arizona desert, far from Lake of Bays, Muskoka, Ontario, where most of his boats were kept.

Wanda III was a Canadian yacht in the Peck collection. The much-publicized buildup of vast steam navies in the early years of this century caused some patriotic boat owners to model their pleasure boats on naval vessels. *Wanda III* has a bow somewhat like that of a battleship, and a ''torpedo-boat'' stern. With a big triple-expansion engine and a Yarrow boiler, she could run 21 miles in an hour. (Edward O. Clark photo)

Naiad, of 1890, was the prize of the Peck collection. In these 1951 photographs she appears as crisp and fresh as the day she was launched. The 80' x 10' yacht was built at Toronto by the Poulson Iron Works. She was of composite construction, with oak planking on steel frames. Peck reports that the slender vessel cruised easily at 14 m.p.h., with her rather small compound engine, 7.5'' & 15'' x 10'', and a hardwood fire in the Scotch marine boiler.

Contrasty and striking varieties of wood, such as bird's-eye maple, walnut, and cherry, set off the teak and mahogany joinerwork. All of the design elements in the superstructure were raked strongly aft, in harmony with the steamer's big funnel. In *Naiad* this comes off well and deserves imitation by some future steam launch designer who is sure he knows what he is doing. (Edward O. Clark photos)

Naval Launches and Torpedo Boats

The first fighting steamer, the *Demologos,* was designed by Robert Fulton, and he supervised her construction for the U.S. Navy in 1814. *Demologos* was many years in advance of other naval steamers designed for battle. She was a grandiose freak, a product of the momentum created by Fulton's commercial success on the Hudson. With a 120-h.p. engine, thirty 32-pounders firing red-hot shot, and wooden sides almost five feet thick, she might have wiped out a becalmed British squadron and changed naval history — but the War of 1812 ended before she was ready.

In 1815 and 1816, with several commercially successful steamers already in service on the Thames, the British Admiralty began to look closely at steamers. Proceeding cautiously, the Admiralty commissioned small-scale tests with *Congo* and *Regent. Congo*'s Boulton and Watt machinery overburdened the small vessel so much that she was slow and unseaworthy, and Marc Brunel reported that *Regent,* a Margate packet, successfully breasted wind and tide, but her 16-h.p. engine did not provide enough reserve power to tow the Navy's large warships.

In 1819, *Eclipse,* an 80-h.p. Margate packet, towed a line-of-battle ship against the tide. Lord Melville, First Lord of the Admiralty, was impressed — and worried. By 1828 he regarded steam power as the principal threat to British naval supremacy.

The first commissioned Admiralty steamer was *Monkey,* of 1821, 212 tons and 80 nominal horsepower. *Lightning,* built at Deptford in 1823, saw action off Algiers in 1824. *Diana* served in the first Burmese War of 1824. The first steamers designed for fighting (and not merely towing or dispatch-carrying) were *Dee, Medea, Rhadamanthus, Phoenix,* and *Salamander.* These were launched soon after 1830. By 1836 there were 27 commissioned steam vessels. At this time the standard list of engine-room stores for the Royal Navy included sweet oil, tallow, lamp oil, candles, spun yarn, oakum, white lead, and red lead. Melted tallow was the all-purpose lubricant, and if a bearing ran hot, it was flooded with seawater.

Between 1815 and 1860 the Royal Navy dominated the seas as no other navy has before or since. The navy that helped defeat Napoleon was a sail navy, with all of its skills and traditions based on wooden walls and sailing tactics. While antiquated broadside concepts dominated tactical planning, even the largest and most powerful steamers were considered inferior. A fast dispatch steamer with one or two guns was not even counted, though she might perform military duties a hundred times more quickly and surely than a becalmed frigate.

All senior officers — and all junior officers who hoped to rise in the Royal Navy — feared and despised steam power. If a few loutish and illiterate "enginemen," indifferent to naval custom and not under naval discipline, could deliver as much effective horsepower as the swarming crew working the sails of a three-decker, what would the world come to? The military officers on steamers were sometimes taken aback when the men hired to propel the ship (who were not sworn fighting men) walked off the job just before an engagement. Prejudice was so far-reaching that in 1827 the Carpenter Warrant Officers passed a resolution complaining that it was degrading for their class to have to command "enginemen and their stores." The dirtiness of steam power was a common excuse (but less commonly the real reason) for the universal prejudice.

Early engineers learned their trade with engine builders and only later went to work on board ship. They wore no uniforms and had no contract with the navy. Eventually some wise heads in the Admiralty saw that the navy should train its own men to be able to run the steamers. By the late 1820s a few enlisted men were apprenticed to engine builders, and by the 1830s and 1840s there was a small cadre of uniformed navy men who understood their ships' plain, conservative power plants well enough. (The standard naval steam pressure was 8 p.s.i. for many years, and the navy found reasons for clinging to slow-turning paddle engines longer than it needed to.) Engineering officers were not accepted as social equals for another hundred years, if then.

After 1860 many powers began to modernize their navies, buying most of their new ships from British industry. The Confederacy's best steamers, for example, were British built, and these in turn spurred rapid advances in steam engineering in the Union navy.

Launches, Cutters, and Pinnaces

Navies were slow to adopt fighting steamships, but they were even more hesitant to employ mechanical power in small craft. With hundreds of muscular young sailors to pull oars, and with no steam plant really suitable for small craft, it seemed foolish to install engines in ships' boats. The realities of the mid-century wars finally showed the advantage of having boats that could out-tow and out-last rowed boats, even if they were not as fast as a well-manned eight-oared cutter. After the Crimean War and the American Civil War, naval schedules included steam launches to be used for picket duty, carrying dispatches, or towing a ship's rowing boats. Close scrutiny of photographs or models of the naval ships of 1870 to 1920 (and a few up to 1945) will usually reveal a steam launch or two among the boats.

In the American navy, the term "steam cutter" supplanted "steam launch" before 1900, and the powerful 50-footers introduced shortly before World War I were termed simply "steamers." In the Royal Navy small, open, steam-powered boats were called steam cutters. Steam launches were utilitarian workboats, usually unarmed, decked, and shore based. Some of these were built as late as 1944. Powerful boats possessing some shelter and equipped to carry armament were known as pinnaces, and after 1890 the larger, more heavily armed pinnaces were called steam picket boats. The Royal Navy purchased more than 1,000 steam pinnaces between 1866 and the retiring of the class in the 1940s.

Most naval powers bought their ships' boats from commercial suppliers, just as they might put pumps or binnacles out on bid. The American navy's steam cutters were an exception, designed by a navy bureau and built, to the last detail, in naval shipyards — boat and power plant were designed as a single unit.

The U.S. Navy Department had very little money

to spend on ships and boats from 1865 until the buildup of the modern steam navy began, a quarter of a century later. In those years, steam launches were built in limited numbers, and to a variety of designs. An order of 1870 first recommended standardization of ships' boats, and a fully standardized series of boat and engine designs was finally published in 1900; another was developed for 1915.

Today, ex-navy engines are numerous in steam launch engine collections in America. They are handsome, they were built by the hundreds in the U.S. Navy's series of 1900 and 1915, and they were surplused out in a manner that permitted a large proportion to be saved by private collectors.

The 1900-series engines had slide valves on both cylinders and light, cast-bronze frames. The latter feature resulted in distinctively handsome but rather flimsy and vibration-prone engines. The 1915 engines were fully engineered and cost-be-damned, and they were manufactured in quantities to suit the scale of the war that was then in progress. The Type-K and other engines of the 1915 series (which were actually designed several years earlier) are for the most part above criticism. Some hobbyists object to the difficulty of maintaining a piston valve on the high-pressure cylinder without machine tools, and the 50-h.p. Type-M engine is laid out "backward." (A normal marine steam engine for a single screw has a right-hand rotation, and the high-pressure cylinder is forward. The Type-M violates the latter canon.)

'The Torpedo-Boat Threat'

Steam launches were a late and minor development in the application of mechanical power to watercraft. They were not part of the main evolutionary stem of seafaring — either civil or naval — with one exception. All destroyers, from the 5,000-h.p. *Havock* of 1893 through the dashing 60,000-h.p. fighting machines of 1945 to the ponderous 7,000-tonners of today have a clear and direct bloodline from certain small steam launches of the 1860s.

During the Civil War, steam launches became useful in patrolling, or "picketing," the perimeter of the many coastal- and river-dissected fields of action. When the need was dire, some screw picket boats were equipped with spar torpedoes and sent out to destroy units of the enemy fleet. The occasional success of these missions resulted from the suicidal bravery of the crews — not from the speed or agility of the launches, which traveled only a little faster than rowed boats.

In 1865 no one believed that small craft could ever travel as fast as ships. The "facts" of naval architecture made this impossible. By 1873 young men in England — notably John Thornycroft — had built a few small launches that exceeded the 14-knot speed of fast river packets or the Confederacy's blockade-runners. Naval planners suddenly realized that a small boat might destroy a great battleship, in daylight, by speed, daring, and a torpedo. In 1868 Robert Whitehead, a naval contractor working in Austria, had constructed a self-propelled torpedo that could sink a ship from a thousand yards.

For a generation, from 1870 to 1895 (a time of great uncertainty and rapid change in naval construction), navies were preoccupied with "the torpedo-boat threat." Throughout the 1870s there was a ferment of development of fast torpedo launches, and 15-, 17-, and 18-knot boats were supplied to the Norwegian, Dutch, and American navies. The Royal Navy's TB #1, of 1877, was a highly evolved weapon capable of 19 knots.

In the same year, Imperial Russia ordered 110 small torpedo boats from several builders, upsetting the naval balance for years thereafter. The possessors of heavy, old-fashioned ships, some still using sails, began to wonder if they had been outflanked. France, Britain, and Germany built frantically during the 1880s. There were more than 1,200 torpedo boats afloat or on order in 1890, and in that year the United States Navy purchased its First Class Torpedo Boat #1, the *Cushing*.

The single-purpose torpedo boats were regarded as light, cheap, and expendable weapons requiring no refinements or accommodations. They were to be all steel, steam, and speed. The first-class boats increased in power in 10 years from 400 horsepower to 1,600 horsepower. They became increasingly complex and expensive but were still not really seaworthy, and they were lightly armed and had limited endurance and vile accommodations. The little steamers, propelled by hundreds of horsepower and with no plates thicker than one eighth inch, were often driven to death in ordinary Force-5 weather if they attempted to accompany the fleet. Advances in quick-firing cannon, searchlights, and heavy machine guns soon made their 20- or 25-knot speed inadequate.

Japanese torpedo boats scored against Chinese ironclads on the Yalu River in 1894, but the first torpedo-boat destroyers — larger, faster, heavier-gunned, and more seaworthy than the torpedo boats — had been launched in 1893. A Russian destroyer soon topped 30 knots, and the torpedo-boat fad withered away. By 1910 most of them had been melted down to make dreadnoughts, the new super-weapon.

Still, the hope for a big return on a small investment persisted, and a few 60-foot, 12-ton (lighter if aluminum) steam launches, rated as third-class torpedo boats, continued to be built. These were carried on the decks of ships or stationed where they might make quick dashes against opportune targets. Torpedo boats didn't accomplish much militarily until they went undersea in 1914, but for a time they exemplified the romance of steel, speed, and power. Before fighter planes and moon rockets, there was no more poignant union of a small crew and a small craft.

LAUNCHES, CUTTERS, AND PINNACES

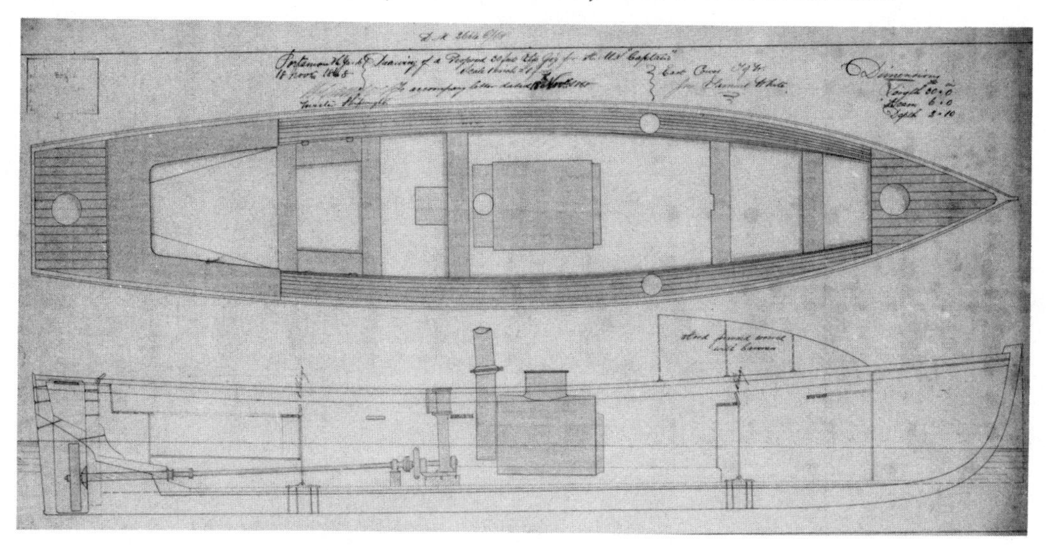

Many Royal Navy steam launches originated in the East Cowes, Isle of Wight, yard of J. Samuel White. This 30-foot life gig for H.M.S. *Captain* was built in 1868. (National Maritime Museum, London)

''Pinnace'' is an old word, now in disuse, deriving from the Latin for ''pine tree.'' It identifies a vessel larger and tougher than a boat, but not quite a ship. This 1896 Royal Navy specimen was as stout and hearty as a naval steam pinnace should be. Naval steam launches did not spend all their time towing liberty boats or taking the marketing party ashore to buy cabbage. That quick-firing three-pounder might have sighted in on slavers off Zanzibar or pirates near Penang. (Crown Copyright. Science Museum, London)

Above: Late in the last century, the Royal Navy's powerful steam pinnaces of about 50 feet in length were standardized as the 56' x 9.5' steam picket boats. This 1895 view is of one such boat, belonging to H.M.S. *Resolution.* She was (smokelessly!) coalburning, with a fire-tube boiler and compound engine. She could get up to 15 knots in smooth water. The boat was fitted with side-dropping gear for two 14-inch or 18-inch torpedoes. Her light armament would have been a 3-pounder quick-firing gun and a Maxim machine gun.

The great steam navies added major ships one after the other in the early years of this century. The new fleets needed to have more boats with great speed and seaworthiness for dispatch-carrying and scouting or picket duty. The Royal Navy's pinnaces performed heroically at Gallipoli, Turkey, in 1915, but they are more picturesquely remembered for their part in suppressing the Indian Ocean slave trade a generation or two earlier. Many a fast Arab slaver surrendered to a steam picket boat off the East African coast. (Imperial War Museum photo) *Below:* This 56-foot pinnace is coaled up for a hard day of picket duty off a naval anchorage in the south of England. The salt on the stack and the deadlighted ports tell what it was like on a heavy, powerful steam launch in the winter of 1915. (Beken of Cowes photo)

Steam launches were essential to the amiable summer gatherings of several emperors and their friends. Great yachts and naval fleets from all Europe used to assemble at Cowes, Isle of Wight, on special occasions between 1903 and 1913. Visits yacht-to-yacht and landings at Cowes and other towns along The Solent were carried out in steam launches such as this 42-footer, built by J. Samuel White in 1899.

It would make titillating reading to report that the boat was carrying Edward VII off on one of his private excursions, the Royal Standard at the prow carried low for discretion's sake. The truth is that this is Edward's old launch in 1923, long after his death and after the violent end to the Edwardian gatherings of friendly emperors at Cowes. The "Sailor King," George V, and Queen Mary (visible in the cockpit) have chosen the flawlessly maintained old steamer to carry them about their sedate and regal duties. (Beken of Cowes photo)

Above: Hundreds of ex-navy steam launches came into private hands as great wartime fleets were reduced in strength. The new owners usually dismantled the boats or did violence to their ''navy launch'' character just as soon as they could get their hands on a wrecking bar and cutting torch. Only one type of navy steamer escaped the wreckers, and this was the Royal Navy's 52.5-foot Harbour Service Launch. These boats were commercially useful as small tugs or tenders. Several hundred were built during both world wars.

R.W. Partis' *Puffin* is the most interesting surviving Harbour Service Launch. She was built in 1919 and retains the plain, masculine personality of a small steam workboat of that period. Her accommodations and the details of her equipment remain close to the original design. On her longest summer cruise, Harwich to Dartmouth and return, *Puffin* consumed six tons of coal in steaming 640 miles. Her engine is shown on page 275. (Courtesy R.W. Partis) *Below:* Peter Collins' Harbour Service Launch, shown here on The Solent, is close to her original state. A 6' x 6' Scotch boiler and an 8'' & 16'' x 8'' compound engine give her 75 h.p. at 250 r.p.m. on 150 p.s.i. (Portsmouth & Sunderland Newspapers, Ltd.)

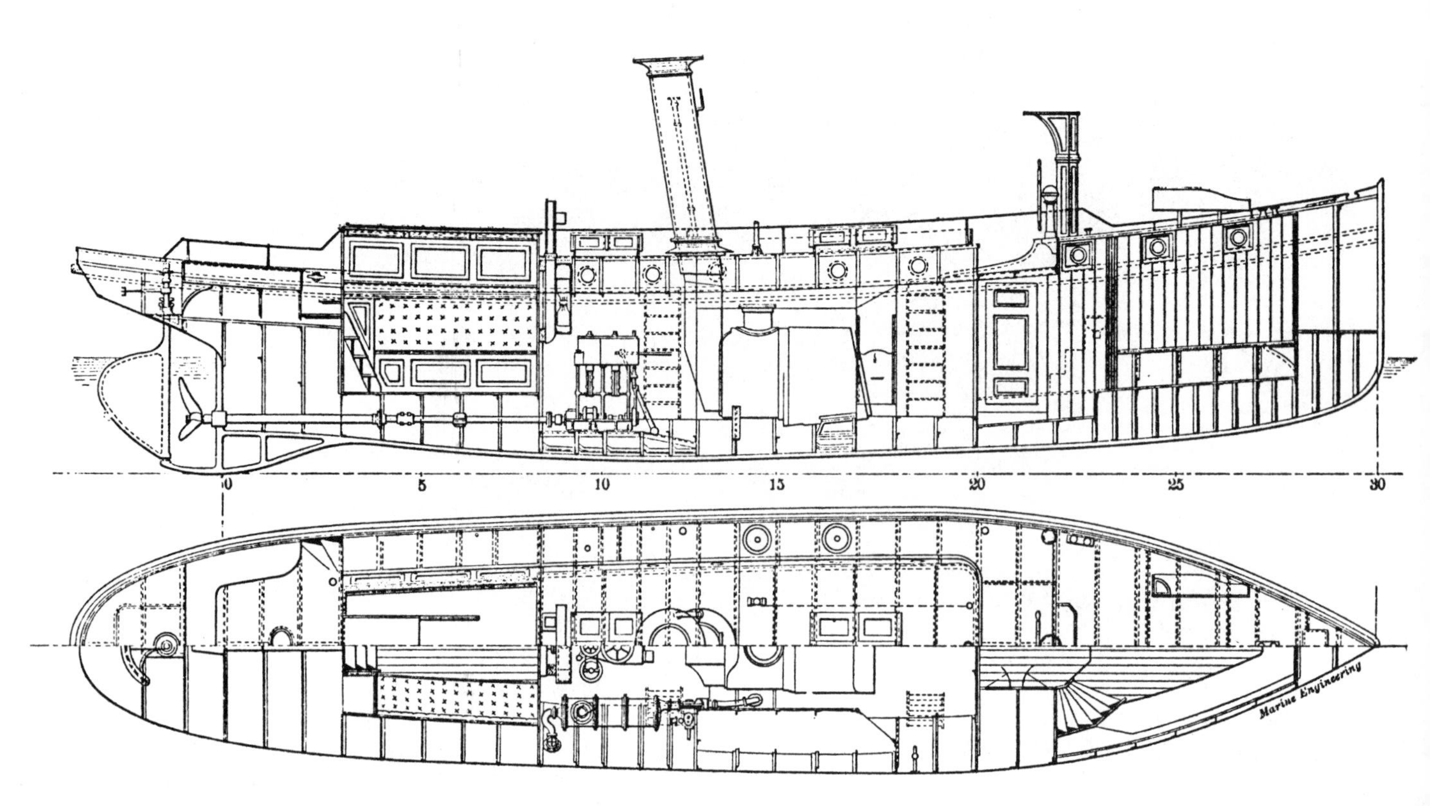

This 47.5' x 10.3' German pinnace of 1901 was designed to be seaworthy in all seasons for the waters around the Imperial Dockyard at Kiel. She was stoutly built ''of good German steel'' by the Sachsenberg Brothers of Rosslau, and she had four watertight bulkheads in her deep and powerful hull. The direct-tube boiler looks surprisingly small in the big, pressurized machinery space. With 154-pound steam, the 5.3'' & 13.8'' x 7.9'' engine indicated 75 h.p. at 350 r.p.m. The stokehold pressure was ½ inch of water. (From *Marine Engineering,* 1901)

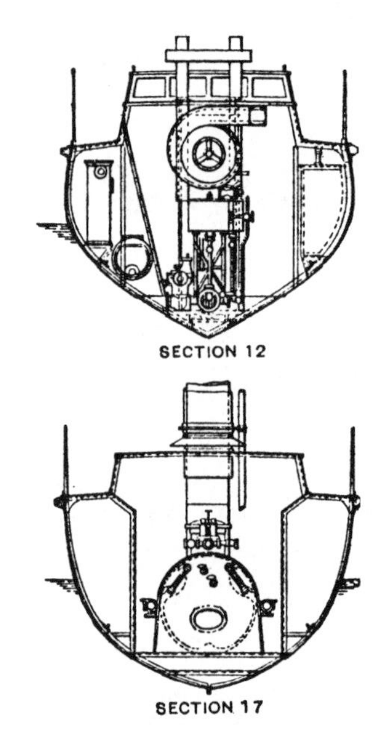

The 10-meter steam cutters of the Regia Marina Italiana were among the most successful and long lived of naval steam boats. The photo above is from 1886; the drawing below is of the 1896 re-design, which remained in service until 1926.

The excellence of design and structure, with full-length flotation chambers and durable, foolproof machinery, could scarcely be improved upon for the service. The compound engine worked at 66-p.s.i. steam pressure, and the pinnace-type boiler was fired from forward. (Drawing and photograph courtesy Artú Chiggiato)

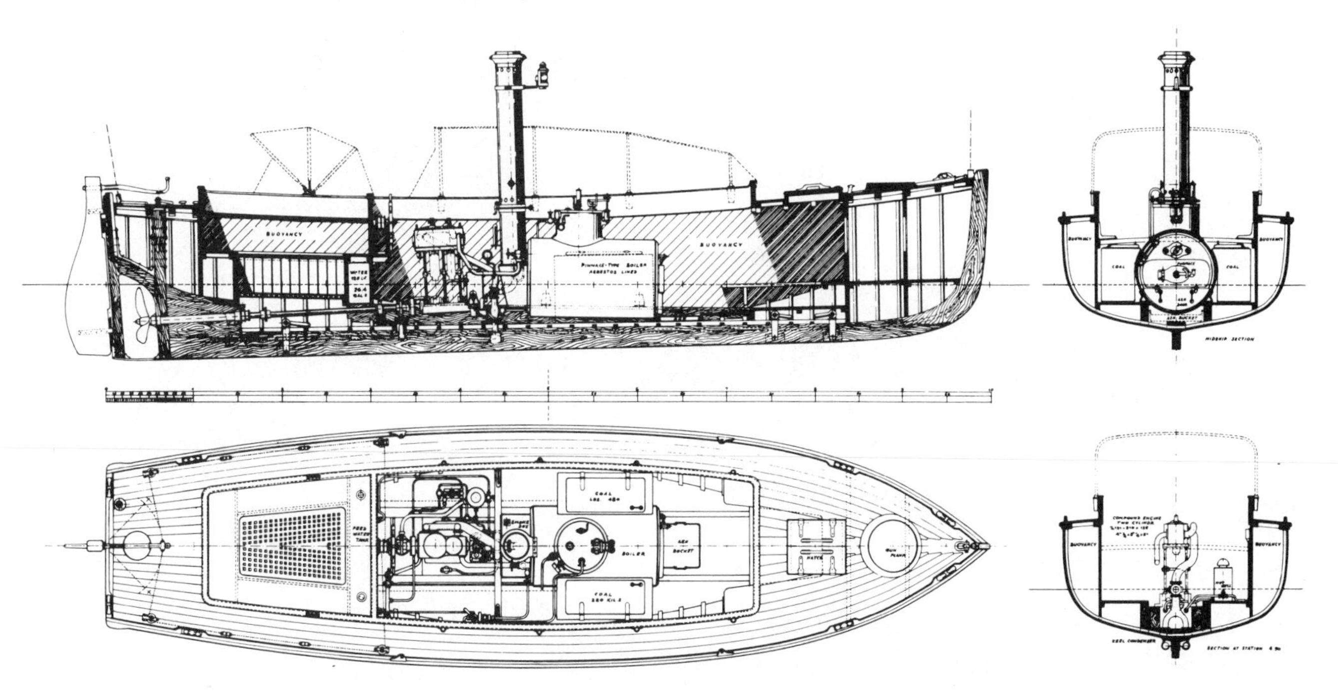

Above: The Royal Italian Navy's 40-foot picket boat in this 1910 photo is still an unprotected open boat. These boats were later given the enclosed cabins and steel-protected crew stations usual in the pinnaces of other navies. (Courtesy Artú Chiggiato) *Below:* "Vedette" (scouting) pinnaces were part of the equipment of every large steam-navy unit before aircraft, radar, and radio took over their functions. This 1908 Italian vedette was fast for her 46-foot length, and she could steam up her coal-fired boiler from cold in 35 minutes. (Courtesy Artú Chiggiato)

Small naval steam launches existed mainly to run errands and to tow all the rowed boats carried by their ship. They were also equipped for some warlike missions, principally as landing craft. This French steel-hulled launch (with pinnace boiler) carries a surprisingly heavy gun. A three-pounder? Landing parties such as this maintained order along thousands of miles of North African, Indochinese, and Madagascar coastline. (Musée de la Marine, Paris)

This gaggle of visitors to a U.S. Navy installation in the 1890s is pictured in a steam whaleboat. These boats were popular with some commanders in the "new steam navy," but they were not adopted as "standard boats," and little is recorded of their design. (From a stereopticon card)

Members of the U.S. Naval Academy class of 1903 man their launches for a sham battle and lessons in signaling and tactics. Nameless 28-foot steam cutters were the "first command" of most of the admirals who earned fame in both world wars.

This 36-foot U.S. Navy steam cutter of 1900 typifies American naval steam boats. Short, stout (in both senses), and formal in her accommodations, she made 7.5 knots on 18.65 indicated horsepower. The exquisite woodwork and finish (*above*) were done to navy standards, with two-tone staving in cherry and birch and planking of clear white cedar.

The boat was given auxiliary sails (*below*) a long lifetime after sails were dispensed with on most civilian steam launches. The loose-footed lug foresail permitted sailing while the smoke pipe was up. The steamers were also equipped with 17-foot sweeps, in case they ran out of coal and the wind died.

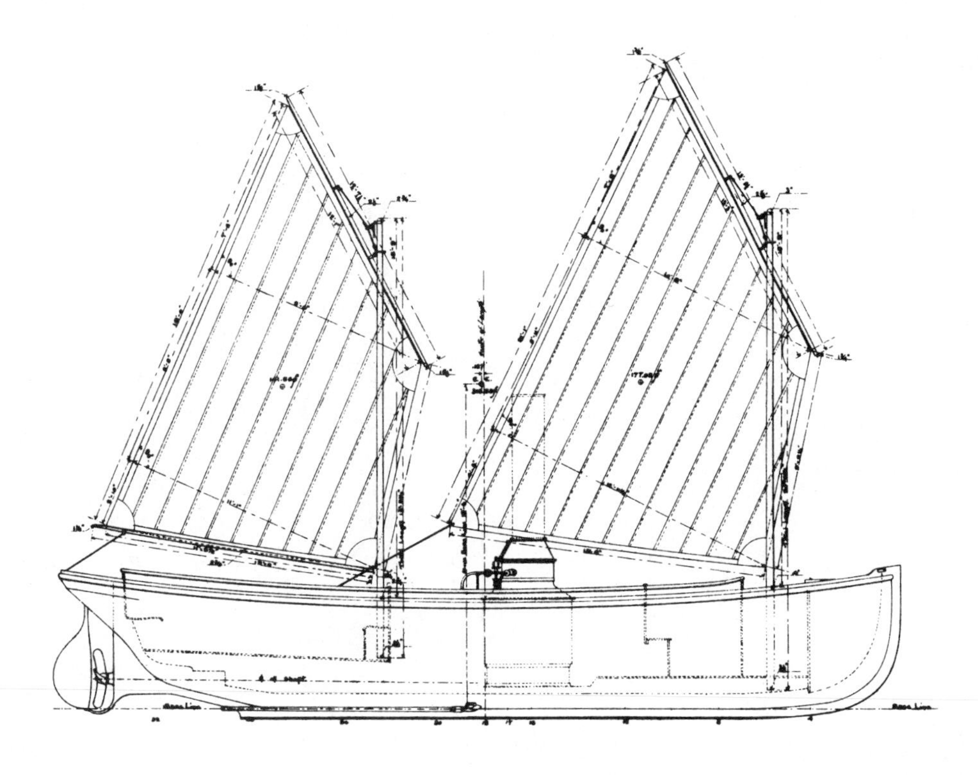

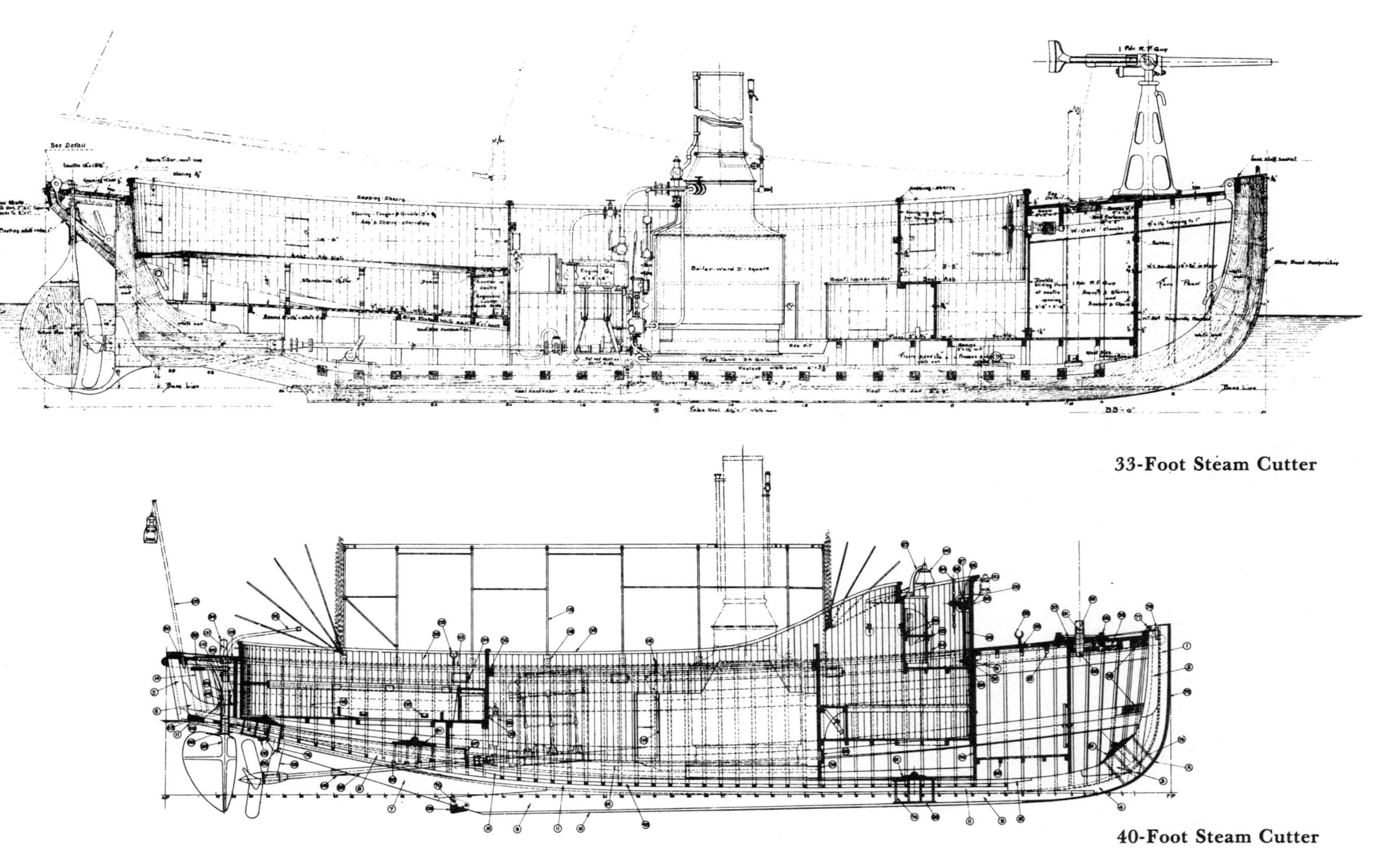

33-Foot Steam Cutter

40-Foot Steam Cutter

A revolution in small-craft design occurred between 1900 to 1915, when planing hulls, hydrofoils, and, most especially, extraordinarily light and powerful engines were developed. The last standard steam cutters built by the U.S. Navy, the 40-footers of 1915, were touched by these onrushing changes, but they remained steamers through and through. The steam cutters of the 1900 series (*top*) had hull lines not very different from those of Roman grain ships or Elizabethan pinnaces. The 1915 boats (*above*) had radically altered lines, to accommodate new levels of powering. Many old salts who operated these boats between 1917 and 1919 are still around to tell about them: "They were *powerful,* boy!" When the 50-h.p. Type M compound engine turned that 38-inch screw, the effect was quite different from a 50-h.p. outboard spinning its piddling 12-inch "prop."

The navy failed to save any of its boats (where were the Historical Officers?), and someday a naval museum will have to spend millions to obtain handmade copies of one or two of these historic steamers.

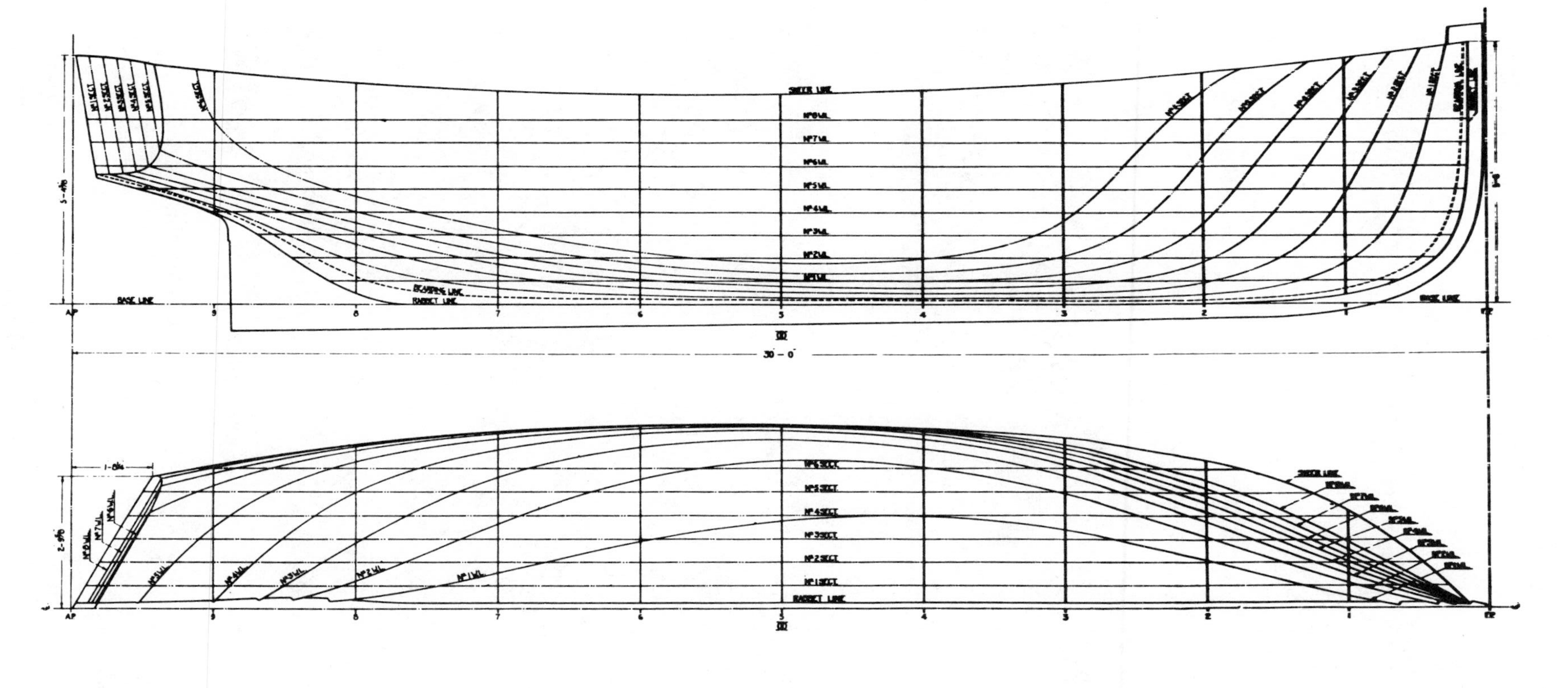

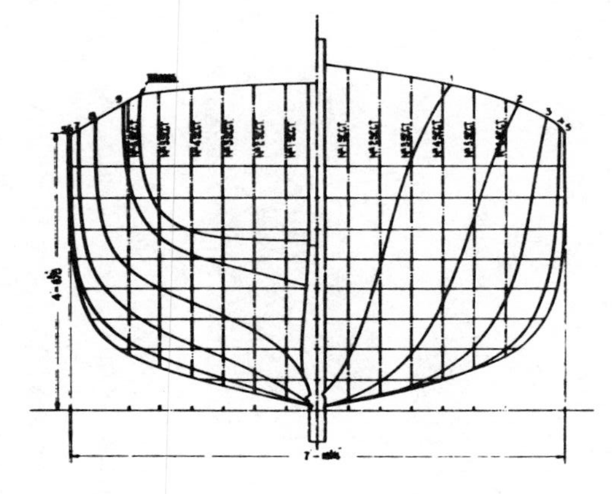

U.S. Navy steam cutters of the 1915 series were designed with painstaking care, including tank-testing of models. The design of the 25-h.p. 30-footer achieved its objective of crowding great power and capacity into a hull of limited dimensions.

The U.S. Navy 50-foot steamers, entering service in 1908, were a pinnacle of American steam launch design. Their 200-h.p. power plants, turning 50-inch screws at 400 r.p.m. for 16 knots, made them internally as fierce and high-strung as a destroyer. The steamers were worthless in a gas-boat world, post-1918, and were sold or given to anyone who would take them. A few 1920s-style motor cruisers are still afloat on the hulls of these 50 footers. (Photos above and opposite top courtesy Steven Pope)

This rare 1893 U.S. Navy engine (the 1892 *Maine* would have had one or two on board at Havana) with Bremme valve gear is described as a "Type F engine, serial #110." The present owner traded a bottle of whiskey for it at a Vermont lake. (Courtesy Francis P. Coughlin, Jr.)

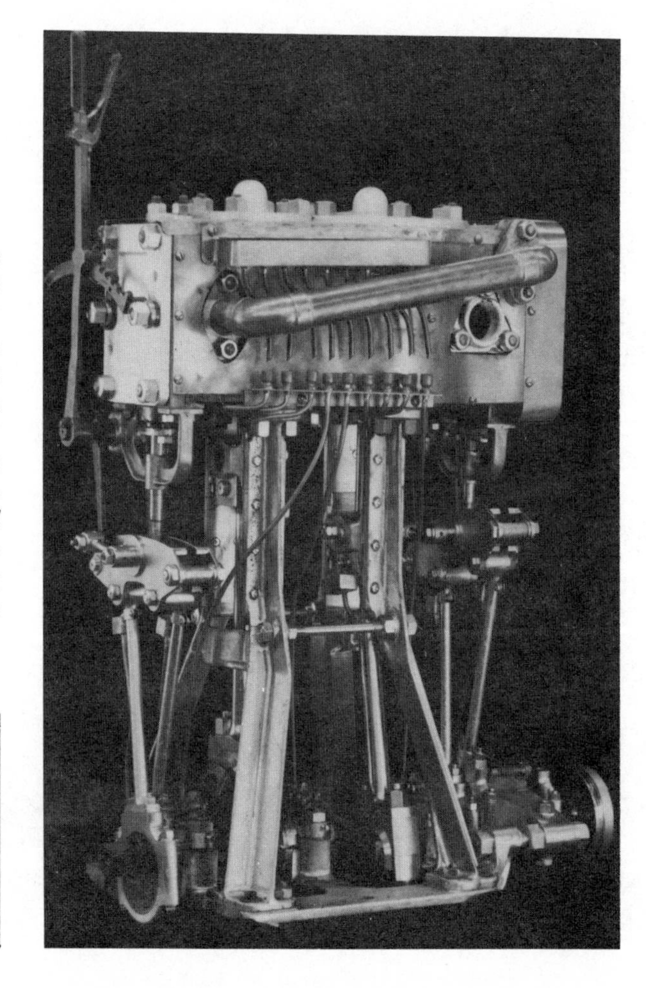

After the fact, anybody can see that the wasp-waisted framing of the 1900 Type B engines lacked rigidity. The "Navy Bs" have a special charm, with their short bedplates and cast-bronze frames, and not many men throw them away because they're not perfect. Gary Mitchell's "B3," built at the Portsmouth (New Hampshire) Navy Yard, was sold in 1980 to Jim Webster for use in his steam launch *Cloud Nine*. Jim rebuilt the engine, restoring it to its original beauty. Bill Shaw's drawing was developed from a rubbing off this engine. (Jim Webster photo)

TYPE B 3 NO. 712
3 ½ X 7 X 6.
NAVY YARD
PORTSMOUTH N. H.

Above, left: This view of the back side of Tom Rick's "Navy K" reveals the perfect proportions of the engine, 3.75" & 7.75" x 6". The base is solid bronze, the cylinder jacket bright brass, and the columns polished steel.

Above, right: U.S. Navy engines of the 1915 series are unmistakable in their perfection of proportions and finish. The 50-horsepower Type M engines are worth only half as much to collectors as the 25-horsepower "K" engines — the designer put the high-pressure cylinder on the wrong end! This is Frank Orr's "M." (Courtesy Everett Arnes)

Left: For American collectors of steam launch engines, no plum is so large and juicy as the 198-h.p. triple engine installed in the U.S. Navy 50-foot steamers of 1908-1918. Francis P. Coughlin, Jr., is the lucky collector who was taking this one home when this photo was taken. Students of the subject detect in this design many details from the last reciprocating-engined battleships and cruisers. (Courtesy Francis P. Coughlin, Jr.)

TORPEDO BOATS

Above: A spar torpedo shown rigged and ready on a steam launch while the crew poses for a portrait. The explosive charge, fastened to the end of the spar, could be projected forward or abeam and lowered well below the waterline of an enemy ship. The charge was exploded by contact or by means of a lanyard. *Opposite:* The Confederate Navy's original *David* was built in 1863 as a private venture by T. Stoney at Charleston, then turned over to the navy to be used in destroying units of the Union fleet that was then blockading the port. The cylindrical boiler-plate midsection of the 50-foot hull was fitted with conical ends of timber, and a horizontal mill engine was bevel-geared to the propeller shaft. The boat floated nearly awash (but should not be confused with the submarines also used in the Civil War). She was capable of 7 m.p.h. but was considered unmanageable and dangerous in a seaway.

David attacked *New Ironsides,* an enormous Union ironclad, on October 5, 1863. When the 60-pound torpedo exploded under the ship's quarter, the geyser of water fell down the launch's stack and put out her fire. Officers and crew escaped by swimming, but Assistant Engineer J.H. Tombs stayed with the boat, relighted the fire, and finally steamed safely away (with the pilot, who couldn't swim and had stayed with the launch out of necessity). The damage suffered by the *New Ironsides* from the attack kept her out of the war for a year.

Several other nameless "Davids" were built, mainly at Charleston. After the war, these were popular trophies, and they were displayed in locations as far north as New York City. The Confederate torpedo boat *(opposite bottom)* that was kept at the U.S. Naval Academy for several years was probably the original (and most successful) *David.* Is her machinery gathering dust in some Annapolis subbasement?

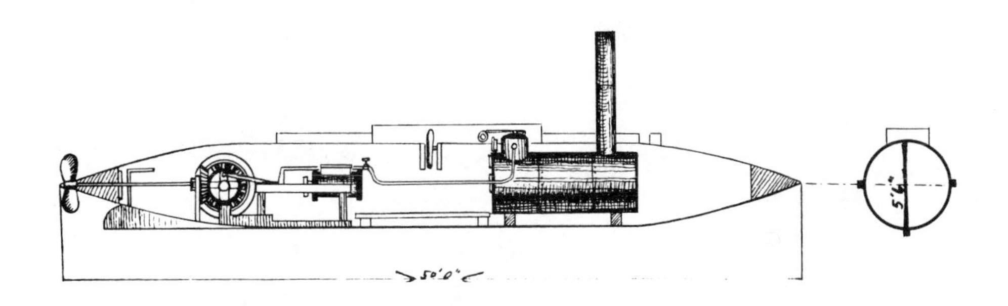
50'0"
5'6"

Only a few popular heroes emerged from the mass slaughter of the Civil War, and the most popular of these was the commander of a fighting steam launch. On October 27, 1864, Lt. William B. Cushing jumped his 36-foot screw picket boat over a log boom at Plymouth, NC, and discharged a spar torpedo against the Confederate ironclad *Albemarle,* sinking her.

The *Albemarle* had been a major obstacle to Union advance, overpowering all Union ships capable of operating on the North Carolina sounds. Lt. Cushing set out to destroy the ironclad with two steam launches carrying torpedoes, 12-pound howitzers, and " . . . 20 men, well armed with revolvers, cutlasses, and hand-grenades." He made his famous dash at 3 a.m., at 6 knots, through a hail of cannister and musket fire.

Twenty-six years later, America's first modern torpedo boat was christened the *Cushing,* and the creation of patriotic literature and art about the exploit continued into this century. (Photograph courtesy Smithsonian Institution)

Yarrow & Co. built this 20-m.p.h. torpedo boat for the French navy in the 1870s. The boat may be the *Torpilleur,* Yarrow torpedo boat #4, of 1875.

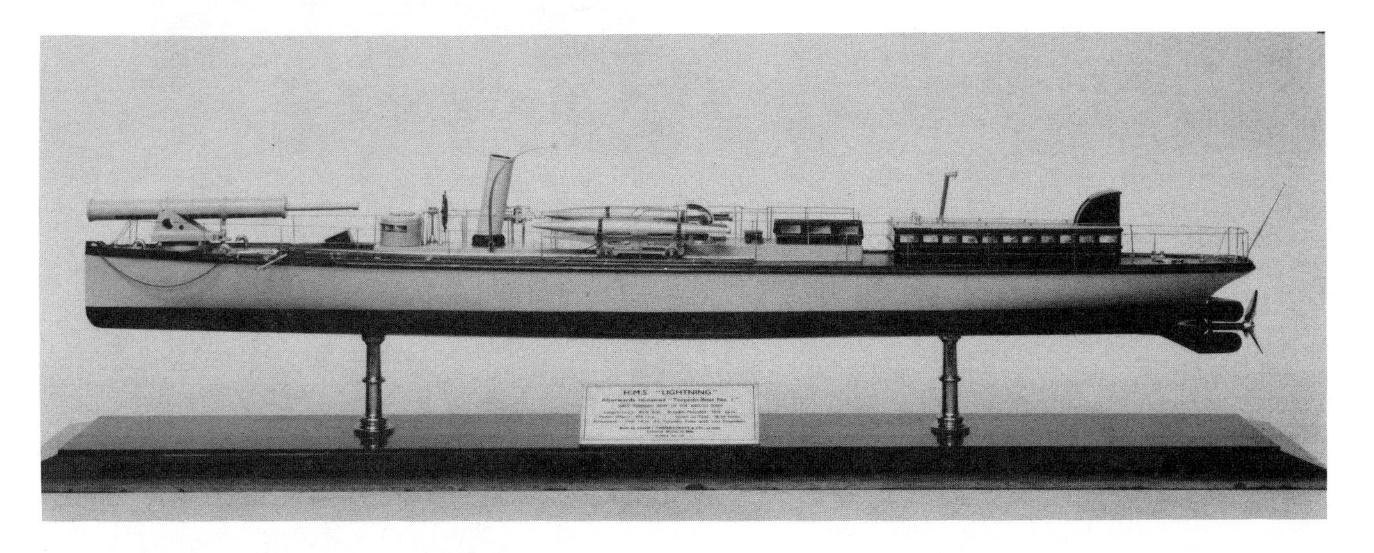

The Royal Navy's steam launch *Lightning* of 1876 was 84.5' x 10.8' and was capable of 18.54 knots with her 478 i.h.p. She carried two 14-inch automobile torpedoes. Built by Thornycroft, she was renamed *Torpedo Boat #1* when it was noticed that she was so fast and carried such deadly weapons that every fighting ship afloat was at risk. (Crown Copyright. Science Museum, London)

During the first six months of 1878, F. Schichau, of Elbing, East Prussia, turned out ten 66-foot coastal torpedo boats (*above*) for the Russian government. Their three-cylinder compound engines (*right*) had a 9.75'' high-pressure cylinder between two 12.62'' low-pressure cylinders. Stroke was 10.25 inches. Locomotive-type boilers supplied 10-atmosphere (147-p.s.i.) steam for 260 h.p. at 380 r.p.m. on a 4-foot screw.

Early in July 1878, the boats were delivered from Elbing to St. Petersburg, Russia, under the charge of the head engineer at Schichau's works. During the voyage, the boats spent a good deal of time scurrying to shelter in stormy weather and begging coals from tramp steamers to replenish their bunkers, but they covered the 630 nautical miles in 53 hours, for an average speed of nearly 12 knots.

The torpedo boats averaged 17 knots in two-hour trials in rough seas. Their performance and the engines' light, taut design were advanced for their time, and they must have set several British builders to sharpening their pencils. (From William Maw, *Recent Practice in Marine Engineering,* 1883)

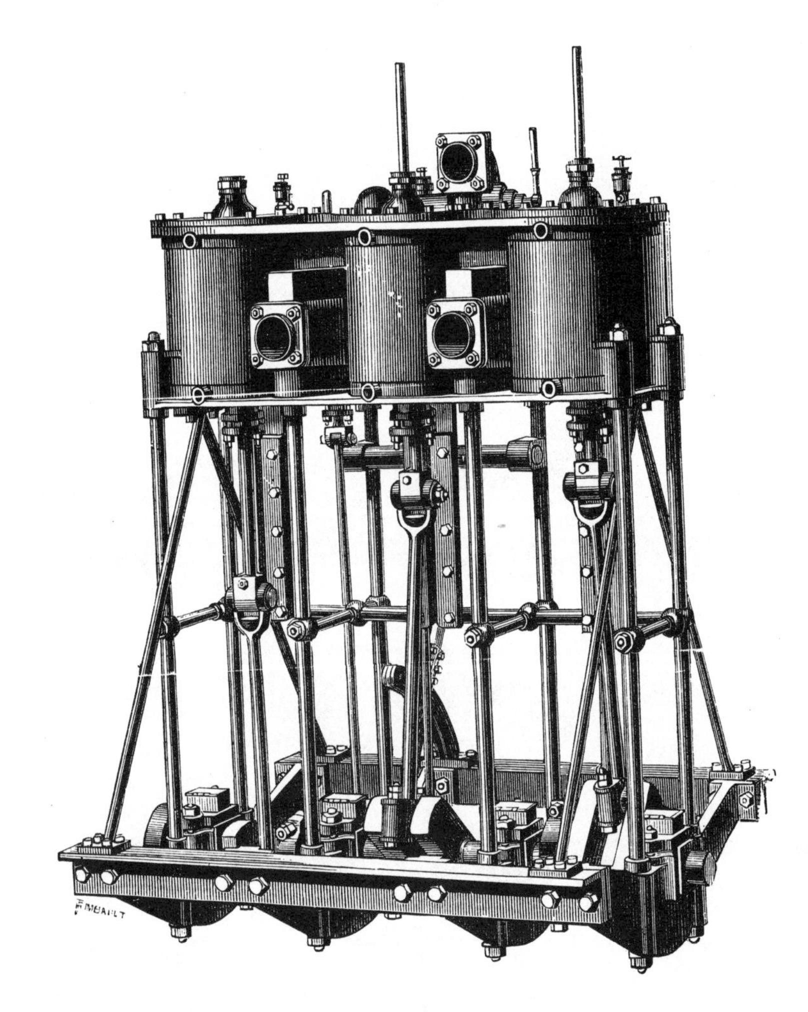

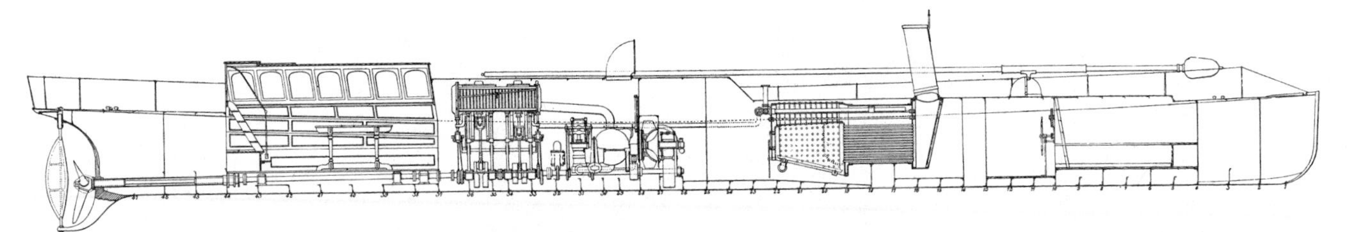

By the late 1870s most of the British colonies had engineering works capable of building steam launches, though it was usually more economical to ship them out from Britain. *Acheron,* 80' x 10', was built by the Atlas Company, Sydney, Australia, in 1879 for the New South Wales government. These second-class torpedo boats were expected to serve frequently as dispatch boats, so they were given more liberal accommodations than was usual in fighting boats.

On her trial run, March 1, 1879, *Acheron* " . . . attained a speed of 16 m.p.h. in a heavy sea, the engines making 330 revolutions per minute . . . . The conclusion was that the vessel was over-screwed (the wheel was 5' x 5'), the engines not being able to get away . . . . A speed of sixteen miles was, however . . . creditable to colonial engineering." The hull was of steel plating, ordered by telegraph from England, on iron frames shaped locally. The 11" & 19" x 14" engine was connected to a forward-extending shaft that drove the pumps and blower. (From William Maw, *Recent Practice in Marine Engineering,* 1883)

Simple Engines

Plain, functional, simple-expansion engines suit plain, functional, steam launches. Simple engines were installed in most of the severely classical launches of the 1870s and 1880s, and they remained dominant in working steam launches and in small pleasure launches (under 10 horsepower or so) until gas engines took their place. Noncondensing simple engines were once the prime movers of locomotives, agricultural machines, sawmills, and factory lineshafts. Today, the operator of a simple-engined boat can feel a profound appreciation for the long, unbroken thread that connects him with the rhythms and values of the 19th-century industrial revolution. Some of the chief virtues of steam power — ruggedness, mechanical simplicity, and adaptability — are exemplified in one-cylinder launch engines.

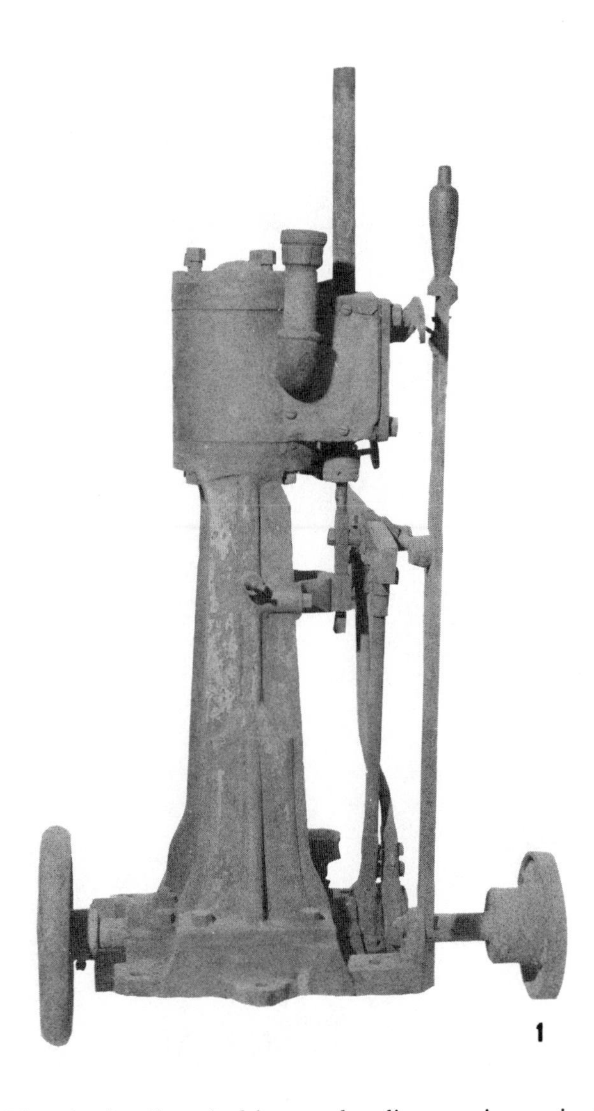

1

2

The basic "vertical-inverted, direct-acting, single-expansion, double-acting, reversible" steam launch engine was perfected about 125 years ago. The design was so "right" for the purpose that it remains fundamentally unchanged today. The old engines here are from the collections of William W. Willock (1), Bob Thompson (2), and Lewis Conant (3). Thompson found his engine standing neck-deep in a backwater of Lake Winnipesaukee. It was in *Mineola* about 1872, and may have turned the first propeller on the lake. (*Nellie,* of the same year, has her own partisans claiming primacy.) With a new piston and valve spindle, *Mineola*'s engine supplies power to grind Thompson's apples each fall.

The modern engines are Fred Semple's standard 5-h.p. launch engine, put on the market in 1955 (4), Roy Anderson's 3-h.p. design of the late 1950s (5), and a 1½-h.p. simple-expansion version of the Blackstaffe-Wood engines of 1962 (6). The Semple model shown here has a wheel reverse; Semple also has a lever-reverse model (see page 82). Bob Kerr of Seattle now holds patterns and rights to the 2.5" x 2.5" Anderson engine. Valve gear parts are precision centrifugal castings in bronze.

3

4

5

6

Above left: Among many types of engines built by H.R. Stickney of Portland, Maine, was this single-cylinder launch engine, which has been completely restored by a Maryland owner. *Above right:* This 3'' x 3'' marine engine was built by A. Brown, of Bedford, England, and is owned by Mike E. Delanoy, of Haddenham, Cambridgeshire. The typically British lubricator on the top of the cylinder is extremely handy if you open the petcocks in the right sequence. (Mike Delanoy photo) *Right:* The old engine in William Willock's miniature tug *Benj. F. Jones* (see page 192) is just right for the job, evoking the crude but sturdy tugboat technology of 100 years ago. The builders of steam tugs crowded as much power into the boats as they could, at the lowest capital investment. Most harbor tugs were simple-engined and noncondensing, blowing their exhaust steam into the atmosphere — notorious coal-wasters, as detailed by Frank Graham in *Audel's Engineers & Mechanics Guide 3.* (W.W. Willock, Jr., photo)

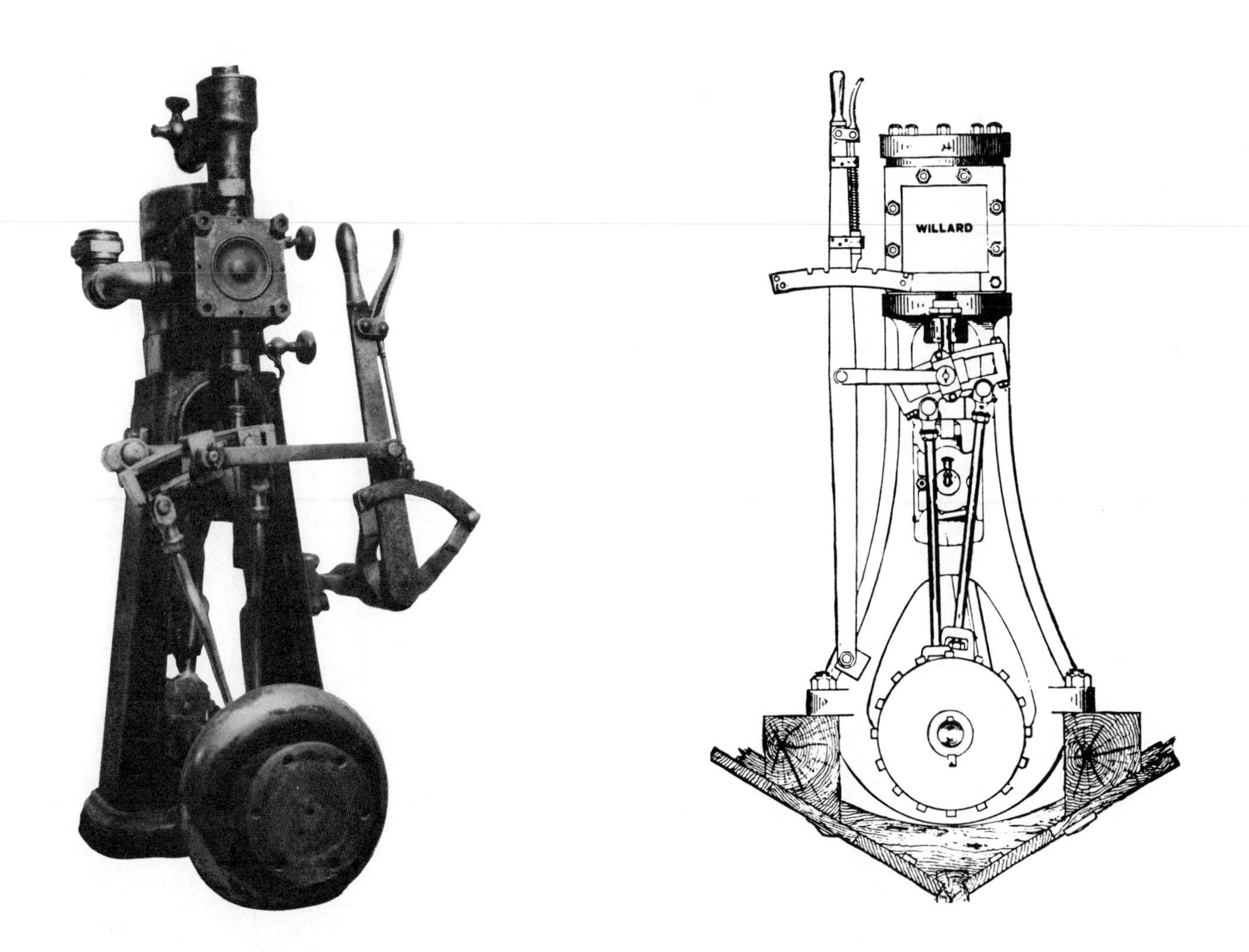

Left: This sturdy 2.5'' x 3'' engine was built for a retired sea captain 80 or 90 years ago. It powered *Lady of the Lake* on Spofford Lake, New Hampshire, for many years, and then was installed in the author's *Lourick* in 1943. Much of the engine and the boiler jacketing were nickel-plated long ago. The engine now powers an unnamed boat owned by Wesley J. Trathen, in Flint, Michigan. *Right:* The Chicago-built Chas. P. Willard & Co. engines of 100 years ago typify the ruggedness and simplicity of mid-American marine engineering. These were the engines of thousands of small tugs, launches, passenger propellers, and excursion boats. Not many specimens survive, because typical simple-expansion workboat engines of 20 or 30 h.p. stood as tall as a person, weighed half a ton, and thus fell outside the scope of most hobbyists' collections. (From C.P. Kunhardt, *Steam Yachts and Launches,* 1891 edition)

Whether you call them "inverted" or "crosshead," marine engines with the crosshead and guides above the cylinder are admired by most steam launch men as an interesting and practical layout. The author is familiar with only two builders, both of Boston, Massachusetts, who specialized in launch engines of this type — John Paine and Edward S. Clark.

The "inverted" vs. "crosshead" verbal controversy harks back to the side-wheel engines of 120 to 200 years ago. In those, the cylinders were in the bottom of the boat and piston rods worked upward toward the transverse shaft that carried the side wheels. When cylinder-above-crankshaft propeller engines were introduced, after 1840, they were called "inverted" engines. This term was used/misused for crosshead-above-cylinder engines at least 90 years ago (by George Whitney, for one). The latter engines were also called "steeple" engines in large steamers, because the crosshead guides towering above the weather deck required a steeple-like enclosure. (This appellation must not be confused with "steeple-compound" engines, which are unrelated.)

A George Whitney 3.5" x 5" inverted engine. (George Whitney collection)

An Edward S. Clark engine, built before 1885. (Lee Burgess photo)

A John Paine engine. This engine was in Bob Thompson's *Rum Hound* for many years, on both Lake Winnipesaukee and Squam Lake, New Hampshire.

The author bought this engine from Nat Goodhue, passed it along to Wesley Trathen in Michigan — and so it goes.

Southern Sweden is a land of lakes and waterways, which were aswarm with steam launches a century ago. *Montala* (1873), 56.5' x 11.5', could carry 106 passengers. Her 8-h.p., 60-p.s.i., noncondensing engine might be described as "Teutonic" in style, but there were several internal niceties beneath the plain iron surface.

A rotating-leaf throttle valve could be positioned to admit steam to both an "expansion-valve" chest and the normal steam chest, to provide full-stroke steam admission for starting. The slip eccentrics were rotated for going astern by a tram-handle reversing lever at the top of the engine, bevel-geared to a pinion gear that engaged a grooved collar on the shaft. The massive engine standard was a hollow casting containing an exhaust feedwater heater. The screw was 3.2' x 6.8' (old Swedish feet, slightly less than the English). (From William H. Maw, *Recent Practice in Marine Engineering,* 1883)

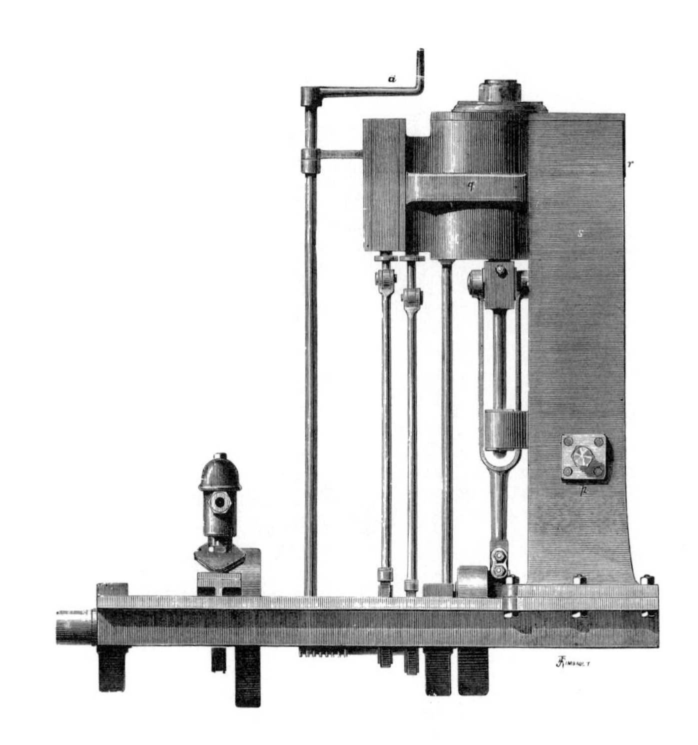

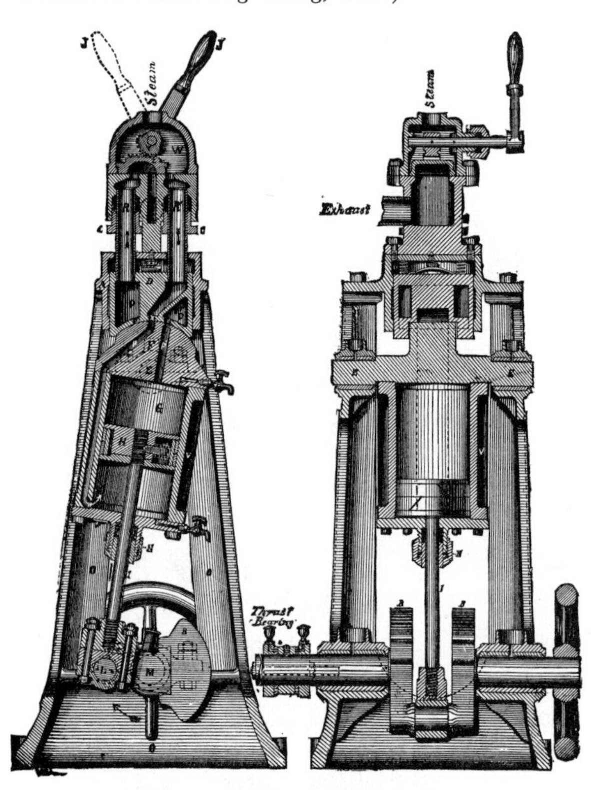

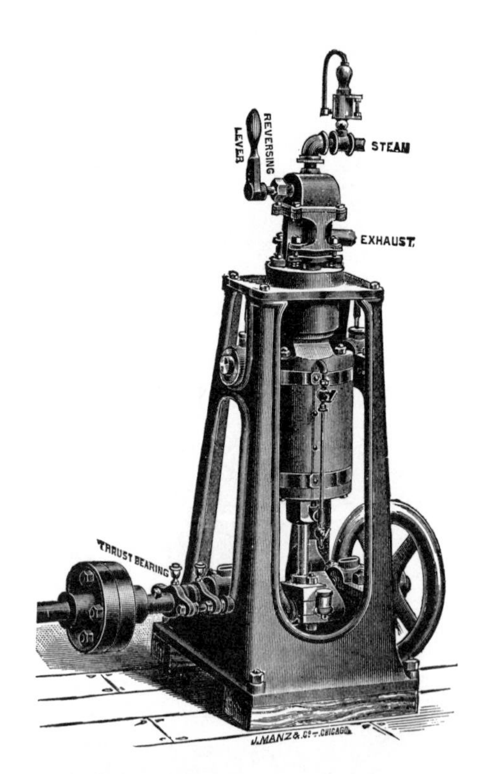

During the 1840s and 1850s, oscillating-cylinder engines were common enough in steam vessels, both propellers and paddlers. With 8- or 10-p.s.i. working pressure, there was little difficulty in controlling loss of steam at moving joints (then usually at the trunnions).

During the 1880s the Kriebel oscillating engine, employing the 60-pound steam popular in that decade, was widely used in small American steam launches. Steam admission in a Kriebel is through ported sliding faces, and the worn Kriebels in service today are sometimes lost in a cloud of leaking steam. (From C.P. Kunhardt, *Steam Yachts and Launches,* 1891 edition)

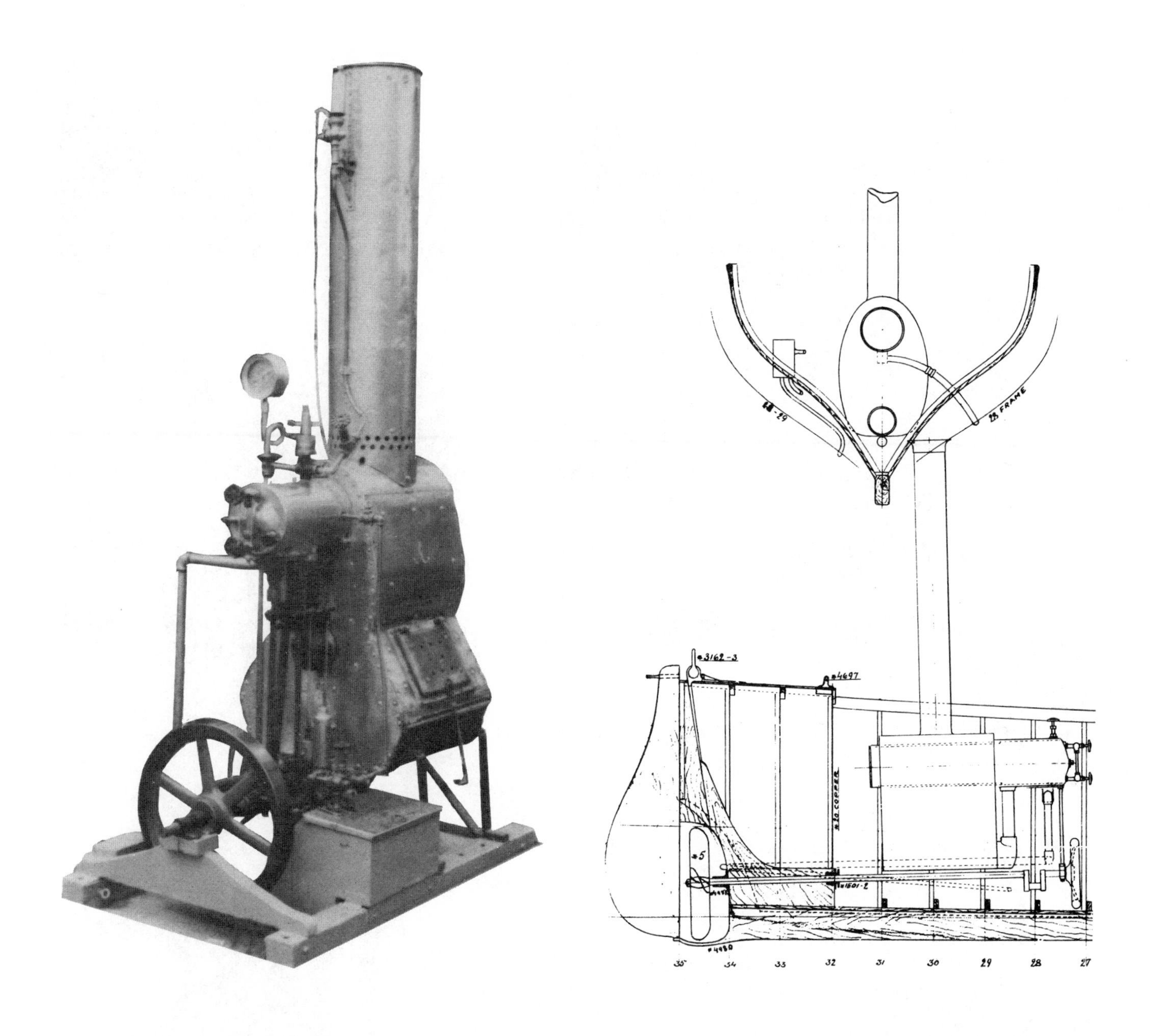

This exquisite 1893 power plant is Herreshoff through and through. The stressing of parts is so precise (unusual in steam launch design) that the machinery appears spindly, even frail. The very long connecting rod is a Herreshoff trademark, and the high finish of the boiler is matched by its light weight. Locating the engine's cylinder inside the boiler's steam drum results in compact dimensions and measurably improved thermal efficiency. (The spoked flywheel is extraneous, representing the size and weight of a boat's propeller.)

Nathanael Herreshoff designed the unit, Herreshoff engine #543, for use in his own light 22-foot launch (stern-area profile and section drawings shown here). Originally fired with kerosene, the boiler was later converted for solid-fuel firing from either side. Henry M. Luther most recently owned the plant and loaned it to the Fall River (Massachusetts) Marine Museum for public display. (Drawings from the Hart Nautical Museum, M.I.T.)

Above left: The 2.625" x 2.625" engine in Gordon Sullivan's *Feeble II* (see page 146) once drove the forced-draft fan in a U.S. Navy 50-foot pinnace. Characteristically, Sullivan has prettied her up with brass nuts, mahogany lagging, copper bands, etc., but nothing could be done to gild the luster of the engine's base and back column — a single beautiful casting of naval bronze. *Above right:* There are innumerable "Stuart 5A's" (from Stuart Turner, Ltd., in England), 2.25" x 2", in hobbyists' little boats. P.J.C. Robinson, a ship's engineer, was used to bigger machinery, but between voyages he looked forward to firing up *Jamie,* which has a 5A, for a quiet spin around Lake Windermere. The engine was modified by Robinson to stand up to the hard work of propelling an 18.5-foot launch. He also built the 120-p.s.i. boiler, which has 32 half-inch copper tubes. Robinson was killed on February 22, 1979, in an engineroom accident at sea. (P.J.C. Robinson photo) *Right:* This handsome little specimen was designed by Ray Hasbrook, of New Paltz, New York, and built by Everett Smith, of Homer, New York. Smith installed it in his launch *Panetelet* (see page 139).

Left: A desire to overcome the drudgery of rowing small boats was widespread for 20 years or more before Ole Evinrude started the outboard-motor revolution. A "rowboat" boiler was likely to weigh 200 pounds, so even careful weight-shaving on a rowboat engine such as this one did not promise much of a future for steam rowboats. The engine is "overhung" — the crankpin is outboard of the main bearings instead of between them. The crosshead is held in alignment both by a guide-rod extending below it and by a crossbar that rubs along the edges of the frame.

Cliff Hills's father used this 2.25" x 3.5" engine in a rowboat on the Ashuelot River, Swanzey, New Hampshire, about 1905. Cliff had the engine in a 21-foot boat (see page 139) with a Stanley car boiler during the 1920s and 1930s. He cherishes the engine now as a memento full of pleasant family associations. *Below left:* This little gem provides power for the 17.5-foot *Soot* (see page 140), owned by Fred Sweetsir.

In the established pecking-order of steam launch engine collections, "bottle-frame" engines rank pretty low. This is because engines of this design were the lowest-priced upright types, and they were much more often put to work in creameries or sawmills than in boats.

Bottle engines are honest laborers, with a longer history and a more stolid and durable character than most flashy "all-marine" engines. Some of the collectors who spurned them in the junkyards 20 years ago are kicking themselves now.

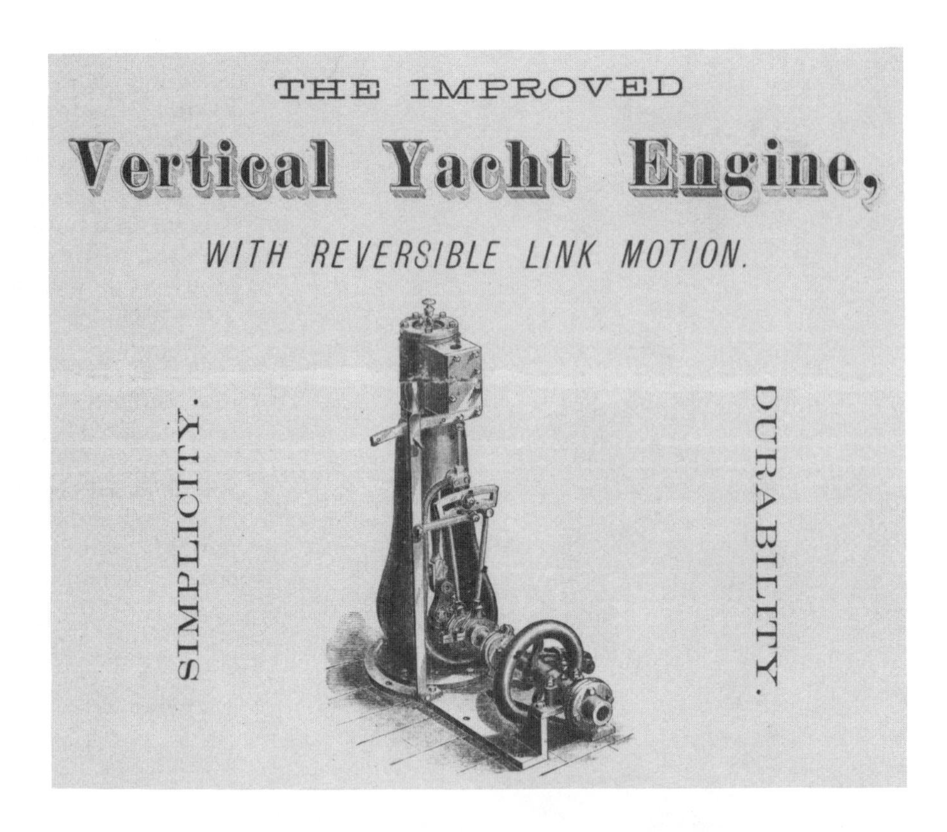

A Ward B. Snyder engine. (Courtesy John S. Clement)

THE PEERLESS MARINE ENGINE.—This engraving shows the style of engine which we call our "Peerless" Engine. The lever for handling the link is very convenient, strongly attached, self-adjusting, and a most satisfactory arrangement in every way.

Our Peerless Marine Engines are among the simplest, most compact, and strongest in the market. The crank, connecting rod, piston rod, valve stem, and pins are made of steel. The main frame of the engine and the slides, as well as the bearing for the crank shaft, are made in one piece, so that it is impossible for the working parts of the engine to get out of line or change their relation to each other. The construction of the engine is such that the action of the piston rod is exactly central, and all overhanging strains are avoided.

We warrant these engines to be economical of steam, thoroughly well made in every particular, and to give full rated power.

We can furnish engines complete, with stern bearings, and any length of shaft required for different length of boats, at a moderate advance on the price of the engine.

A Chas. P. Willard & Co. offering of the mid-1880s.

Above left: Sawdust Sally's 3.5'' x 4.5'' engine was built in the Boston & Albany Railroad foundry many years ago. Marine steam engines do not require flywheels (the propeller weighs enough!), but most owners like to have a handwheel to position the single crank throw where they want it for maneuvering. *Above right: Nellie*'s engine. *Nellie,* the oldest steam launch in America, is shown on page 155. (Francis P. Coughlin, Jr., photo) *Left:* Frank Fuegeman's engine was a common bottle-frame industrial type, but his expert conversion has made it a truer steam launch engine than many rare antiques misused by the inexpert. Fuegeman's launch *Ada* is shown on page 52, and the boat's Fuegeman-built Clyde boiler on page 305.

Shipman Engine Company

Boston Model (above left)

H. P.	Floor Space.	Height.	Marine Weight.	Revolutions.	Size Cylinder.	Shaft.	Price.
1	19 x 35 in.	30 in.	475 lbs.	450	2¼ x 3 in.	1⅛ in.	$175 00
2	20 x 45 in.	34 in.	806 lbs.	400	3 x 4 in.	1½ in.	275 00
4	20 x 48 in.	35 in.	870 lbs.	400	3½ x 4 in.	1½ in.	375 00
6	21 x 59 in.	42 in.	1365 lbs.	350	4½ x 5 in.	2 in.	550 00
8	34 x 65 in.	43 in.	1476 lbs.	400	4½ x 5 in.	2 in.	700 00

Rochester Model (above right)

H. P.	Floor Space.	Height.	Marine Weight.	Revolutions.	Size Cylinder.	Shaft.	Price.
1	31 x 20 in.	23 in.	300 lbs.	450	2¼ x 3 in.	⅞ in.	$165 00
2	43 x 27 in.	31 in.	625 lbs.	400	3 x 4 in.	1⅛ in.	235 00
3	43 x 27 in.	33 in.	750 lbs.	350	3½ x 4 in.	1⅛ in.	300 00
4	48 x 30 in.	48 in.	1000 lbs.	300	4 x 5 in.	1⅛ in.	375 00

Launch Outfits

For the convenience of a large number of correspondents who desire prices on launch outfits, with a view of building their own hulls, we quote the following prices, which include engine, boiler, iron propeller, shaft, stern and thrust bearings, stuffing box and whistle:

1 H. P. Boston Model $198 00	1 H. P. Rochester Model $188 00
2 " " " 304 00	2 " " " 264 00
4 " " " 409 00	3 " " " 330 00
6 " " " 596 00	4 " " " 409 00

Racine Boat Company

Horse Power.	Cut-Off.	Pressure.	Cylinder Diam. Stroke.	Revolutions per Minute.	Floor Space.	Height Boiler to Dome.	Approx. Weight.	Price.
1	1/4	100	2 1/4 x 3	300	17 x 30	34	350	$ 175 00
2	3/8	100	3 x 3	300	19 x 34	39	550	250 00
3	1/4	100	3 1/2 x 3	350	19 x 34	43	750	325 00
4	1/4	100	3 5/8 x 4	325	25 x 40	48	850	400 00
6	3/8	100	4 x 5	325	31 x 48	50	1100	525 00
8	1/4	100	5 x 5	350	31 x 48	62	1250	600 00
10	1/4	100	5 x 6	350	42 x 54	70		675 00
12 1/2	1/4	100	6 x 6	300	42 x 54	80		750 00
15	1/4	100	6 1/2 x 6					850 00
20	1/4	100	7 x 8					1,050 00
30	1/4	100	8 x 8					1,350 00

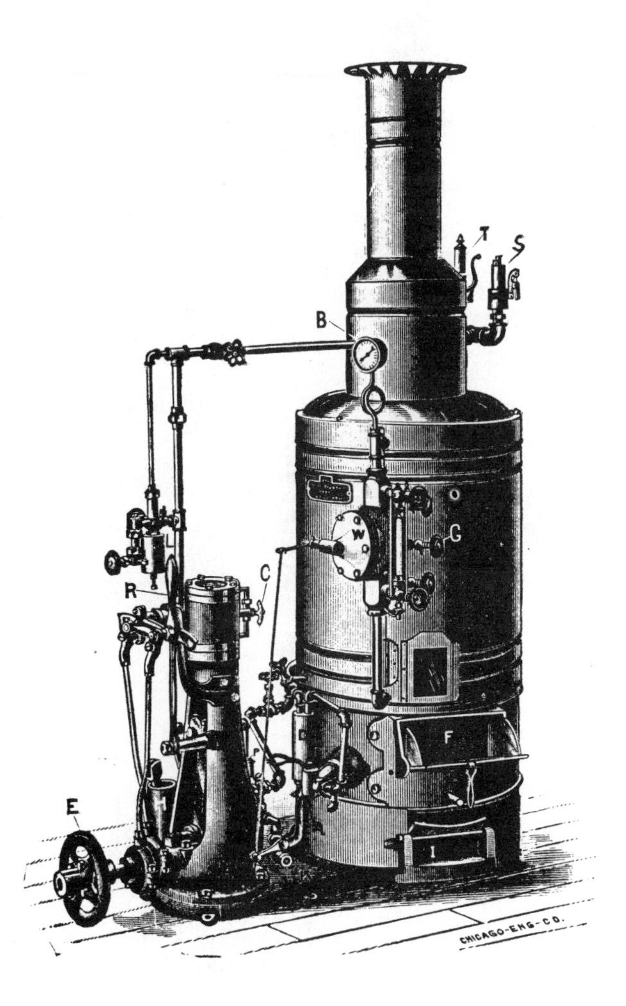

RACINE AUTOMATIC ENGINE AND BOILER.

The best Yacht Outfit on the market for the following reasons:

It is self-contained, engine and boiler on one base, consequently can not get out of line.

It is automatic, requiring very little attention.

It is arranged with a combination fire box, permitting the use of wood or coal in connection with oil for fuel.

It has been on the market for eight years, which speaks for itself, when you take into consideration that there are nearly 2,000 of the Racine Automatic Engines now in operation, which in itself is a sufficient guarantee as to its practicability.

Space prevents us from publishing a long list of testimonials which we have in our possession.

We invite comparison with any other outfit on the market.

By 1895, with naphtha, electric, and gasoline boats pointing the way to clean and easy powerboating, simplified operation became a principal selling point for steam launches as well. The Racine Boat Co. and the Shipman Engine Company were major competitors for a growing market among steam launch owner-operators. The lightweight, unitized power plants portrayed in their sales brochures had automatic water-level control, as well as automatic fire control when on liquid fuel. These designs could have led to foolproof steam launches for Everyman within a few years, but history decided otherwise. The development of gasoline engines was so rapid after 1895 that "automatic" steam launch plants died in mid-career. They are rare curiosities today.

Double-Simple Engines

A two-cylinder high-pressure — or "double-simple" — engine has twice as many parts as a single-cylinder engine, and it may cost twice as much and take twice as much space in the boat. It promises none of the economy and little of the distinction of a two-cylinder compound engine. However, double engines were the standard for locomotives, steam cars, hoisting engines, and stern-wheelers, and they were installed in many oceangoing ships of the 1850s and 1860s. They are positive-starting, instantly reversible from any crank position, smooth-turning at low r.p.m., and capable of throttling down to 5 or 10 r.p.m. Quite a few steam launch men prefer these engines to all others, even if they have to shovel a little more coal.

Above: This 3" & 3" x 4" double-bottle-frame engine drove a boat on Lake Winnebago, Wisconsin. It is full of design eccentricities and historical mystery. (Courtesy The American Precision Museum, Windsor, Vermont) *Above right:* David Thompson had this double engine in his *Annie E. George* for a time. When a double engine is exceptionally massive or plain, there is a suspicion that it came out of the basement of an 1880s office building. Pre-electric-motor elevators were driven by double-simple steam engines with Stephenson-link reversing gear, and some of these ended up in boats. *Right:* The unfamiliar details and Victorian-Gothic style of this 2" & 2" x 3.25" launch engine suggest that it was built in the 1870s, or possibly the 1860s. (William W. Willock, Jr., photo)

Above left: When *Lady Woodsum*'s engine outlived its usefulness in the 1920s, it was given a decent burial in a pit on the shore of Lake Sunapee. Ozzie Woodward (shown at right in the photo), a lifelong resident nearby, remembered the boat and the burial, and dug up the engine 40 years later. He completely rebuilt the big Paine return-connecting-rod engine in his basement shop. *Lady Woodsum* can be seen on page 22. *Above right:* The original engine in Bill Willock's *Clermont* was built by H.W. Petrie of Toronto. The double-simple engine has cylinders of 2 inches by 3 inches. (William W. Willock, Jr., photo) *Left:* Hugh Cawdron and his *Bubbly Jane* (see page 165) on the River Thames. Merryweather boiler, double-simple engine. (Courtesy Hugh Cawdron)

Compound Engines

The advance from simple-expansion to compound-expansion engines was fundamental to the triumph of steamships over sail, which occurred during the 1870s. In steam launches, compound engines are more like the frosting on the cake — not necessary to a successful boat, but certainly adding interest and distinction, and perhaps even saving some coal.

Steam launches reached their peak of economic importance and interest to the public in the 1880s, but steam launch machinery design peaked a decade or two later. Serious students of launch technology and aesthetics are torn between admiration for the classic "iron age" machinery of the 1870s and fascination with the exactly elaborated engines of the 1890s.

Most steam launch men are engine men. They would give an arm to possess a perfect compound launch engine with all the trappings — such as mahogany lagging, brass drip-oilers, and polished steel frame — but may be willing to settle for a nondescript boat or boiler. The possible saving of fuel with a compound engine is not of great moment to modern enthusiasts. The main interest is the beauty and appropriateness of the mechanism.

Fitzhenry engines are gemlike miniatures, suited to small yacht tenders or highly finished day launches for lakeside estates. This 2'' & 4'' x 3'' specimen, built about 1910, is in the author's collection. Most of the engine's parts are bronze, and the design is exceptional in having the crank throws 180 degrees apart and in having both valves controlled by a single set of eccentrics and links. Perhaps the only other builder that made compound engines with cranks at 180 degrees was Murray & Tregurtha, of South Boston, Massachusetts. The tracing was drawn by W.U. Shaw from the low-pressure cylinder cover of this engine.

When admiring an old launch engine, there is added pleasure if its source and original use are known. Most antique engines have been handed along from collector to collector until their true history is lost or blurred.

This engine, now in a museum, is believed to be one of three built for identical launches that carried passengers from downtown Chicago to the World Columbian Exposition grounds in 1893. James Kloosterman, of Ocqueoc, Michigan, has another of the engines in his launch, and the third is at the bottom of Lake Michigan. Slip eccentrics to drive the valves of both cylinders are at the low-pressure end of the engine. (Collections of Greenfield Village and the Henry Ford Museum, Dearborn, Michigan; courtesy James Kloosterman)

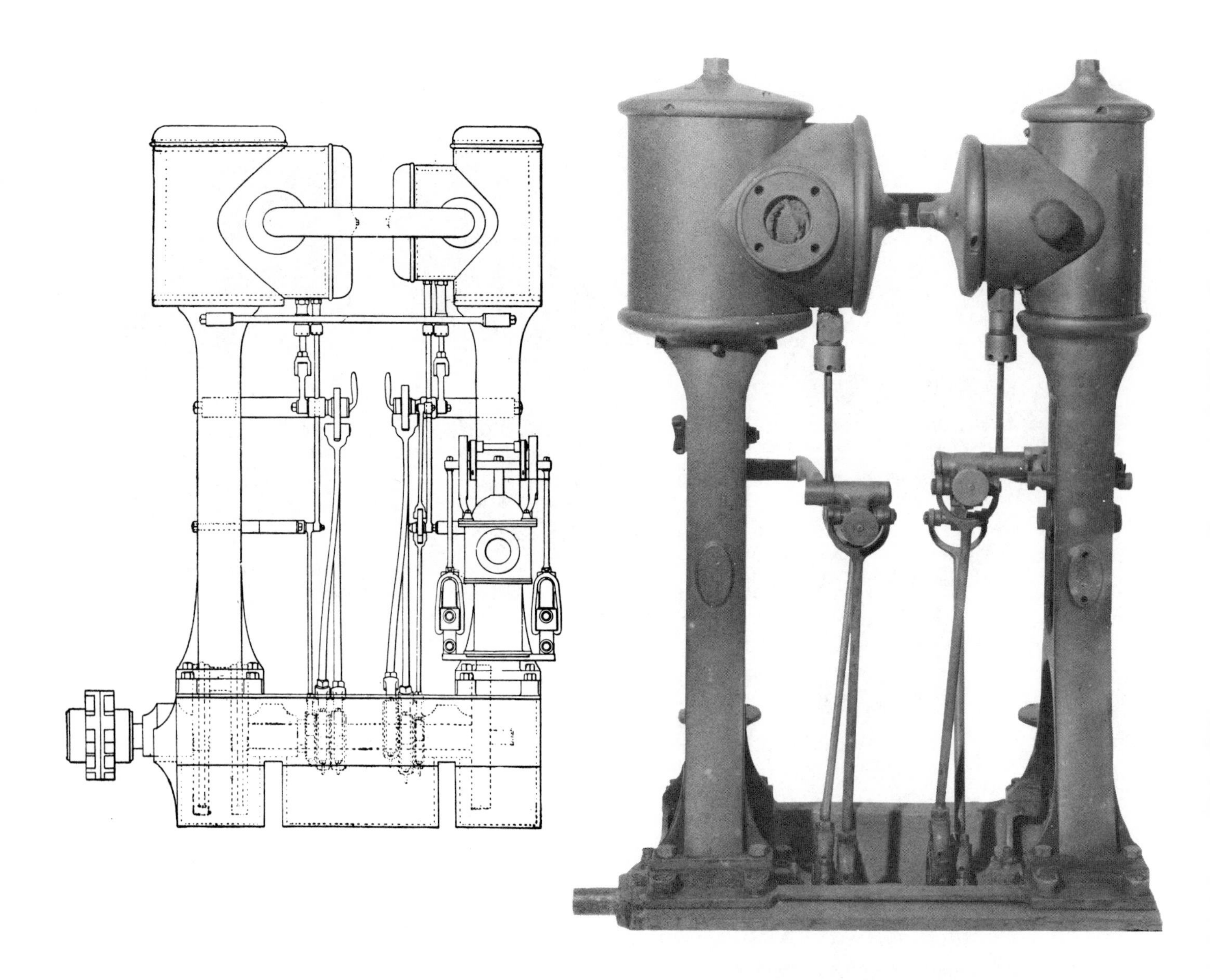

The early engines of the Herreshoff Manufacturing Co., Bristol, Rhode Island, did not look much like other steam launch engines. This nonconformity was the result of Nathanael Herreshoff's absolute confidence in his own skills and understanding. If other engine builders met design requirements in other ways — they had to be wrong!

This 2.5'' & 4.25'' x 5'' all-bronze 1885 engine, Herreshoff #120, is in the collection of W.W. Willock, Jr. The two cylinders' reversing controls are not interconnected. (William W. Willock, Jr., photo; drawing from C.P. Kunhardt, *Steam Yachts and Launches,* 1891 edition)

Everett Arnes found a stack of glass-plate negatives recording some of the marine engines built by the Sumner Iron Works, Washington Territory. These are excellent examples of the sort of rawboned "country" engines that could be produced by any shop able to pour iron and turn steel. (Courtesy Everett Arnes)

OPPOSITE PAGE

Left: The 10-h.p. V-compound Semple engine would power a boat up to 30 feet in length. Since the high-pressure cylinder is the same size as the single cylinder of Semple's 5-h.p. model, it can be fed by the same Semple boiler — the F-40 VFT. Twice the power is obtained by reusing the steam in the 5-inch-bore low-pressure cylinder. (Courtesy Semple Engine Company)
Right: Elmer Brooks designed and built the boiler (see page 309) and this fine 3'' & 6'' x 4.5'' engine for his steam launch *Artemis* (page 201). (Courtesy Ric Shrewsbury)
Bottom: Dr. Henry Stebbins' *Zephyr* (see page 158) cruises the New England coast, so the emphasis is on sturdy components that will keep working, not on polished brass. (Photo by Judy Buck, *Boston Record-American;* courtesy Dr. Henry Stebbins)

THIS PAGE

Top: *Panatela*'s engine, built in 1965 from castings supplied by the Reliable Steam Engine Co. of New Baltimore, Michigan, has cylinders of 2.5-inch and 3.75-inch bore and 2-inch stroke. This photograph shows Peter Moale's uncluttered installation. The entire launch, including hull (see page 147) and boiler (page 312), is Moale's own work. (Courtesy Peter Moale)
Left: *Puffin*'s Plenty & Son engine indicates 75 h.p. at 250 r.p.m. with 120-pound steam from a single-furnace Scotch boiler (see page 233). (Stephen Partis photo)

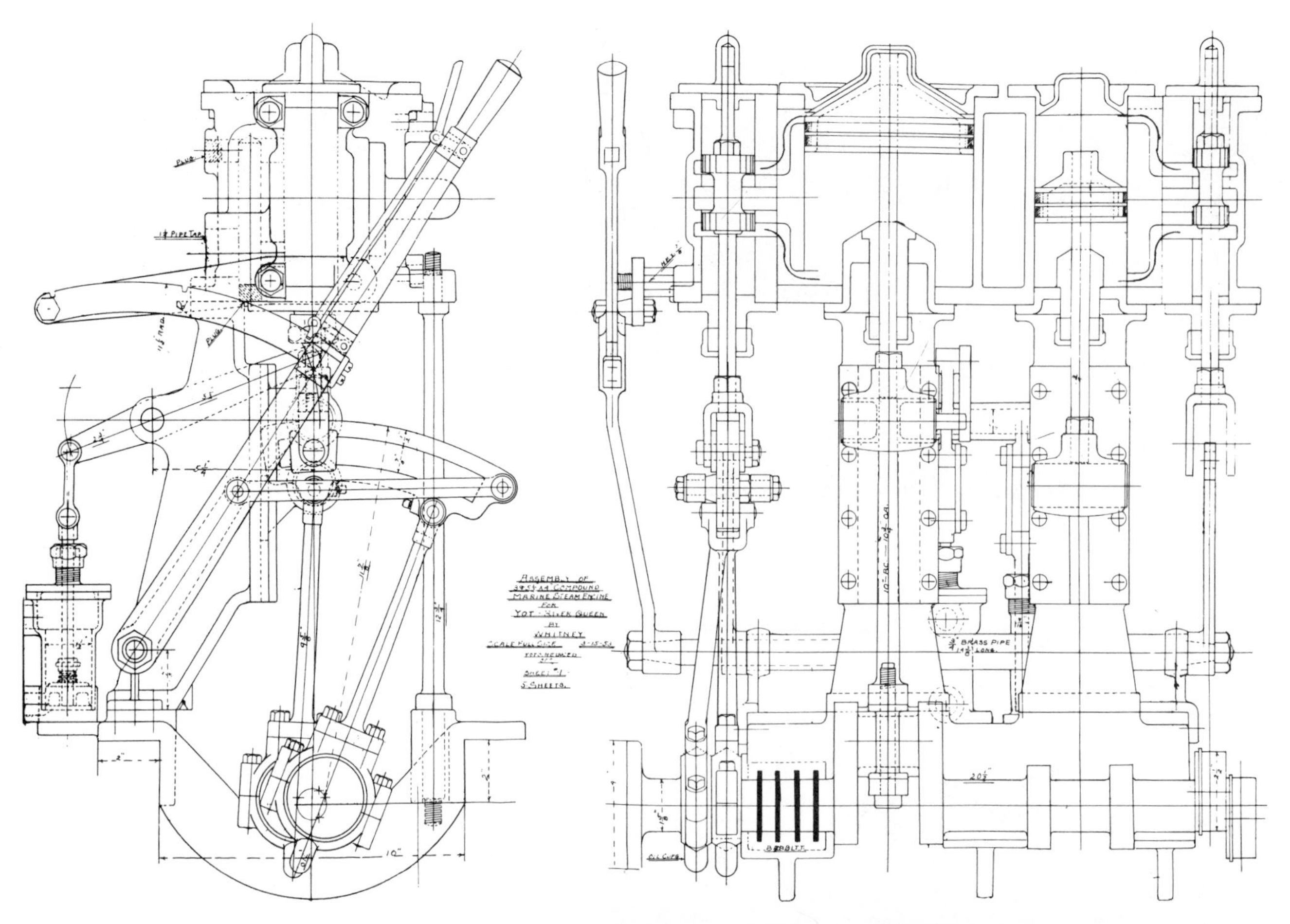

Most of George Eli Whitney's engines were compounds, and most of these were unorthodox in one way or another. Two Whitney steamers were given triangle compound engines, perhaps the only two of this design ever built for marine use. That of the *Montauk* (*right*) was 7'' & 14'' x 9''; the *Ida F*'s (*opposite left*) still exists, and its motion defies description. The latter engine has boiler-water-jacketed cylinders. The 4'' & 7'' x 6'' specimen of 1888 (*opposite top*) had Marshall valve gear and very light, diagonally braced framing.

When Whit was past 90, with 65 years of engine-building experience behind him, he sold Ted Middleton two 3'' & 5.5'' x 4'' engines. One of these (*opposite bottom*) was eventually installed in the *Lucky Star,* owned at the time by Frank V. Robinson of Mill Valley, California. Whitney built several engines of this design during the 1950s, and these have become famous as ''the last launch engines from an old-time builder.'' The ''River Queen'' engine (never installed in the author's *River Queen*) is another of these, and one of Whitney's drawings for this engine is reproduced here (*above*). (*Lucky Star* engine photo by Frank Robinson; 4'' & 7'' x 6'' engine photo from George Whitney collection)

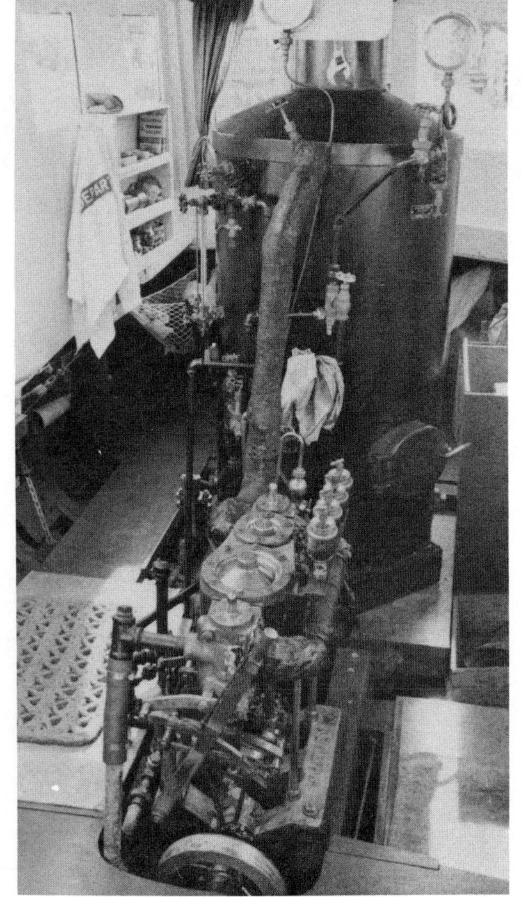

Left: "Steeple" (tandem) compound engines were widely used in larger vessels, especially tugs, but were little used in launches. A few of launch size were built, however. The fine example shown here was manufactured in 1898 by Fifield Bros. of Augusta, Maine.

Steeple compounds do not have a large following among steam launch hobbyists. (When Cliff Blackstaffe designed his initial 2-h.p. marine power plant around a steeple engine, he was besieged with appeals: "But when will you offer us a *real* compound, a fore-and-aft compound?") This Fifield Bros. engine is said to have been built for fishing-boat use. Apparently fishermen were more interested in low initial cost and compact dimensions than in having a beautiful little engine "just like a ship's." Hobbyists don't claim to base their preferences purely on economic considerations, and that's OK, too.

Above: Frank Lord, the proprietor of Mercer & Langmaid, Inc., Bath, Maine, built three identical 5-h.p. steeple engines sometime between 1880 and 1890. One was sold, Frank kept one, and one (shown here) he gave to his brother Fred. Fred used his in a launch on Maine's Kennebec River, where it performed long and well, finally outlasting the boat. The exact year is not known, but sometime before 1920 Fred reverently crated the engine and stowed it in his barn. Doug Lee found it there as a boy and bought it for $10. Doug in his turn had the engine for a score of years before he lent it indefinitely to the Maine Maritime Museum, in Bath, where the author snapped this photo in 1979.

Above left: This old laborer was the first engine used by Richard Hovey in his launch *Skookum Jack* (see page 191) at Manchester, Massachusetts. *Above:* The standard Blackstaffe-Wood compound, 1.5'' & 3'' x 2.5'' — handsome, sturdy, well built. Cliff Blackstaffe and Eric Wood built between 25 and 30 engines to this design. (Courtesy Everett Arnes) *Left:* This is *Phoebe*'s original engine, which has been brought out of storage to replace the Poulson compound that powered the boat (see page 193) for many years. The 65-h.p. Davis compound engine turns a 36'' x 48'' propeller at 150 r.p.m. to give the *Phoebe* a speed of 10 knots. (Courtesy Frederick G. Beach)

Good steam-engineering practice is much in evidence in the power plant of the 31-foot *Victoria.* The 3.25'' & 6'' x 5'' compound (10 indicated horsepower) was built in 1898 by J. Samuel White & Co., Cowes, Isle of Wight. The Merryweather boiler was new in 1946. M. Gilbert, of Reading, England, installed the steam plant in the 80-year-old hull in 1959-60; the present owner is J.D.L. Drower, who operates her on the Thames River. (Courtesy M. Gilbert)

GAS ENGINE & POWER CO
AND
CHARLES L. SEABURY & CO
CONSOLIDATED
BUILDERS
MORRIS HEIGHTS ON THE HARLEM NEW YORK CITY

Above: Seaburys are possible the most numerous — and also among the plainest — antique American steam launch engines. In small Seaburys, the steam passage between the cylinders — the ''receiver'' — is cast into the cylinder head. This compound engine is in the collection of W.W. Willock, Jr. The nameplate tracing was made from this engine by W.U. Shaw. (William W. Willock, Jr., photo)

Right: Without doubt the most refined participant in the steam launch revival in England is the 2.5'' & 4.5'' x 4'' compound engine produced by the Beaumaris Instrument Company, Ltd. Known in steam launch circles either as ''the Beaumaris compound'' or ''the Mills open compound'' (Victor Mills designed the engine in 1972), the 4.75-h.p. (on 100-p.s.i. steam) engine has an initial cost of about $1,000 per horsepower if ''exhibition finish'' is desired. (Courtesy H.A. Jones)

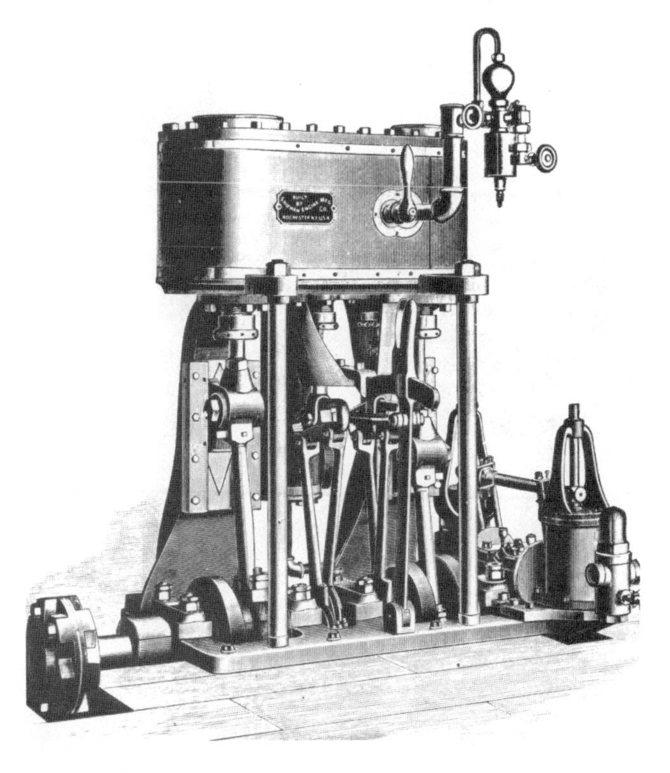

The Shipman Engine Company's 22-horsepower fore-and-aft compound, 4.5'' & 9'' x 6'' ''counterbalanced and vibration free,'' stood 32½ inches high and weighed 900 pounds. Engine, feed and vacuum pumps, and boiler together cost $1,350.

John Samuel White & Co., Ltd., Cowes, Isle of Wight, built this compound engine, which now has an honored place in Neptune Hall at the National Maritime Museum in England.

Above: Al Giles's *Crest* (see page 184) has a fore-and-aft compound engine, 3.5'' & 6'' x 4.5''. Of unknown build, the engine is counterbalanced and smooth-running. (Al Giles photo) *Left:* This 4.5'' & 9'' x 5'' Murray & Tregurtha engine, built in South Boston, came out of a 40-foot New Jersey launch named *Fox.* For many years the engine was owned by Frank D. Graham, author of two of the books in the Audel series of pocket guides to technology. Graham's books on steam engineering have a premium value for steam launch men, as the author often used his own steam launch interest and knowledge, or details from his steamer *Stornoway,* to illustrate points of engine design or economy. Frank Graham's widow sold the engine to Steamtown in Bellows Falls, Vermont. (Frank D. Graham photo)

This rugged and nicely proportioned compound was built recently by David H. King of Norwich, England. It has cylinders of 3'' & 6'' x 4'' and would go very well in a 20- to 25-foot launch. (Courtesy David H. King)

Thomas Thompson's *Firecanoe* (see page 137) has steamed enough miles to crisscross America three times and then some, and she's done it with this Thornycroft compound, 3.5'' & 6'' x 3.5'', built about 1905. Durability? You bet.

Above left: Frank C. Fuegeman, a German-born machinist living in Berlin, Maryland, takes a back seat to no one when it comes to fine machine work. The engines that he has built and displayed at steam meets all over the East — including this lovely Reliable compound — have captured the admiration of all who have seen them. *Above right:* This compact little compound of unknown make powers Bill Kimberly's *Polly* (see page 146). *Left:* Neat, functional, and new (Semple 10-h.p. compound engine). John Tiffany's *Polly* (page 174). (Courtesy John Tiffany)

Stickney engines, from Portland, Maine, often achieve an especially happy balance of lightness and openness together with sturdy marine proportions. The 3" & 6" x 4.5" engine with the remarkably handsome teak lagging (*right*) was originally in *Thalia,* on Sebago Lake, Maine. Evers Burtner saw it in a barn by Sebago Lake, purchased it, and has since given it to the author. The tracing was made by W.U. Shaw. The Stickney in *Gemini II (below)* is the same size as *Thalia*'s engine and also came from Sebago Lake. The neat and compact installation includes a boiler designed and built by Pete Levitre. *Gemini II* can be seen on page 148.

MADE BY
H.R.STICKNEY
38 UNION ST.
PORTLAND.ME.

Triple- and Quadruple-Expansion Engines

Triple- and quadruple-expansion engines are scarce in steam launch sizes, but this scarcity is not cause for concern among steam launch operators. Although they are all familiar with the multiple-expansion principle and with the beauty and historical significance of marine triple engines, they agree that one- or two-cylinder engines best suit small launches. The technical-minded are likely to argue that the extra surface radiation and condensation and the added rubbing surfaces may make small triple-expansion engines *less* efficient than compounds. Still . . . .

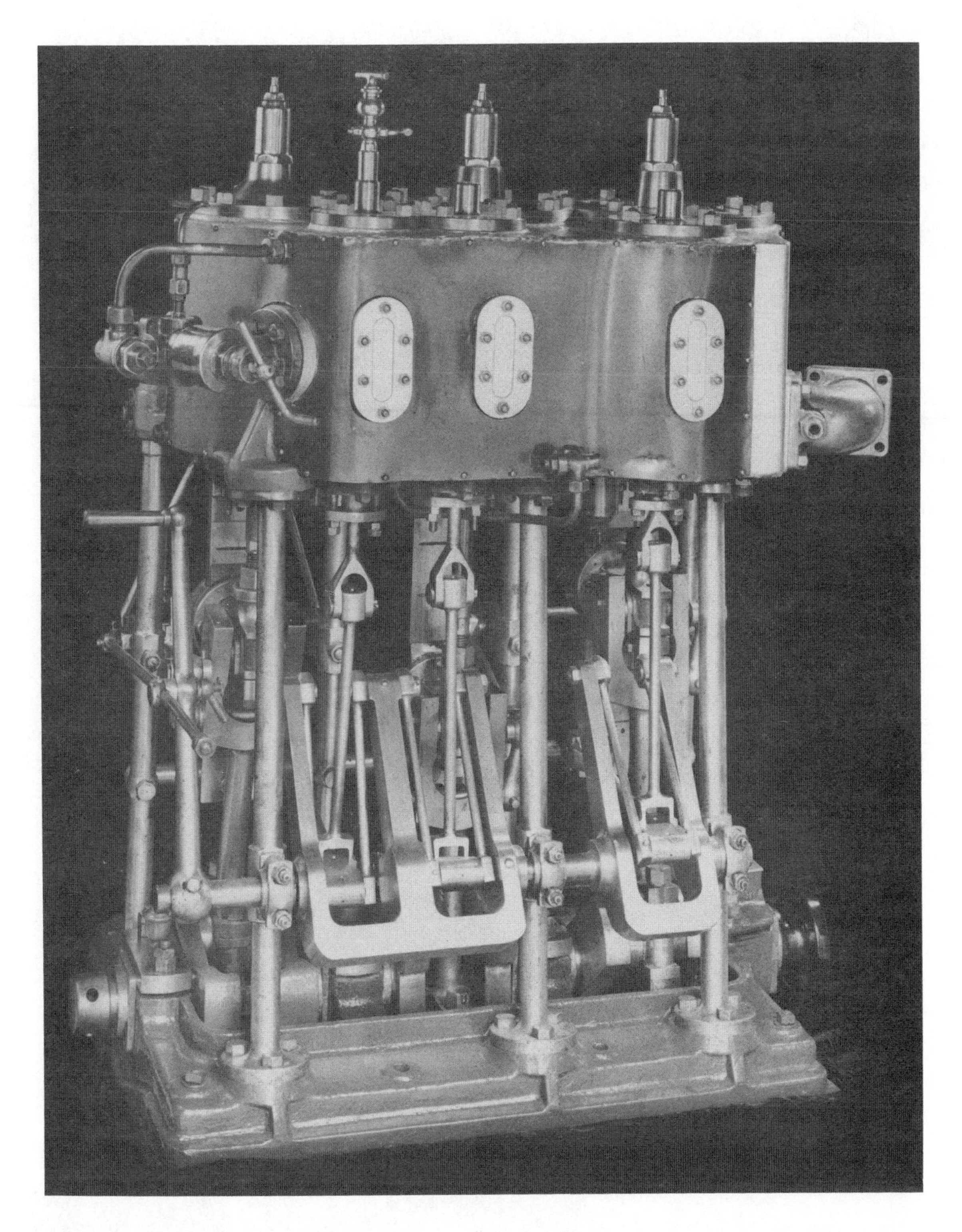

There is nothing in the photo to show the size of this engine, but it would be beautiful in any size. Built by W. Sisson & Co. (engine #612), in Gloucester, England, in 1901, the 2,000-pound triple, 6.12'' & 8.38'' & 11.19'' x 8'', develops 75 h.p. at 450 r.p.m. and 200 p.s.i. It was installed in the Thames River passenger steamer *Reading,* operated by Salter Bros. of Oxford. Several engines of this size once powered 50-foot excursion boats on the Thames. In 1962 the engine was removed from the boat and crated. Robert Maccoun purchased it and shipped it to his friend Frank Coughlin of Massachusetts, to sell on commission. In 1978 Jim Webster, of Webster, New York, bought it, in anticipation of finding a suitable hull. (Len Rosenberg photo)

There are numerous private collections of steam launch engines, but few have a centerpiece to match the miniature ship's engine owned by Francis P. Coughlin, Jr. The engine was constructed between 1917 and 1919 at the Revere Rubber Co., Revere, Massachusetts, for the plant manager. The plant was tooled for the manufacture of machine parts and belting, and the manager wanted a triple engine for his steam launch on a New Hampshire lake. It was a happy marriage of a desire and the means to achieve it — but the company's board of directors had to be kept ignorant of the affair. Whenever the "brass hats" came around, the engine was shoved under a staircase and hidden behind oil drums. Fred Semple found the engine, packed in grease in a box covered with mouse nests, in a northern New Hampshire antiques shop. The engine shows no evidence of ever having been under steam. Maybe, some day . . . .

A large proportion of American "marine" engines were built by Mid-westerners who had never smelled salt water. They were not quite true to Atlantic-basin traditions but often beat the world for price and practicality. Shown here is a corn-belt yacht engine of 1890, built by the Davis Boat & Oar Co., Detroit, Michigan.

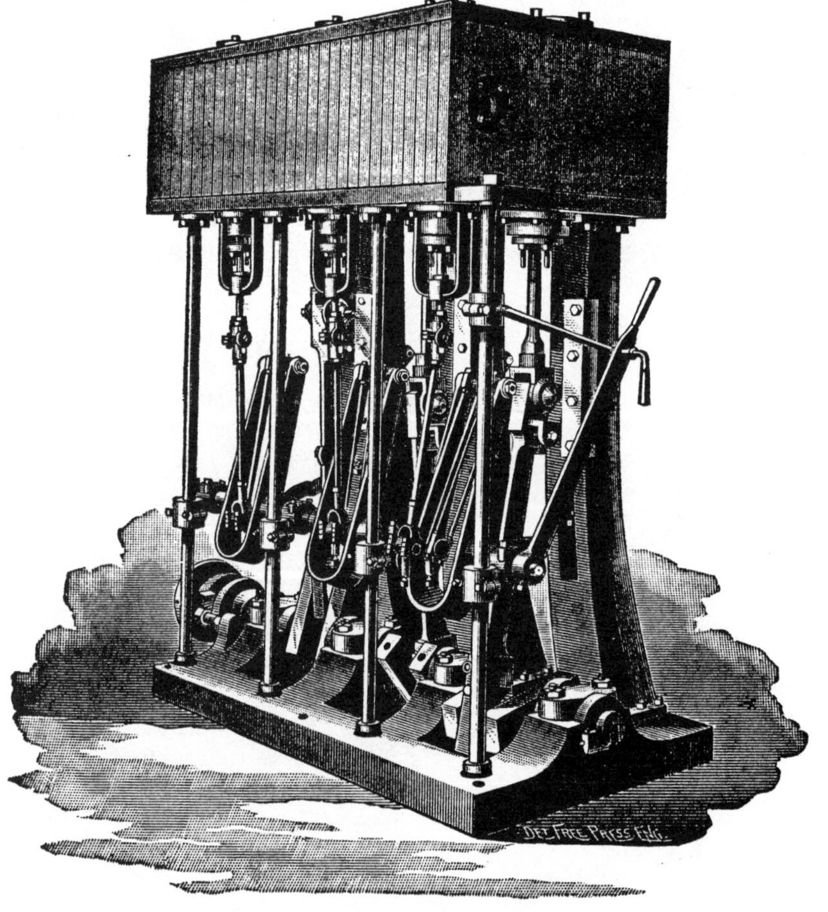

Steam yacht engines of one hundred to a few hundred horsepower were always beautiful specimens, and not too large to set aside in a garage or shed after a fine yacht was scrapped. Quite a few are still around, in museums or in private hands. This typical Herreshoff triple was one of two in *Navette,* a 114-footer built for J.P. Morgan in 1917. The engine's lines are as austere and perpendicular as those of a Gothic cathedral. *Navette* is said to have been the last of the steam commuters, designed to run her owner to his office each day in greater comfort than that provided by the Long Island Railroad. (Courtesy Mrs. Muriel Vaughn)

This 1899 Herreshoff engine was installed in *Scout,* one of eight 20-m.p.h., 81-foot steamers designed to serve as tenders for the New York Yacht Club "Seventies." *Scout* was owned by August Belmont and served his *Mineola.* The engine is preserved at Mystic Seaport.

GEO. LAWLEY & SON CORPN.
BUILDERS
SOUTH BOSTON MASS.
No 25. 1902.

Fred Semple owns this tall, businesslike Lawley triple, engine No. 25, built in 1902. Bill Shaw made the nameplate tracing from the 95-h.p. engine. (Courtesy Fred Semple) *Above right:* Lawley engine No. 38 is this 175-h.p. triple, which, like Lawley engine No. 25, was built in 1902. J.S.L. Poeckel, who was for many years Lawley's chief engineer, had previously been employed by the Herreshoff Manufacturing Co., so many Lawley engines closely resemble their Herreshoff counterparts.

Number 38 powered George Armstrong's *Gilnockie,* believed to have been the fastest boat on Lake Winnipesaukee in her day. (Edward Hilt photo)

Right: Almost small enough to be a model, this very attractive triple was built by a Mr. Levitt, a power-plant engineer in Newburyport, Massachusetts. It was to have gone into a 21-foot hull, but Levitt did not live long enough to realize his dream. The nuts for the cylinder heads were never made, and the cylinders were never lagged. His children sold the engine after his death. The present owner could name his own price, but the engine is not for sale. The author knows — he tried.

TOP AND CENTER:
Steamship engines lose one thing but gain another in the translation to miniature size. An engine as tall as a house, with gleaming steel parts that weigh tons and are dancing up and down, is above all *big*. Miniatures have a different appeal, especially the crisp detailing and the new insights made available when a great edifice can be examined at close range. Cliff Blackstaffe is a model engineer who is comfortable with mechanisms of any size, having served as first engineer on a Canadian survey steamer with two 700-h.p. triple-expansion engines. He assembled this 1.5'' & 2.25'' & 4'' x 2.5'' engine to please himself, using standard Blackstaffe-Wood engine components for the most part. The engine turns 500 to 900 r.p.m., employing 150- to 180-p.s.i. steam. Blackstaffe has made elaborate provision for thermal efficiency, with three stages of feedwater heating to boost the feedwater temperature to 240 degrees Fahrenheit. He is studying the maximum efficiency possible with a conventional triple-expansion marine engine on the scale of 2 or 3 horsepower. The engine has been installed in Blackstaffe's *Sparrowhawk* (see page 144). (Top photo courtesy Cliff Blackstaffe; center photo courtesy Everett Arnes) *Bottom:* The author's principal recollection of this rare launch engine is that he could have bought it for $35 in the 1950s. There are piston valves on the high-pressure and intermediate-pressure cylinders; the low-pressure cylinder has a slide valve. The manufacture of the three expansions — three ''engines'' in the old terminology — as separate units suggests that the engine was built in a small shop with small machine tools in a small town. Reversing is by slip eccentric.

Commander (later Rear Admiral) Lauren S. McCready, U.S.M.S., proudly displays the engine and boiler that he installed in his 26-foot *Little Effie* (see page 39) in 1953. On a voyage up the Hudson in 1954, retracing the *Clermont*'s route, Commander McCready put in at Catskill, New York, to visit George Krum, who built the powerful boiler and largely hand-fabricated triple-expansion engine about 1935. (Courtesy Rear Admiral Lauren S. McCready, U.S.M.S.)

This 1912 power plant appeared at the end of the steam launch era. The enclosed and balanced high-speed Simpson, Strickland triple receives 250-pound steam from an oil-fired water-tube boiler. The boat is David Kyle's *Pussy Cat* (see page 134). (David Kyle photo)

Quadruple-Expansion Engines

Simpson, Strickland & Co., of Dartmouth, England, specialized in multiple-expansion engines of 3 h.p. and upward. The smaller units were made especially light and compact to suit yachts' hoisting launches. The firm's "Kingdon" patent quadruple-expansion, tandem-cylinder launch engine was an advanced design in the 1880s, with twice as many expansions and considerably higher working pressure and rotative speed than any offered by other launch-engine builders. It remained current and widely admired for 20 years, and the company boasted, "We have built more quadruple engines than all the other builders in the world put together, although of course ours are relatively of small size." The high-pressure and first intermediate cylinders were placed on the forward crank, and the second intermediate and low-pressure cylinders on the after one. The unbalanced 2-crank "quad" design was not suitable for the high r.p.m. demanded of steam launch engines in competition with gas boats after 1905, so the company shifted its emphasis to triple engines that could be balanced for running as high as 800 r.p.m.

Above left: This Simpson, Strickland Model "A," 2.5" & 3.5" & 4.5" & 6" x 3.5", found its way to Meaford, Ontario, in 1920. It was installed in a 28-foot boat, was stolen, recovered, sold, overhauled.... Jim Cooper is the present owner of this now-rare engine, rated at 10 indicated horsepower. The nameplate photograph *(left)* came from this engine. *Top:* This slightly larger Simpson, Strickland model, 3" & 4" & 5.25" & 7" x 4", indicated 14 horsepower. *(Above left:* G.A. Black photo; courtesy James Cooper. *Left:* Courtesy J. Cooper. *Top:* Crown Copyright. Science Museum, London)

Unusual Engines

Steam launch engines were often fashioned in small shops by local artisans who reinvented each mechanism they built. All of these engines were unique from birth. Over the years, even the more standard models have become rare by attrition. Today, any steam launch engine is unusual enough to be a collector's item.

Pointing out these facts, critics will say the title of this section is redundant, and they'll get no argument from this quarter. Deciding which among the unusual are more unusual than the others has its element of arbitrariness, but you will see the reasons for the selections. Now, let's examine these sundry oddities.

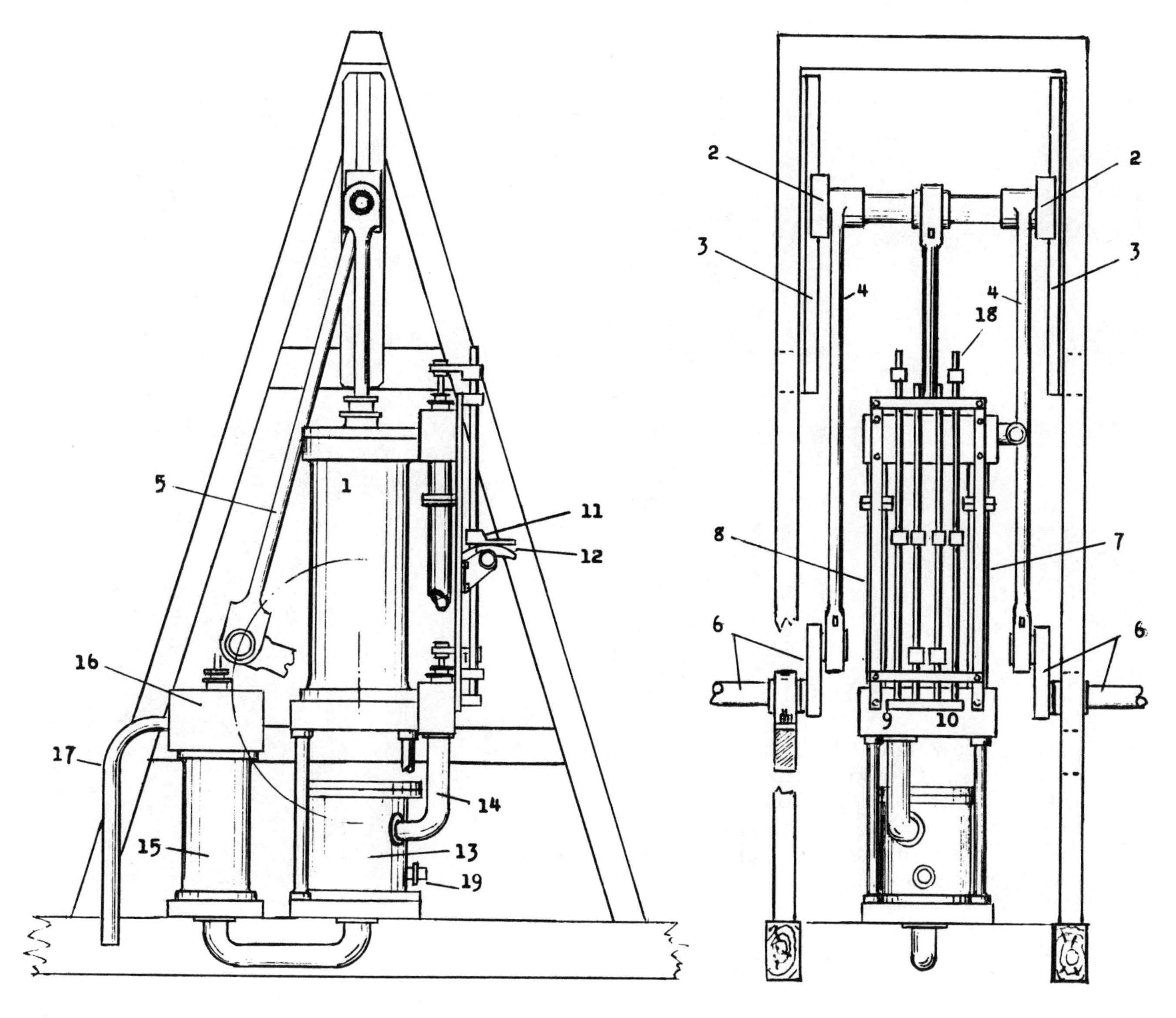

1. Cylinder
2. Crossheads
3. Crosshead guides
4. Connecting rods
5. Connecting rod — side elevation
6. Cranks and main shafts
7. Steam manifold
8. Exhaust manifold
9. Exhaust valve chamber
10. Steam valve chamber
11. Lifting toes
12. Wipers
13. Condenser
14. Exhaust to condenser
15. Air pump
16. Hotwell
17. Waste pipe
18. Valve lifting rods
19. Condenser water supply

From rough sketches and descriptive information, Bill Shaw of Thornbury, Ontario, came up with this drawing of Henry Wiegert's model of the side-wheeler *Norwich*'s unique engine. The original *Norwich,* of 1836, had an engine of 40-inch bore and 120-inch stroke. Wiegert's scale model of *Norwich* (see page 125) is 1/8 size, and her engine is 5'' x 15''. (Courtesy Bill Shaw)

Above: This "inverted" three-cylinder simple engine was built by John Paine, of Boston, Massachusetts. The size of the cylinders is about 6" x 8". Dan Weber took the photo in the Henry Ford Museum, in Dearborn, Michigan, in 1963. *Top right:* In the 25 years that the author knew Clifford Harris, of Gill, Massachusetts, Cliff built or converted five steam launches, including their engines and boilers. This is one of his products, probably the only authentic marine walking-beam paddle engine of its size since the U.S. Navy training boat *Sweetheart* was launched in 1836. Harris used the engine to push his 20-foot quarter-wheeler *French King Belle* (see page 127) on the Connecticut River. *Right:* Cliff Harris converted a high-pressure (single-expansion) "V" engine for marine use, with the result seen here. The engine's cylinders previously powered an automobile.

Above: The 2.125'' x 3'' Rochester-model Shipman engine (see also page 266) is unusual in having two pistons in its single cylinder. The engine has outside steam admission and inside exhaust. The specimen shown here was once owned by the author, who sold it to Richard Dickey. For a time it powered Dickey's *Soot* (see page 140). One summer, Dickey took the engine apart and mounted the pieces on a board to exhibit them to good advantage.

Gas engines have been adapted to power mopeds, golf carts, and snowmobiles. Pedal propulsion has been used for bicycles, submarines, and airplanes. Is it any wonder that a few steam-canoe engines were built? *Above left:* Tommy Thompson's Simpson, Strickland steam-canoe outfit. A fire-tube boiler supplies steam to a 1.5'' & 3'' x 3.5'' steeple compound engine. (Thomas G. Thompson photo) *Left:* Tom Bonomi, of Quincy, Massachusetts, owns this British-built canoe engine, 1.75'' x 2.5''.

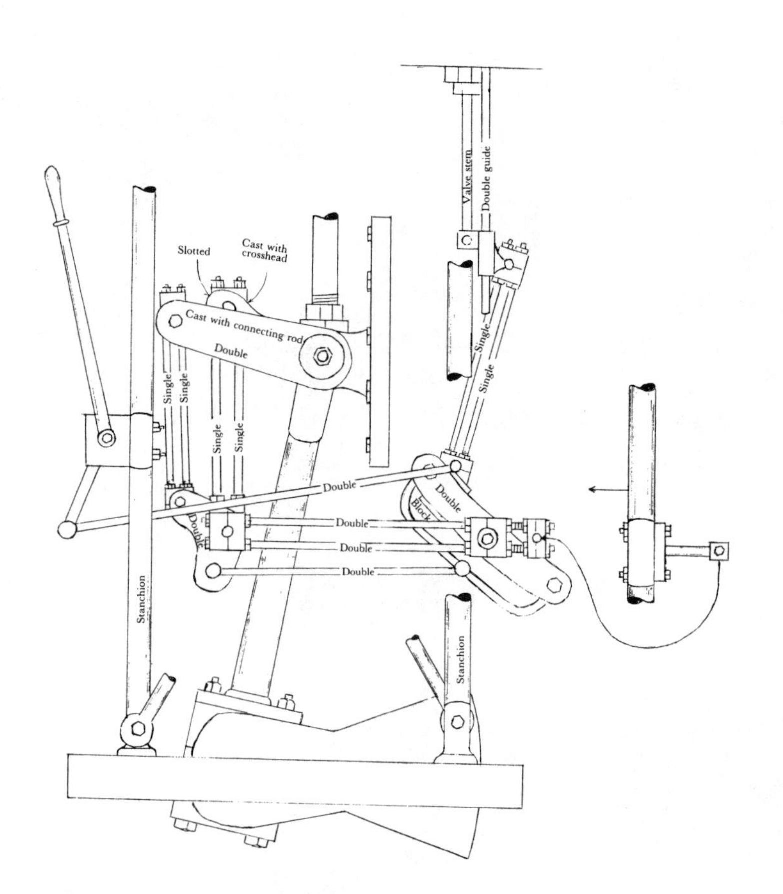

More than 100 different varieties of valve gear were employed on reciprocating steam engines, each having some claimed advantage over the others (although it appears that some ingenious motions were patented out of sheer delight in mechanical innovation). Most marine engines were equipped with the familiar Stephenson-link motion. Others had "radial" gear, driven from the engine crossheads instead of eccentrics on the shaft. This permitted a close-coupled engine, saving space on small yachts.

The valve motion on *Kestrel*'s (see page 224) engine (built by the Fore River Engine Co., Weymouth Landing, Massachusetts, in 1892) is said to be the only one of its kind. The gear has so many components that to watch its action is to witness absolute confusion. The 5-foot-high compound (6" & 12" x 10") so impressed airline flight engineer Bernard Denny, of Chappaqua, New York, that he built this 1½-inch scale model, 0.75" & 1.44" x 1.25". The author made the drawing of *Kestrel*'s valve gear. (Lewis R. Brown, Inc., photo)

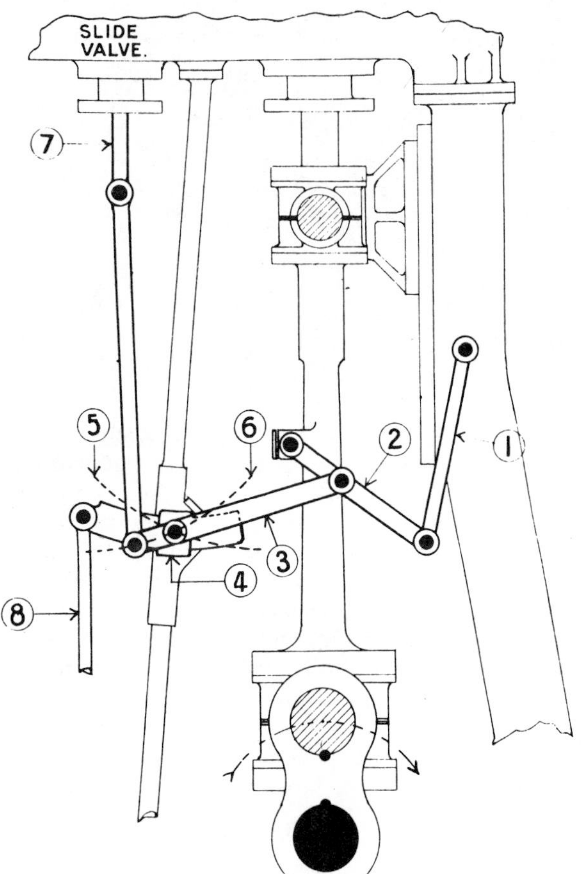

The Fore River Engine Co., builder of the *Kestrel*'s power plant, apparently made a habit of unusual valve gears. Both engines shown here were given Joy valve gear. The smaller specimen *(above left),* about the right size for a 20- to 25-foot launch, is on display at the Hart Nautical Museum, Massachusetts Institute of Technology, Cambridge, Massachusetts. This is the only small engine the author has ever seen that incorporated this type of valve motion. The larger engine *(above)* presently resides in a barn in Camden, Maine, but the author remembers it in operation, pushing the yacht *Swallow* (see page 223) on Lake Winnipesaukee. The schematic drawing shows the Joy valve gear in detail.

Joy Valve Gear

1. Suspended link
2. Compensating link
3. Lever connecting compensating link and valve rod through quadrant block 4
4. Quadrant block (traveling)
5. Angle of quadrant for "ahead"
6. Angle of quadrant for "astern"
7. Slide valve spindle
8. Reversing engine rod

Top left: Several hobbyists, including machinist Walter Kleinfelder of Summit, New Jersey, have adapted small steam engines to outboard-motor housings. Each such installation is unique; Walt has done a superb job with his. The little copper-tubed boiler is fired by propane gas, and with four square feet of heating surface it supplies steam to the 1.5'' x 1.25'' Stuart-Turner engine for a speed of 3 or 4 m.p.h. (Courtesy Walter Kleinfelder). *Top right:* The only steam plant ever seen by the author in which both engine and boiler were hung outboard was in this boat, the *Hingey,* built by Mahlon Lamoureux of Shaw Island, Washington. *Above:* This steam conversion of a Johnson outboard motor was accomplished using a #9 Stuart cylinder bored out to 1$\frac{9}{16}$ inches. A belt-driven pulley drives the feed pump through an approximate 3-to-1 reduction. This was a project of Jerry Heermans, of Tigard, Oregon, in 1965. (Ray Rogers photo; courtesy Jerry Heermans)

Some steam engines resemble gasoline engines. These are ''single-acting'' engines, in which the steam pushes against only the top of the piston, so no crosshead or piston-rod gland is required. The enclosed crankcase permits splash lubrication of all parts, and the steam valves are poppets working off a camshaft at the side of the engine.

This 1903 engine from the Dieter Steam Engine Co. of New York delivered 20 horsepower at 600 r.p.m. (steam pressure not given). The engine weighs 165 pounds and was originally offered for $200. (Courtesy Tom Rick)

Boilers

Buying or building a steam launch boiler is not really difficult, but the process may be fraught with uncertainty, hard decisions, or monotonous handwork. What type is best for my boat and uses? *Must* I pay so much for that uninteresting mass of steel and firebrick? Can I do any of the work myself (and visions of coil, flash, and porcupine boilers dance in the amateur's head)?

Nearly all boilers require some welding or brazing to "pressure vessel" standards. It is foolhardy to have this done by anyone not qualified for the work, but if the design is selected with care, pressure-vessel welding can be kept to a minimum. Most of a water-tube boiler's structure can be made up of threaded fittings, bolted flanges, and bent or coiled tubes that can be assembled by anyone capable of doing the piping for a steam launch. Even with a fire-tube ("tank") boiler design, it may be possible for the amateur to do the tube-rolling, fitting, and lagging himself.

The rock-bottom cheapest and most durable boiler that can be bought over-the-counter is nearly always a vertical fire-tube (VFT). These have been the commonest small utility boilers for more than 100 years. Although most VFTs are not entirely suitable for steam launch service, no other design is as quickly and easily available to the steam launch builder who wants to get up steam next week.

Sometimes horizontal fire-tube boilers are built. Of these, the Scotch boiler, the most traditional and "marine" of all boat boilers, is the most popular choice.

Water-tube boiler designs allow free rein to the creative impulse, and skilled amateur welders sometimes design and build water-tube boilers as showpieces. These may contain many more small tubes and welded joints than necessary. The same tube arrangement and surface area was achieved with the Roberts-type threaded-joint design 90 years ago, and it can be duplicated with pipe unions and close bends in today's much-improved iron pipe. Copper tubing and bronze tube or pipe fittings make the job easier.

Some water-tube boilers consist almost entirely of bent tubes. Others have straight tubes that run between headers, and some feature pressure-resistant shells as large as that of a fire-tube boiler, with only short internal water tubes. British steam launch builders in recent years were fortunate in having a supply of World War II surplus Merryweather boilers. These had been coupled to steam pumps for firefighting during the German blitzkrieg. The ability of a Merryweather to generate 10 pounds of steam per square foot of heating surface per hour on natural draft attests to the benefits of copper tubes.

Porcupine ("thimble-tube" or "hedgehog") boilers require no more than a central standpipe or pressure drum into which short pipes are fitted. Other central-standpipe designs that lend themselves to amateur construction employ vertical tubes or helical coils in various arrangements around the standpipe.

The Ward boilers employed in most U.S. Navy steam cutters consisted of a central array of porcupine "quills" descending from a bell-shaped steam drum, surrounded by a water-wall of tubes radiating from the lower rim of the drum.

A water-tube boiler design often includes lengths of bent or coiled tubing that are regarded as separate from the boiler's "generating" section. A feed heater, or "economizer," may be located at the top of the boiler, on one or both sides of the steam drum. A "superheater" or "steam dryer" may add heat to saturated or wet steam after it has left the boiler proper.

Even during the Golden Age of steam launches, the engines and boilers were a varied lot — steam power allows a greater range of individual expression than internal combustion. The boilers that have powered the steam launch revival are no less varied, and they provide fertile territory for fine craftsmanship, experimentation, frustration, and endless debate and interest.

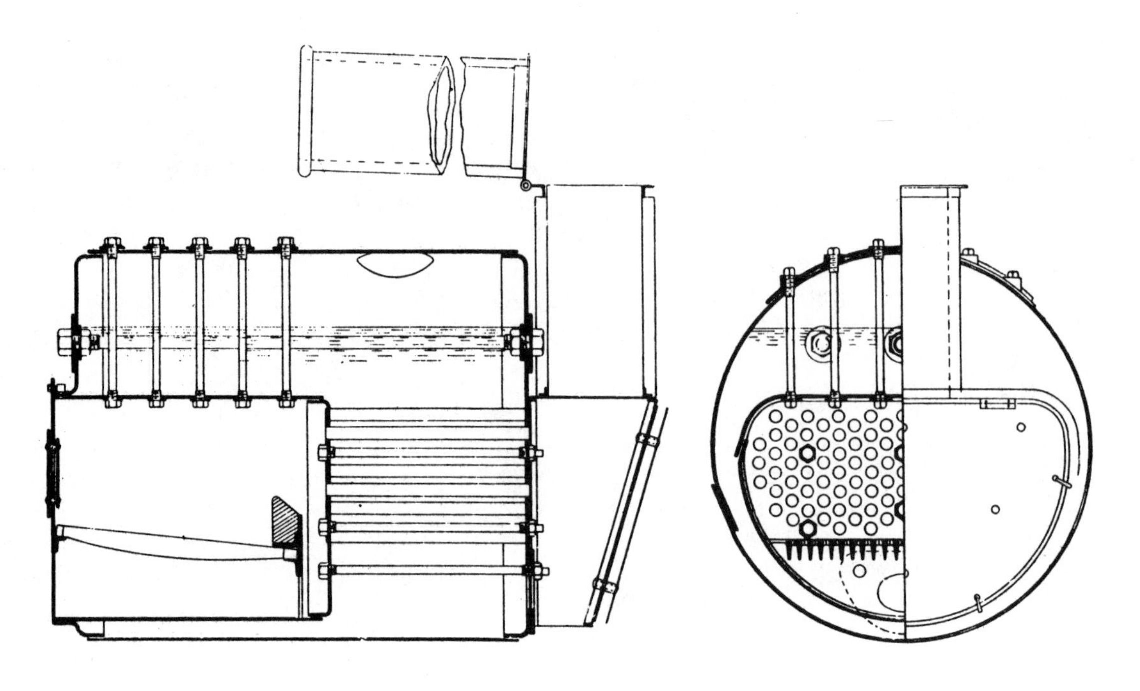

Direct-tube ("gunboat") boilers offer a low center of gravity, the steady-steaming characteristics of all fire-tube boilers, and a long tradition, but their great bulk and weight per horsepower make them one of the least likely choices for a steam launch today. By the turn of the last century, Simpson, Strickland and Co. recommended their horizontal boilers only "for work abroad, where no great speed is required, and the boiler has to be placed in the hands of persons used only to this type." (From the facsimile edition of Simpson, Strickland and Co.'s Catalogue No. 6)

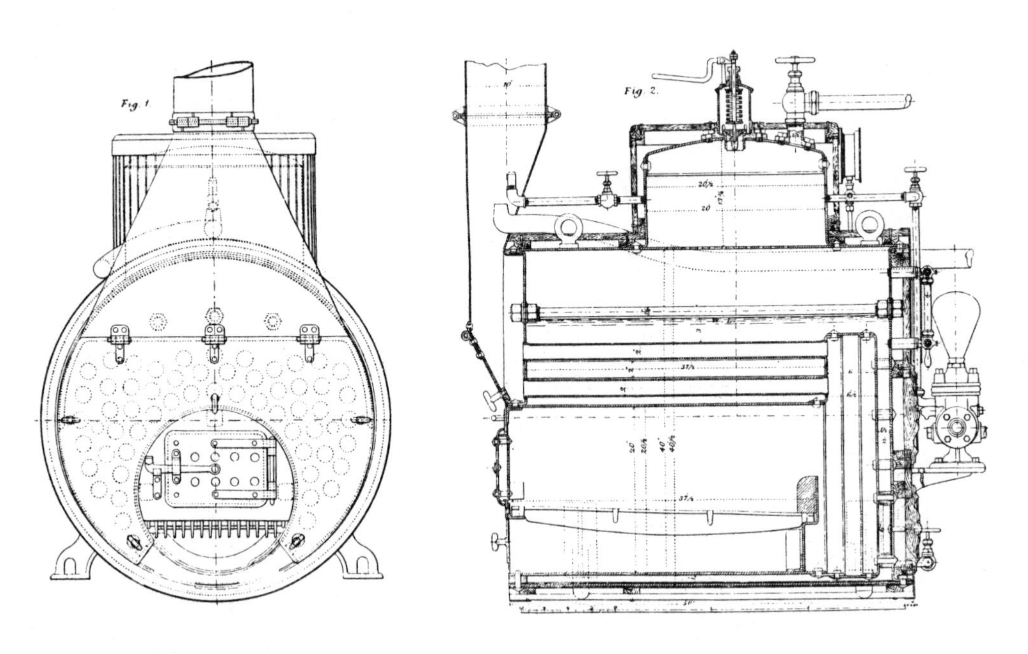

During the 1870s, the U.S. Bureau of Steam Engineering designed heavier and more powerful machinery for the U.S. Navy's existing 33-foot steam cutters. The new power plant included this handsome little Scotch boiler, 3.4 feet in diameter and 4.2 feet long, which carried 125 square feet of heating surface and weighed 2,350 pounds with the donkey pump mounted on the back head. The boiler supplied an 8" x 8" engine that stood four feet tall and weighed 550 pounds. (From William Maw, *Recent Practice in Marine Engineering,* 1883)

Left: John Winters' *Soar Point* (see page 175) has ample stand-by steam capacity with this oil-fired Scotch boiler built by Winters in 1973. Fifteen inches in diameter and 18 inches long, the boiler has nine square feet of heating surface. (J. Harvey photo) *Below:* When it's all put together and ready to steam up, the beauty of a steam launch power plant depends on the taste and skill of the builder, not on having particularly rare or costly components. Frank Fuegeman's launch *Ada* can be seen on page 52, and *Ada*'s engine is shown on page 265. For his boiler, Fuegeman built a replica of an ordinary return-tube boat boiler of 1885, and it is perfect as such.

A Clyde boiler such as Fuegeman's differs from the "true Scotch" in having a "dry back," that is, in the absence of a water leg between the combustion space and the back head. The tubes of return-tube boilers can be easily brushed, and, in contrast to VFT boilers, rust and scale can be readily removed. *Bottom:* A 19th-century marine engineering text used this drawing to illustrate the placement of a locomotive-type boiler in a hull. Locomotive-type (or "firebox") boilers provided somewhat more power relative to their weight than other horizontal tank boilers, and they routinely operated at higher pressures. They were widely used in high-speed launches and torpedo boats before express WT boilers began to supersede them.

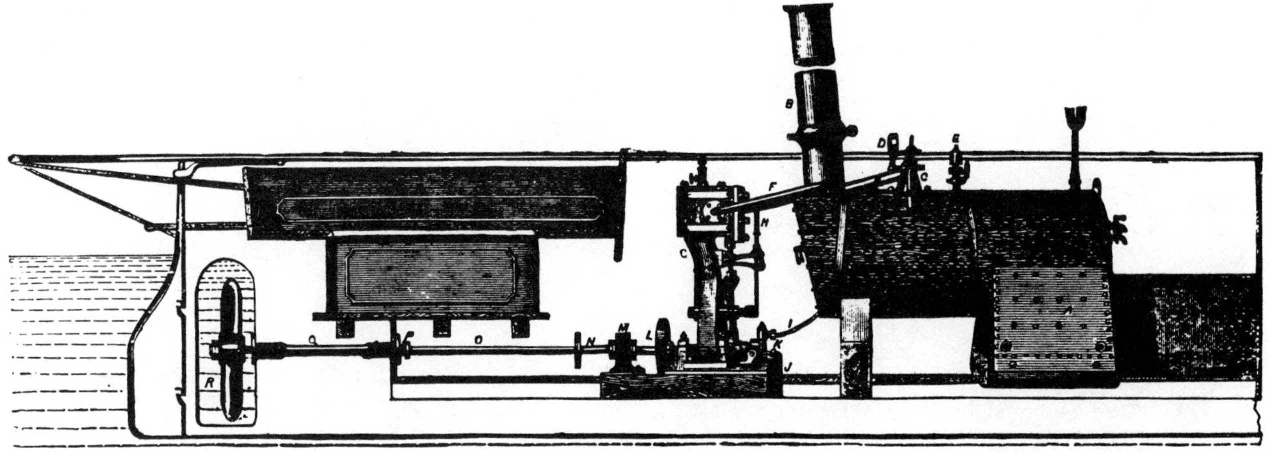

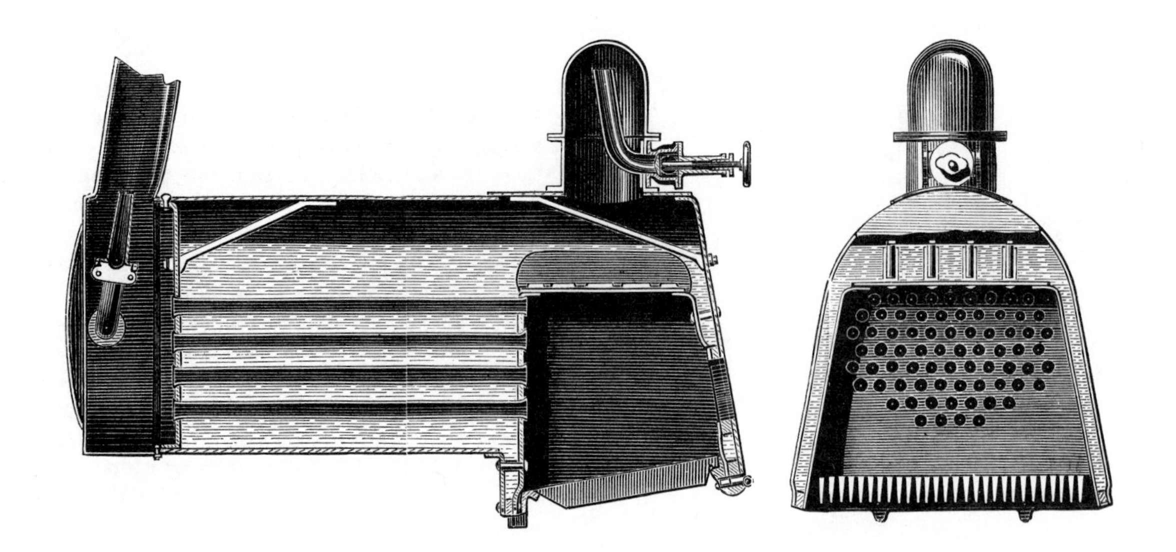

Above: The coal-fired locomotive-type boiler in John Thornycroft's *Ariel* (1863) had a barrel 2.1 feet in diameter, containing seventy-two 1½'' tubes for 85 square feet of heating surface. The firebox walls provided another 12.5 square feet of heating surface, and the grate area was 3.7 square feet. (From William Maw, *Recent Practice in Marine Engineering,* 1883) *Left:* An 1884 photograph of the VFT boiler in *Olive,* George Eli Whitney's third steamer (and the first he built for his own use). The boiler was 26 inches in diameter by 40 inches high, and the inverted-crosshead engine was 4'' x 6''. Working pressure was probably 70 to 80 p.s.i. Every detail was a lovingly considered decision by a man who lived for steam launches, and the resolute piping is as positive a statement of faith as the Brooklyn Bridge. *Olive* is shown on page 42. (George Whitney collection)

The Simpson, Strickland and Company's Kingdon VFT boiler is shown in transverse section and as part of a unit also including their two-crank quadruple-expansion engine. The company claimed that the boiler would not prime and would steam well "with natural or nearly natural draft," and that its lower tubeplate was entirely accessible for maintenance. (From the facsimile edition of Simpson, Strickland and Co.'s Catalogue No. 5)

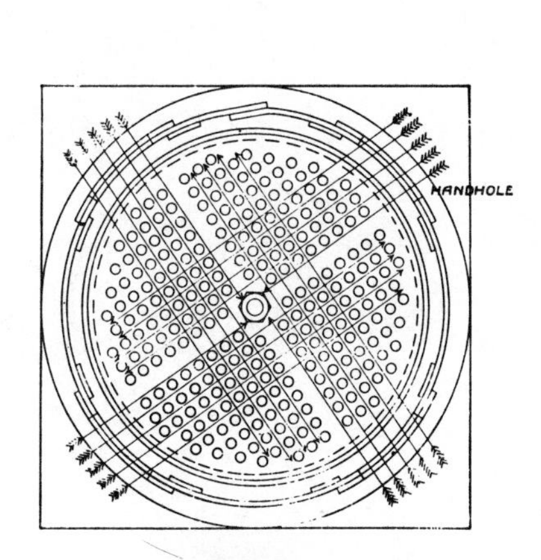

The F-40 VFT boiler has been distributed by the Semple Engine Co. since 1956. With 40 square feet of heating surface, it supplies steam for a 5-h.p. simple engine or a 10-h.p. compound. (Courtesy Semple Engine Company)

Rauchen Verboten's 8-inch boiler has model-shop fittings, an ice-cream freezer can for a firebox, and a stove lid drilled full of holes for a grate — but it is packed full of copper-tube heating surface and steams like Billy-O.

The oil-fired VFT boiler in *Ramona* (see page 161) is over 40 years old, but under Lou De Young's care, it shows not a blemish. There are fifty ¾-inch fire tubes inside the ⅜-inch shell. Operating pressure is 80 to 100 p.s.i. (Courtesy Lou De Young)

Main features of the Roberts boiler design are a single large fore-and-aft steam drum, large downcomers forming the four vertical edges of the frame, mud drums (the company called them "side pipes") running fore and aft on either side of the grate, and upflow coils. Each upflow coil incorporated five 180-degree turns in its run from a mud drum to the steam drum. A feed-water heating coil nestled either side of the steam drum, and a superheating coil was placed on either side of the boiler between the upflow coils and the casing. Each downcomer had a blind "mud-pocket" at its foot. Sediment collected in these pockets, where it could be discharged through blow cocks. Patents for the Roberts boiler were obtained in the United States between 1887 and 1893. (From C.P. Kunhardt, *Steam Yachts and Launches,* 1891 edition)

Above: This homemade water-tube boiler is of Babcock & Wilcox persuasion in the inclined straight-tube generating sections, and Roberts-like in the fore-and-aft steam drum and the vertical water-wall tubes rising to high headers. Elmer Brooks built the boiler for his *Artemis* (see page 201).

Hundreds of different WT boiler designs have been proposed, both by professionals aiming for lowest manufacturing and maintenance costs and by amateurs who have hit on a novel idea in tube arrangement. The design often reflects the builder's preference in manufacturing techniques, and this builder obviously prefers welding to tube-bending or pipe-threading. Repairing a leaky tube or joint would require high skills in cutting and welding. *Below:* Two Roberts-type WT boilers under construction years ago by George Whitney, one for Lester Hill and one for Dick Mason.

SIZES AND PRICES.			
Width.	Length.	Grate Surface.	Price.
20 in.	20 in.	1½ sq. ft.	$150
23 in.	23 in.	2¼ sq. ft.	200
26 in.	26 in.	3 sq. ft.	250
29 in.	29 in.	4 sq. ft.	300
32 in.	32 in.	5 sq. ft.	325
35 in.	35 in.	6 sq. ft.	375
38 in.	38 in.	7½ sq. ft.	400
41 in.	41 in.	8½ sq. ft.	450
44 in.	44 in.	10 sq. ft.	525
47 in.	47 in.	11½ sq. ft.	600
54 in.	54 in.	14½ sq. ft.	700
5 ft.	5 ft.	19 sq. ft.	950
5 ft.	6 ft.	23 sq. ft	1,300
5 ft.	7 ft.	27 sq. ft.	1,500
6 ft.	6 ft.	28½ sq. ft.	1,500
6 ft.	7 ft.	32 sq. ft.	1,800
6 ft.	8 ft.	36½ sq. ft.	2,100
7 ft.	7 ft.	39 sq. ft.	2,[illegible]00
7 ft.	8 ft.	45 sq. ft.	2,500
7 ft.	9 ft.	51 sq. ft.	3,000
8 ft.	8 ft.	52 sq. ft.	3,000
8 ft.	9 ft.	60 sq. ft.	3,500
9 ft.	9 ft.	70 sq. ft.	4,000

THE TRECURTHA BOILER
PATENTED
APR.8.1890. SEPT.1.1891.
BUILT BY
MURRAY & TREGURTHA
SOUTH BOSTON MASS
WROUGHT IRON
Nº [illegible] [illegible]

The Murray & Tregurtha Company offered a pipe boiler in several sizes, the design of which was similar in broad outline to the Roberts boiler, but different in detail. A row of side pipes rising on either side of the grate from a mud drum to a steam drum formed water walls for the firebox. From each side pipe a concentric series of U-shaped "circulating" or generating coils protruded over the fire. The mud drums were below the grate, and since the circulation within each side pipe and its generating coil was largely self-contained, water flow through the mud drums was gentle, allowing sediment to settle there. As in a Roberts boiler, each generating coil could be removed independently for maintenance. Bill Shaw made the tracing from a name plate owned by John Clement. *Below:* The (Roberts-type) Davis water-tube "safety" boiler in *Phoebe* (see page 193) burns wood, fired through two doors. (Courtesy Frederick G. Beach)

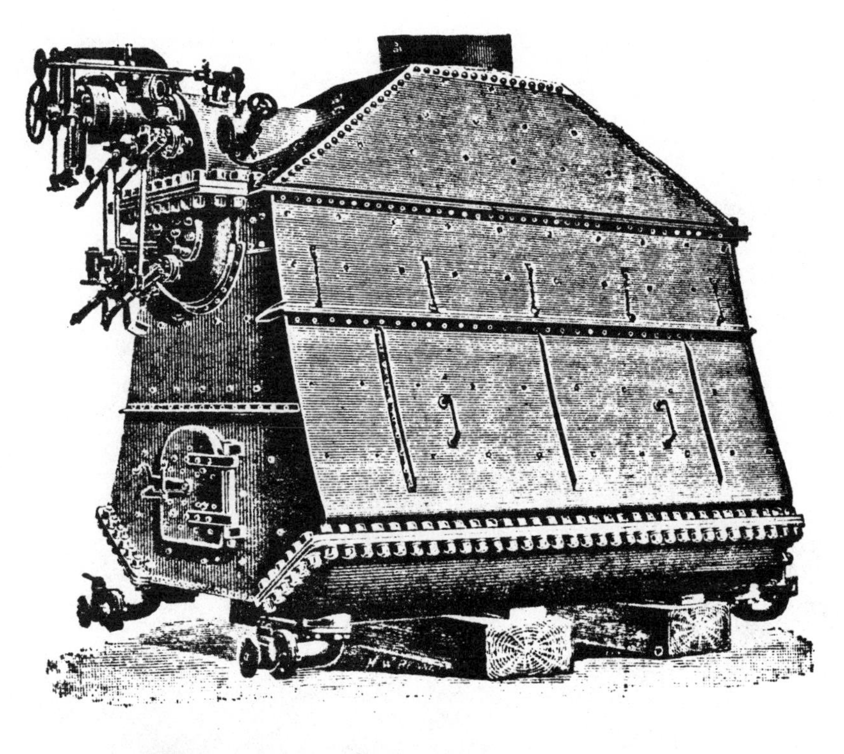

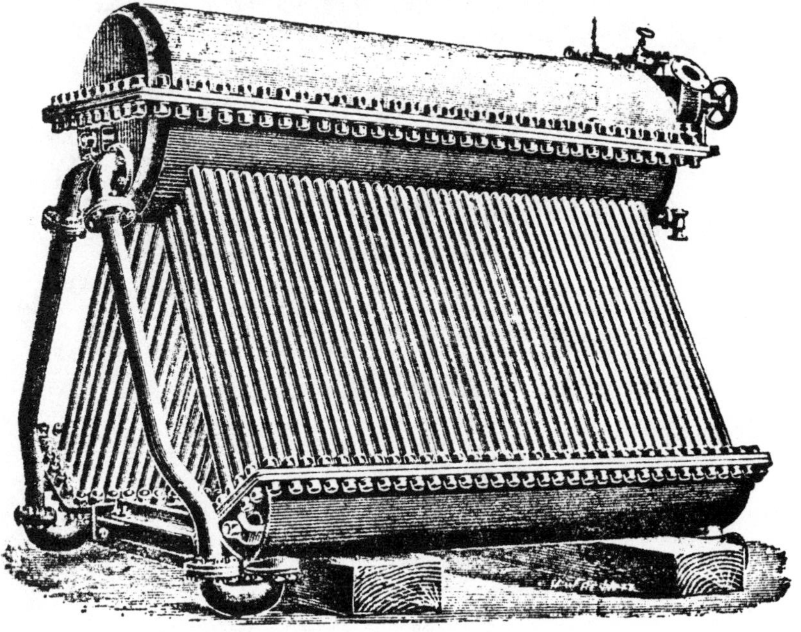

Alfred Yarrow's continuing efforts to lighten power plants led him into a decade of experimentation with WT boilers, culminating in 1887 in his version of the 3-drum express boiler shown here. Normand, a French engineer, and Yarrow's fellow countryman Thornycroft also developed 3-drum boilers about that time. The reliability of these boilers was much increased by the commercial production of solid-drawn steel tubes for bicycles, which began about 1886.

Above: The express boiler (U.S. Navy Type A configuration) manufactured by Horace Nelson for the *Island Queen* (see page 195). *Top right and right:* Peter Moale describes the all-copper, 3-drum express boiler that he built for his *Panatela* (see page 147) as "similar to a Seabury boiler but with two large (1½-inch) downcomers at each end." The woodburning boiler is 24 inches in diameter, 36 inches long, and contains 27 square feet of heating surface. Weight with water is 340 pounds. (Courtesy Peter Moale)

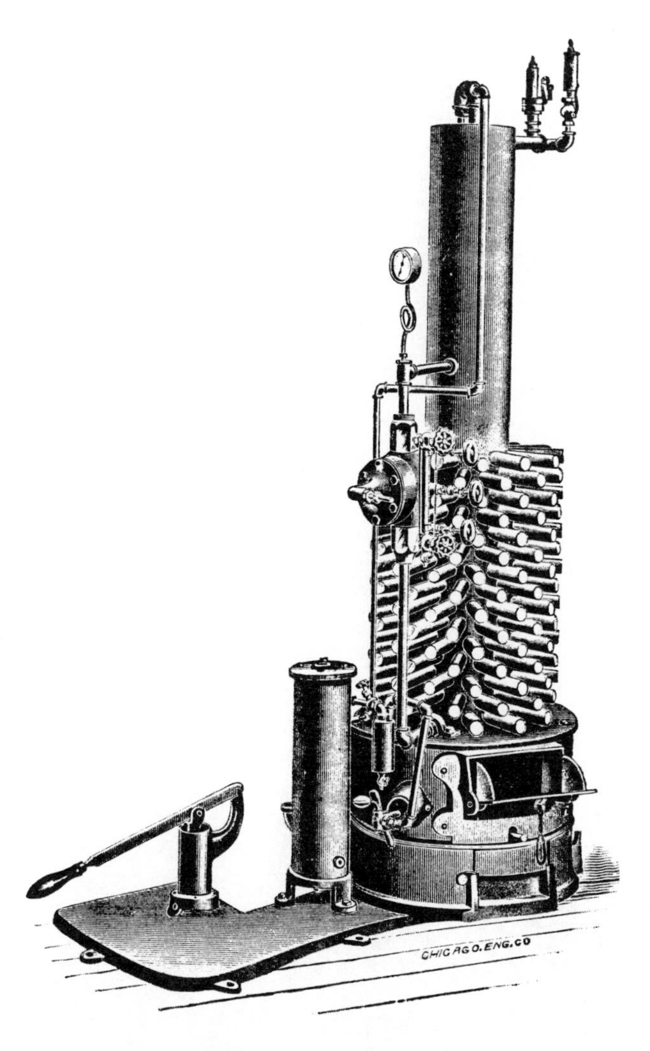

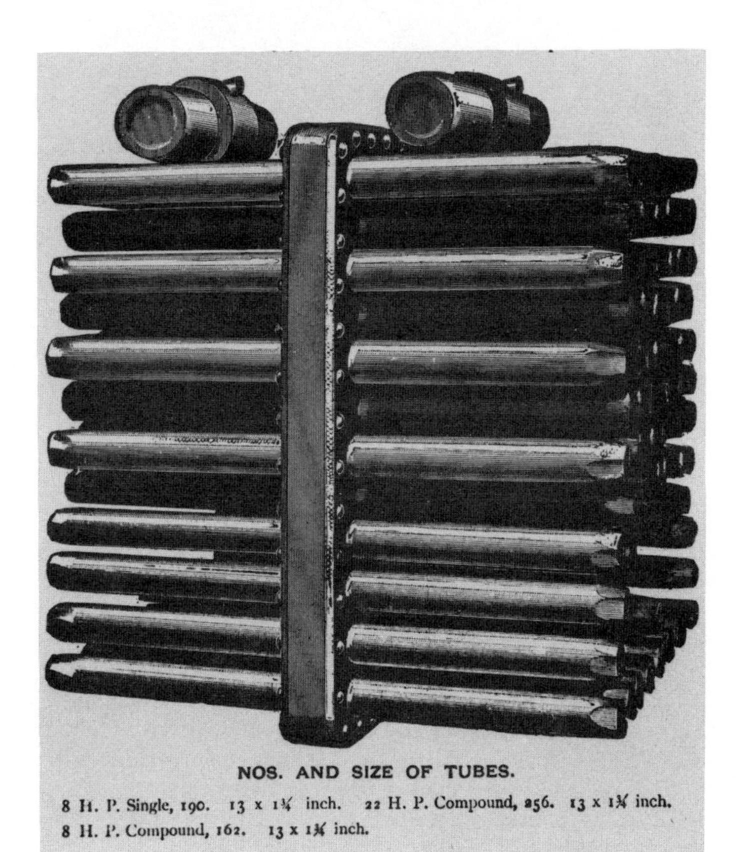

Above: The kingpin of the Racine Boat Manufacturing Co.'s lightweight, unitized power plant was this porcupine boiler, approved by the U.S. Board of Supervising Inspectors in 1888. Like its Shipman Engine Co. counterpart, the boiler had automatic water-level control and automatic fire control on liquid fuel.

For the steam launch enthusiast who lacks welding experience and cannot afford to buy a boiler, building a porcupine is a practical alternative. All he needs is determination, a strong back, and a sense of direction. Room for innovation is almost limitless. Some builders have even used 55-gallon drums for casings.

Fired by "kerosene oil, petroleum, or natural gas fuel," Shipman Engine Co. steam plants were made "automatic," or self-regulating, by the incorporation of a pressure-regulated steam-atomizing burner and a float-regulated feedwater bypass valve. The "single boiler," shown without casing (*above right*), was included in the smaller plants. Porcupine quills were screwed into the rectangular header. The "double boiler" (*right*) that powered the 8-h.p. Shipman simple engine and the 8- and 22-h.p. compounds was similar, but it had a thicker header, with quills and a fire on either side. The 2-h.p. Shipman plant (3" x 4" engine) turned an 18" x 30" propeller at 400 r.p.m., weighed 475 pounds, and occupied six square feet of floor space.

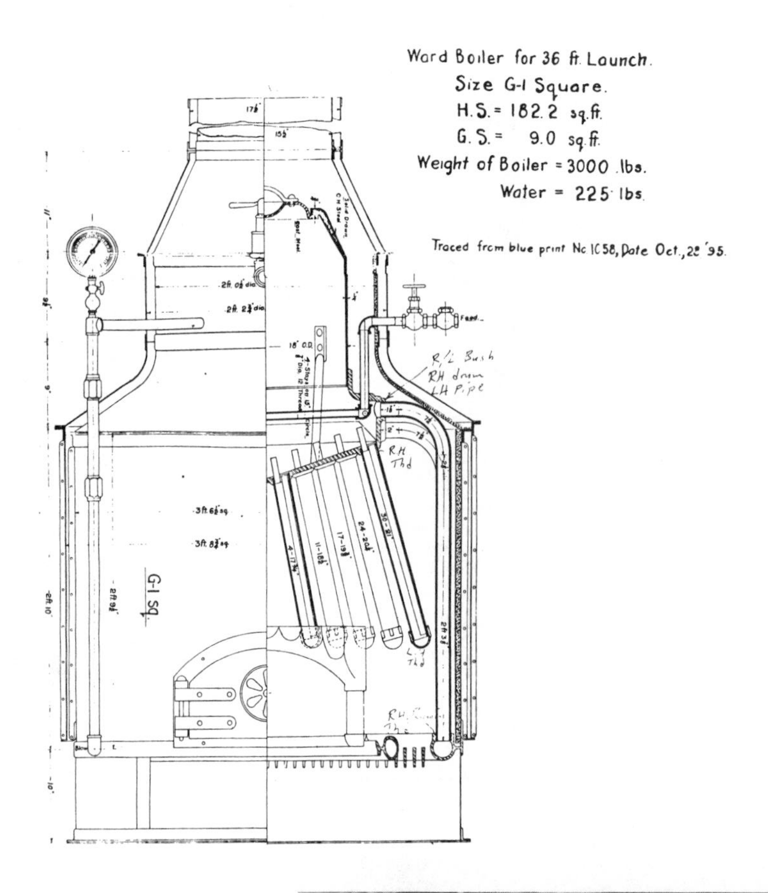

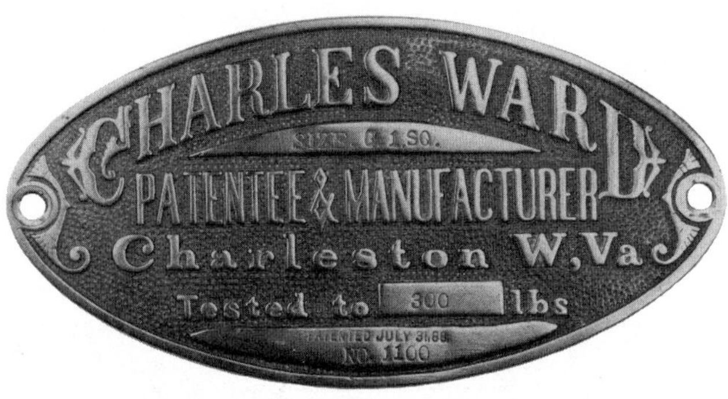

Bill Viden poses years ago with the Ward G-1 boiler that he and Bob Bracchi once installed in *Staurus* (see page 200). Charles Ward built WT boilers of both round and square design. These were used extensively in coal-fired U.S. Navy launches well into this century, but very few found favor with private steam launch operators. The Ward G-1 square boiler had 182 square feet of heating surface and nine square feet of grate surface, and it illustrates the main features of Ward boiler design. A water wall of tubes radiated downward from a bell-shaped drum. Within the water wall, "drop" tubes or "spider legs" absorbed heat directly above the fire.

These boilers were fast steamers, but they had shortcomings. The spider legs were extremely difficult to drain or blow down, and their capped ends came so close to the grate that a very shallow fire was mandatory.

The builder's plate was removed from a boiler once owned by O.D. York.

Many steam launches in England are powered with rugged Merryweather Size "B" WT boilers, but Al Giles's *Crest* (see page 184) is one of the few American boats that have one. *Above:* With the outer drum removed, Giles's boiler shows the pattern of water tubes within the inner drum. The water reservoir between the inner and outer drums gives the Merryweather some of the virtues of a fire-tube boiler. *Above left:* This photograph by Pat Haskett shows *Crest*'s boiler installed. Lagging is composed of glass wool covered by mahogany staving, with stainless steel banding. The Merryweather boiler was patented in the 1850s and was manufactured for nearly a century without a major design change. (Courtesy Al Giles) *Left:* Cliff Blackstaffe's horizontal WT boiler evolved from considerable experience with WT boilers in small sizes. The design incorporates 10 square feet of heating surface, one square foot of superheating surface, and one square foot of economizer surface in its 16.5" x 16" casing. Following are comments Blackstaffe made in *Steamboats and Modern Steam Launches,* November-December 1963, on his then-new boiler: "The boiler supports its casing, yet weighs only 80 pounds empty. The side panels of the casing come off [readily] for tube cleaning . . . . The dome in the stack ensures dry steam, and the boiler has enough water to be solid fuel fired if wanted. A plain atomizing burner with a flame deflector midway gives a very muffled sound and no smoke." (Courtesy Cliff Blackstaffe)

Epilogue: The Future of Steam Launches

by Bill Durham

For more than 20 years I've tried to shape in my mind a concept of steam launches that would let me see them clearly among the myriad recreations, products, avocations, technologies, and other distractions available in an affluent age. I was working aboard low-pressure steamships when I was 18, and I didn't learn about steamboat hobbies until 15 years later, so for me the little steamers have always been connected to a larger world.

I have wanted an expansive, evolving future for steam launches, but I note that each year there are more ways to spend one's time and money, and why should steamboats have a more important future than drag boats or whitewater canoes?

Steamboating shares many of the values of the classic-boat movement that is currently gathering strength, and it holds craft or collecting interests similar to those of live-steam railroad modeling or gun collecting. In its thoughtful use of energy and materials it is comparable to designing "solar" houses or building electric automobiles.

The future of steamboating holds numerous other facets and possibilities, many of them unseen by me, I'm sure. The best potential for growth will continue to lie among people seeking a relaxed, individualistic kind of boating, remote from the uniformity and overconsumption of the sailboat and heavy cruiser fleets. As pleasure steam launches become more numerous, the atmosphere of being always "on show" will fade away, and this will be a good thing for design. When some owners lose their concern for astonishing the spectators, they will be able to give more thought to sweet-running comfort and quiet grace.

By 1970 a new and promising group of recruits to steam launch development began to appear. Proponents of "decentralization" and "intermediate (or appropriate) technology" began to notice that small steam-power applications could be as independent of the mass economy as a log cabin or a raft. Many of these people are attracted by the philosophy of steamboating but want no part of its whistle-

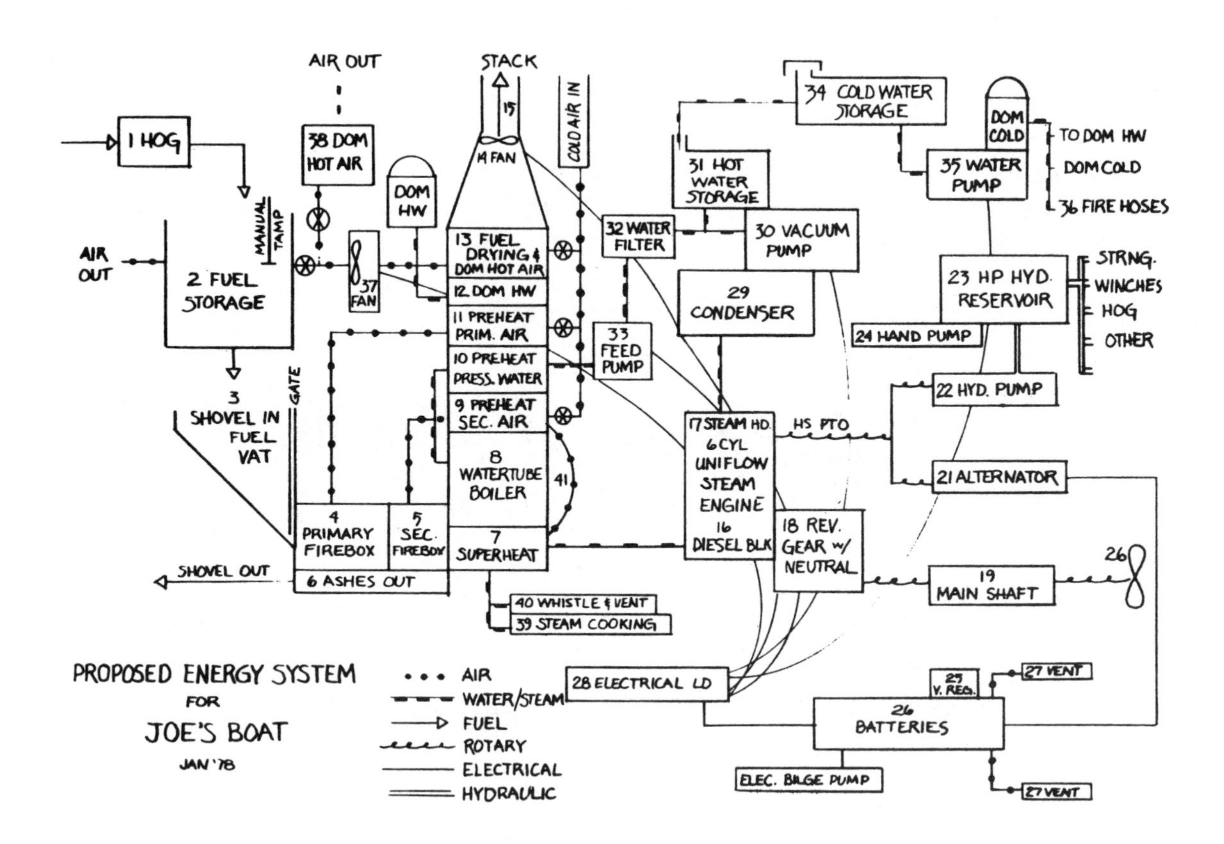

In the fall 1979 issue of CoEvolution Quarterly, *Joe Upton and Bill Smith proposed a wood-chip-fueled, 50-h.p. fishing boat for use in a wood-rich area such as southeastern Alaska. The creators are evidently unfamiliar with conventional marine steam design, and for this reason their concept is full of new slants on old problems.*

blowing, character-boat, cultist aspects. Most of the early inquiries I received from "alternative society" people concerned proposed paddle boats for Alaskan and other rivers — not a very promising field — but by 1980 these people were becoming increasingly knowledgeable about "hard" energies. Current interest in steam co-generation of electricity (with the same heat source used for space heating) will eventually benefit steamboat development. (A few old-time steam hobbyists were co-generating their household electricity 40 to 80 years ago. It's fun to observe the reinventing of the wheel.)

It seems likely that in the future more and more thought will be given to the development of useful and practical employments for small steamers. The steamers' ability to fuel up with anything that will burn is sure to have increasing appeal as the Petroleum Age wanes in the next century. Through most of their history, steam launches made no use of petroleum or electricity. Animal and vegetable fats made fair lubricating and illuminating oils, and coal and wood were universal fuels; early attempts to introduce petroleum fuel seemed contrary to nature and good business sense.

There are several industrial activities where the use of steam launches could be justified, despite the many advantages of internal-combustion power. Why should snag-clearing and debris-collecting boats (or roadside-clearing trucks, for that matter) burn petroleum — a complex, organic marvel best reserved for its values as a chemical, not as a fuel — to haul quantities of steam fuel to a place where it is burned as waste?

"Intermediate technology" steamers might

become economic in some fishing and towing applications, especially if granted tax advantages for using non-petroleum fuels. Steam excursion boats could restore some civility to this form of recreation, which is now characterized by ugly boats, mechanical din, and a vulgar atmosphere. A steam excursion launch could set up business next door to any American excursion-boat business I know of, charge twice as much, and get all the quality business.

Around numerous river and lakeside communities, roaring, high-powered motorboats became intolerable and had to be forbidden. Silent, mannerly steamboats could be reintroduced to these waters, where they could provide pleasant recreation for many more people than the speeding motorboats used to accommodate.

In some locations steam water-buses could incinerate much of a city's burnable wastes while transporting commuters to work. The agencies in several countries that are now studying ways to get commuters and private cars off crowded highways should consider the not-so-old design of the Venetian steam water-buses. They were designed with great care, to perform a useful service at a high level of safety, efficiency, and gentility.

The operator of a steam launch soon learns that his boat is a remarkably effective educational device — almost a whole technical college in itself. Since everyone, everywhere, has become in some degree dependent on work done by heat engines, it is desirable for everyone to know something about how these engines work. The simplicities of burning wood to boil water to make steam to push the piston up and down, turn the propeller, and make the boat go remain the most vivid lesson. Many schools should operate steam launches, including apparatus for generating both AC and DC electricity, to permit generations of students to see and measure inputs, outputs, conversions, and efficiencies in an interesting setting.

Modern steam launches have completed only part of the long journey back from the ebb years. Most new steamers are improvisations or conglomerations; very few are purpose-designed as a whole, with all parts well matched to one another. Most of the steam launch-related "businesses" that have appeared in recent years grew from small and private visions, indifferent to broad markets and to social and economic uses beyond the sphere of the hobbyist. I think the greatest value of this book will be in getting more enthusiasts to contemplate beauty and excellence, appropriate technology, and the various possible paths of future development for a technology that is already 200 years old.

In Chapter Four, "The Naval Architecture of Steam Launches," I have endeavored to explain why steam launches have or should have certain shapes and characteristics. Since almost all modern steam launch designs have been tainted by gas-boat experience and gas-boat expectations, most of the existing designs that are admirable or at least appropriate for steam power date from 80 to 110 years ago. There have been only a few recent efforts to build either "pure" steam launches designed from basic principles, or replicas of pristine classics.

The most successful modern design that I know of is Weston Farmer's *Diana,* which is especially valuable for its structural detailing *(Rudder,* December 1975). When Mr. Farmer sent me drawings of *Diana,* I began to sketch out a boat that I think is more firmly based on steam launch first principles and would have a broader market potential today. Mr. Farmer admired many of the qualities of steamers, which he had observed as a child, but professionally he was a son of the 20th century and the gas-boat era. *Diana* provides a lot of boat on the waterline length, makes no particular effort to keep hull resistance low, and is as heavy-built as a 1910 gasoline tug. (I recognize the short, knuckled counter, however, from the stubby little hoisting tugs that the 1898 Navy called "steam cutters.")

I would design a steam launch that would make good use of a moderate amount of heavy horsepower — not one that could float a large quantity of heavy horsepower. Never forgetting that a horsepower-hour requires a big armload of firewood (instead of a cupful of gasoline), I would want her to be most efficient at a power density less than half that of heavy gas boats. I would give her the deep belly and level shaft of a steamer rather than perch the machinery, as in gas boats, way up above the waterline. (This was made necessary at first by the large flywheels on slow-turning gasoline engines of 70 to 90 years ago, and later by the shallow form normal in all planing boats.) I would make a marketable steamer large and powerful enough for confident family cruising,

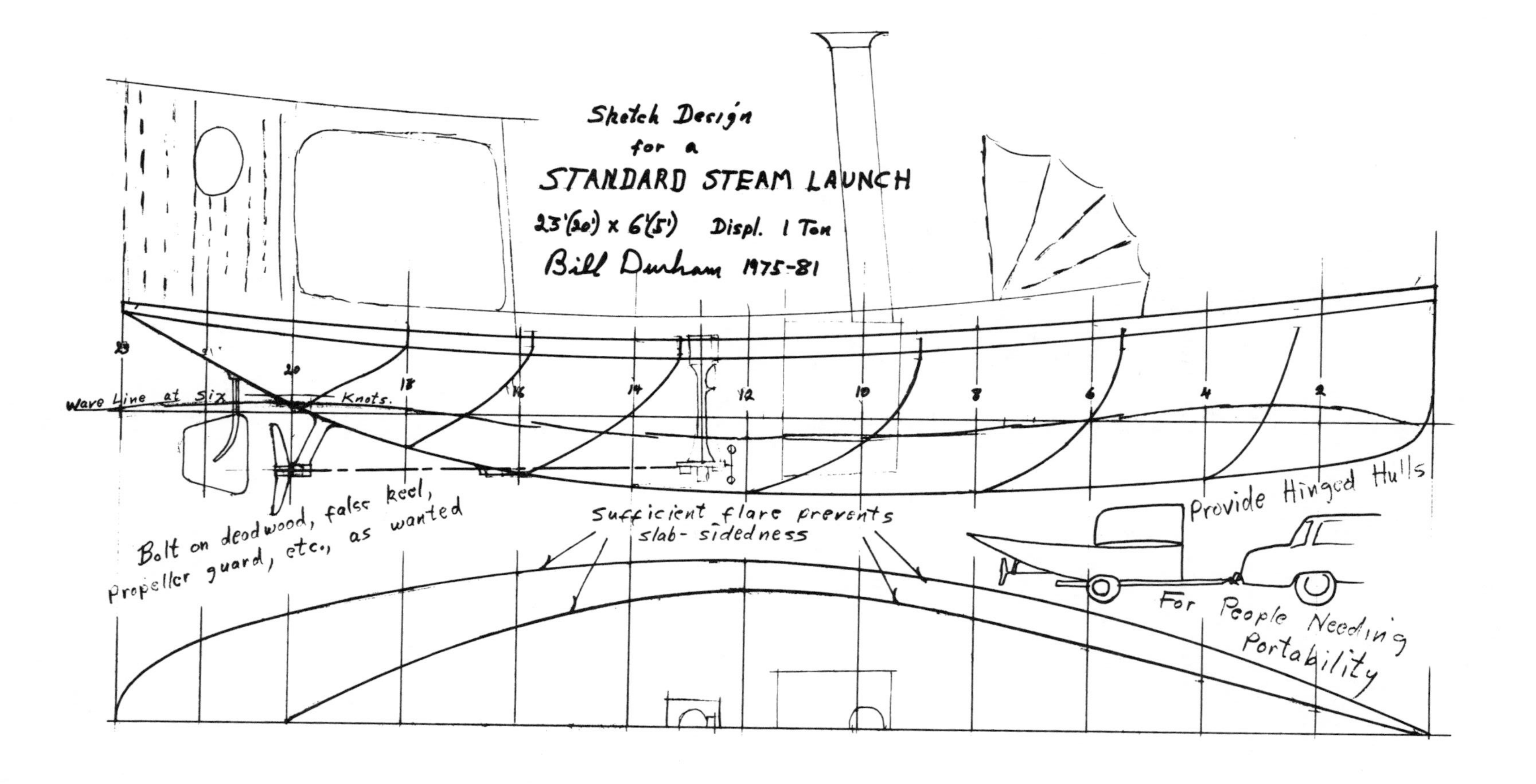

Bill Durham sketched his design (above) *as a light, low-powered alternative to Weston Farmer's* Diana (below). *Durham's 23-footer would have a loaded displacement of a little over a ton and provide accommodations equal to a 16- to 18-foot motorboat — seating for eight as a day cruiser, floor sleeping space for two or three. The boat is big enough to carry a canopy or even a stand-up steering shelter with grace. Giving the hull added flare amidships and forward makes it both drier and more lively in appearance than the old "slabs." The underwater profile is cut away fore and aft like that of a modern sailboat, but this won't show when the boat is afloat.*

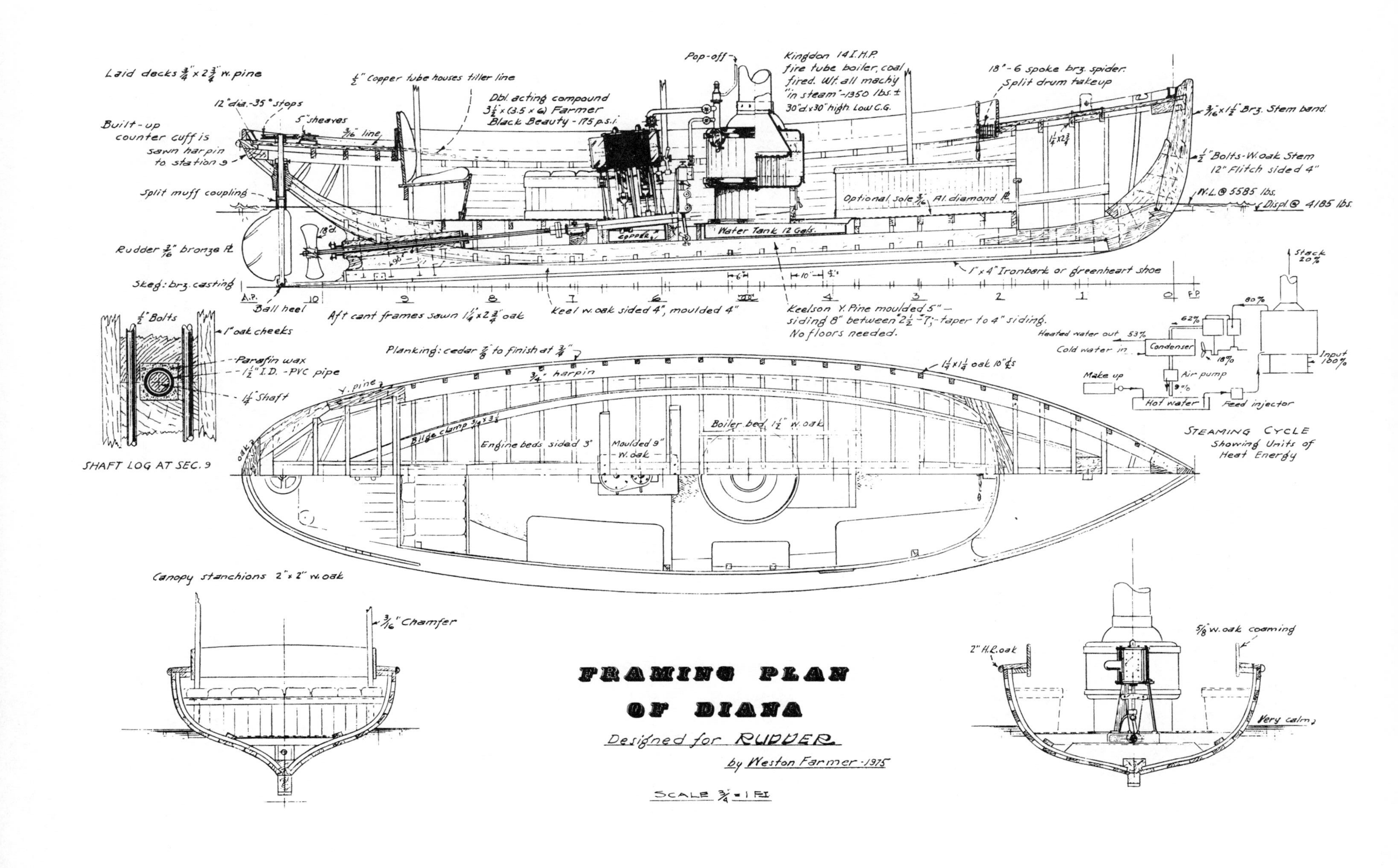
Laid decks 3/4" x 2 3/4" w. pine
1/2" Copper tube houses tiller line
12" dia.-35° stops
5" sheaves
3/16" line
Built-up counter cuff is sawn harpin to station 9
Split muff coupling
Rudder 3/16" bronze Pl.
Skeg: brz. casting
Ball heel
Dbl. acting compound 3 1/2 x (3.5 x 6) Farmer Black Beauty - 175 p.s.i.
Pop-off
Kingdon 14 I.H.P. fire tube boiler, coal fired. Wt. all mach'y "in steam" - 1350 lbs. ± 30" d. x 30" high. Low C.G.
Water Tank 12 Gals.
Optional sole 3/16" Al. diamond Pl.
18" - 6 spoke brz. spider. Split drum takeup
3/16" x 1 1/2" Brz. Stem band.
1/2" Bolts - W. oak Stem 12" Flitch sided 4"
W.L. @ 5585 lbs.
Displ. @ 4185 lbs.
1" x 4" Ironbark or greenheart shoe
Keel w. oak sided 4", moulded 4"
Aft cant frames sawn 1 1/4" x 2 3/4" oak
Keelson Y. Pine moulded 5" - siding 8" between 2 1/2 - 7, taper to 4" siding. No floors needed.
1/2" Bolts
1" oak cheeks
Parafin wax
1 1/2" I.D. - PVC pipe
1 1/4" Shaft
SHAFT LOG AT SEC. 9
Planking: cedar 7/8" to finish at 3/4"
3/4" harpin
Y. pine
Bilge clamp 3/4 x 3 1/2
Engine beds sided 3"
Moulded 9" W. oak
Boiler bed 1 1/2" w. oak
1 1/4 x 1 1/4 oak 10" c's
Stack 20%
80%
62%
Heated water out 53%
Cold water in
Condenser
18%
Air pump
9%
Make up
Hot water
Feed injector
Input 100%
STEAMING CYCLE Showing Units of Heat Energy
Canopy stanchions 2" x 2" w. oak
3/16" Chamfer
2" H.R. oak
5/8" w. oak coaming
Very calm
FRAMING PLAN OF DIANA
Designed for RUDDER
by Weston Farmer - 1975
SCALE 3/4 = 1 FT

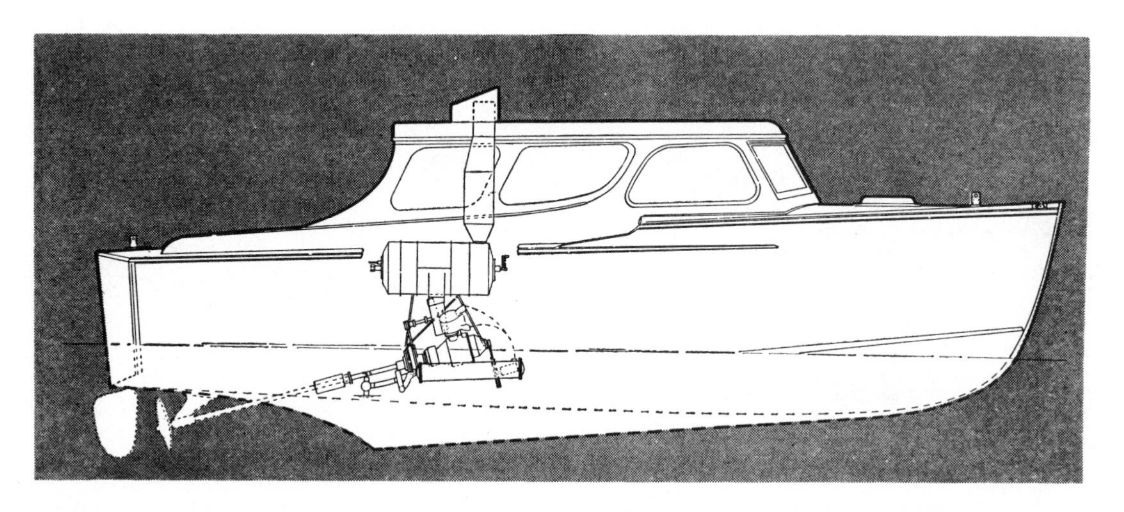

There have been numerous attempts since 1900 or so to adapt steam car technology to boat propulsion. The best-realized of these was a U.S. Navy 28-foot personnel boat of the late 1950s, which made 16 m.p.h. with a 115-h.p. Besler steam plant. With steam at 650 p.s.i. and 750 degrees Fahrenheit, oil-fuel consumption was 0.95 pound per horsepower-hour. The 3'' & 5.25'' x 4'' V-compound engine and coil boiler were very similar to those of the successful Besler steam airplane of the 1930s, which was descended from the Doble steam automobile design of 60 years ago. High-steam, high-efficiency steam launch plants designed to exploit solid fuels remain unexplored territory, with some future promise. (From Bureau of Ships Journal, *January 1960)*

but small and cheap enough for trailering and for competing with outboard runabouts in the marketplace.

Diana's structure provides an illustrated essay on how early 20th-century builders achieved sturdy and durable hulls through the skillful assembly of large quantities of good timber. Nevertheless, ''wooden boat'' nostalgia is not always related to good boat design, and sometimes I worry that the very charm of old-fashioned manufacturing methods will mislead and confuse some of the people who are gathering ideas for their steam launch.

I would design a hull on organic principles, for the highest efficiency and best economy of manufacture, while aiming to take advantage of whatever virtues and benefits steam propulsion may offer. I would install a power plant and stern gear as plain and functional as I could make them — these would turn out to have much in common with the perfected steam plant of 1885. The engine and stern gear would be big and rugged, but hull, gear, boiler, and engine for a 23-foot boat should weigh under 1,200 pounds dry.

I would want a 2- to 4-h.p. engine standing 30 inches tall and turning about 450. Propellers 16'' x 18'', 18'' x 20'', or 20'' x 24'' would suit two, three, or four horsepower at this rate of turning. Two horsepower would not be quite enough to reach six knots and would provide no reserve for adverse weather on broad waters — where this boat has no business anyway. The boiler would be water-tube — because this type provides complete freedom of choice of external dimensions and firebox shape. I would want it narrow so that a passenger could easily step around it, and long enough to accept 24-inch sticks of wood. It would be mainly copper, to escape hauling around a lot of excess steel and water, but it would hold enough water to cause boiler events to occur in a slow and methodical way. A heating surface of about 20 square feet should permit easy steaming on natural draft at two or three horsepower.

I will feel flattered if some rich manufacturer of megahorsepower planing boats borrows these ideas for a trial run of ''something altogether different'' in the mass market. Take it. It's yours! The manufacture and marketing of complete, ready-to-fire-up modern steam launches would be the best thing that could happen to the hobby, at once infusing it with a host of new people, talents, and ideas, and reducing prices all around by economies of scale. There have been dozens of ventures in this direction since 1910, and it is clear by now that special problems hamper the birth of this useful small industry.

Like everyone else who has ever built a small steamer or dreamed about steamer design, I have speculated on how a recreational steamer might garner some of the abundance of disposable wealth

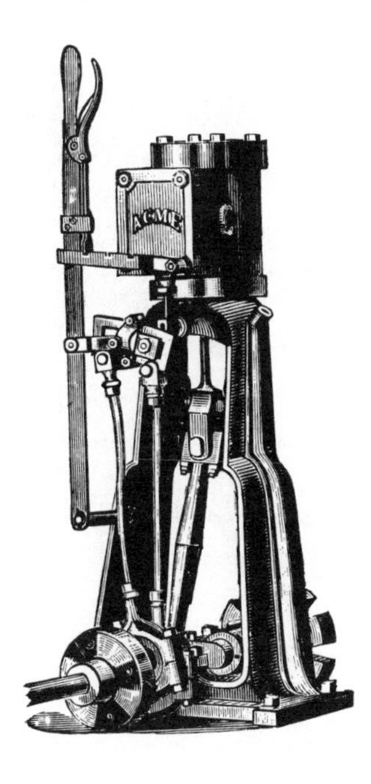

Rugged, easily available copies of the plain launch engines of 100 years ago are needed to help broaden the industry today. This 2-h.p. 3" x 4" Willard of 1885 weighed 150 pounds and stood 30 inches tall. A modern version, with better materials and closer tolerances, would weigh only 100 pounds and develop up to 5 horsepower, while retaining all the important steam launch virtues.

that comes to mind whenever one visits a boat show. (One-tenth of one percent of a five-billion-dollar pleasure-boat market would be five million dollars, and that's worth designing one small boat for.)

Clearly, a complete steam launch is not a very demanding manufacture, since thousands of beautiful and commercially successful boats were built a century ago by tobacco-chewing yokels with no knowledge of thermodynamics, metallurgy, or naval architecture. There are no very difficult design or manufacturing requirements in a steam launch — just a lot of them. Each of the 10 or 12 elements to be considered in a steam launch requires as much planning — and perhaps capital investment — as the entire concept for such recent successes as plastic canoes, inflatable river rafts, or wooden rowboats.

A stock steam launch would have to be introduced to markets unfamiliar with it, and it would have to be manufactured by facilities out of phase with steam launches. For such an enterprise to succeed, each of the following items will require hundreds of hours of professional attention: market analysis, design concept, hull design, hull manufacture, boiler design and manufacture, licensing problems peculiar to steamers, insurance problems peculiar to steamers, engine design and procurement, propeller design, propeller manufacture, promotion and marketing, fuel-burning arrangements, fuel sources, and field tests. Most of the hardware should be imported from countries where simple technologies of cast iron, steel, and bronze are still economic, so importing expertise is also required.

To revive the real virtues of small steam engines, a manufacturer could do worse than simply copy one of the plain, workaday engines of 1875 to 1900. The designs of the new engines introduced since 1920 have been influenced by the dimensions of hobbyists' machine tools — "everything done on a six-inch lathe" — and by a modern bias favoring smallness — "no bigger than a gas engine of the same power."

There will of course be a continuing need for elaborately detailed new compound and triple engines to replace the collectors' specimens as they go out of service (and into museums). The devotees of deluxe equipment don't need any encouragement from me. Their perfectionist passions will lead them to invest large quantities of time and money in building the most beautiful engines they can imagine.

During the 1950s and '60s, the needs of most steam launch builders were filled by ordinary welded-steel vertical fire-tube (VFT) industrial boilers in America and war-surplus Merryweather water-tube boilers in England (they powered portable fire pumps during the blitz). Since then, steam launch owners have reported their experiences with a variety of other boiler types. Most future steam launch boilers will come closer to the needs of the service.

There is room for much improvement in firing and ash-handling arrangements (top-firing of pelletized fuels, for one thing) and in the preparation and marketing of fuel. I can visualize some people making a spare-time business of harvesting orchard prunings, logging slash, or industrial scrap wood and preparing it neatly for sale to passing pleasure steamers.

Every facet of steamboating holds potential for various, even contradictory future development: commercial vs. recreational interests, function vs. surface appeal, classicism vs. sideshow, tradition vs. novelty. This diversity nourishes a self-renewing freshness. Steam launches are immune to the narrowing, specializing trends that overtake many avocational fields. The potential for an interesting and useful future is certainly there, and it remains almost as underexploited now as it was in 1780. Some year late in this century there may be more small steam launches afloat than in the past. It is in the nature of a vigorous civilization that most things are improved, refined, and honed a little sharper year by year. We may see steamers of such grace and elegance, such silentness and warmth as has never been enjoyed before.

Bibliography

Books

Baker, W.A. *The Engine-Powered Vessel.* New York: Grosset and Dunlap, 1965

Barnes, Eleanor. *Alfred Yarrow, His Life and Work.* London: E. Arnold, 1924.

Benjamin, D. Park. *Appleton's Cyclopedia of Applied Mechanics.* New York: D. Appleton and Co., 1881.

Blackstone, Edward H. *Farewell Old Mount Washington.* Providence, Rhode Island: Steamship Historical Society of America, 1969.

An illustrated book on the history of steamboating on Lake Winnipesaukee in New Hampshire. Primarily covers public passenger steamers but pictures of small boats do appear.

Blaisdell, Paul H. *Three Centuries on Winnipesaukee.* Concord, New Hampshire: Rumford Press, 1936. (Reprint. Somersworth, New Hampshire: New Hampshire Publishing Co., 1975.)

Interesting history of Winnipesaukee commercial steamers.

Bureau of Naval Personnel. *Principles of Naval Engineering.* 1970. (Reprint. Washington, D.C.: Superintendent of Documents, 1980.)

Burgh, N.P. *Modern Marine Engineering.* London, 1867.

Couling, David. *Steam Yachts.* Annapolis, Maryland: Naval Institute Press, 1980. London: Batsford Ltd., 1980.

Crabtree, Reginald. *The Luxury Yacht from Steam to Diesel.* New York: Drake Publishers, 1974.

Croil, James. *History of Steam Navigation.* Toronto: William Briggs, 1898.

Donaldson, James. *The Practical Guide to the Use of Marine

Steam Machinery and Internal Management of Small Steamers, Steam Yachts and Steam Launches. London: Norie and Wilson, 1885.

Down East Magazine Editors, ed. *Maine Lakes Steamboat Album.* Camden, Maine: Down East, 1976.

An illustrated book covering some of the steamboat history on the lakes of Maine.

Durant, Kenneth. *The Naphtha Launch.* Blue Mountain Lake, New York: Adirondack Museum, 1976.

A small booklet covering some of the history of the naphtha launch and its use in the Adirondack Mountains of New York State.

Durham, Bill. *Standard Boats of the United States Navy.* Seattle: Bill Durham Publications, 1963.

A booklet on small naval steamers and their relationship to the fleet. Excellent illustrations.

Durham, Bill, ed. *Steamboats and Modern Steam Launches.* San Diego: Howell-North Books, 1981.

This is a collected reprint of the magazine of this title, which was published from 1961 to 1963. The finest example of "how to" pertaining to the steam launch. Profusely illustrated with American and British launches showing hulls, machinery, and fittings. A must for all, whether active or armchair engineers.

Farmer, Weston. *From My Old Boat Shop.* Camden, Maine: International Marine Publishing Co., 1979.

Flexner, James T. *Steamboats Come True.* Boston: Little, Brown, and Co., 1978.

Graham, Frank, and Schank, Kenneth. *Audel's Questions and Answers for Engineers' and Firemen's Examinations.* 3rd ed. Indianapolis: Theodore Audel and Co., 1979.

Herreshoff, L. Francis. *Capt. Nat Herreshoff.* White Plains, New York: Sheridan House, Inc., 1974.

Herreshoff, L. Francis. *An Introduction to Yachting.* 1963. (Reprint. White Plains, New York: Sheridan House, Inc., 1980.)

Hervey, Harcourt, and Hervey, Ellen. *North American Steamboat Register.* South Pasadena, California: North American Steamboat Register, 1980.

A book giving region-by-region coverage of operating United States and Canadian steam launches, complete with owners' names and addresses and details of hull, boiler, engine, and auxiliaries. A must for every steam launch buff. An updated edition is in preparation.

Hillsdon, Brian, ed. *Steam Boat Index.* Vol. 1. 3rd ed. Ashford, Middlesex, England: The Steam Boat Association of Great Britain, 1980.

An illustrated index with complete details of every known steam launch in Britain.

Hochschild, Harold K. *Adirondack Steamboats on Raquette and Blue Mountain Lakes.* Blue Mountain Lake, New York: Adirondack Museum, 1962.

A booklet on the history of the area giving detailed coverage of the steamboats and the men who operated them. Also includes a history of the railroads and the fishing and hunting camps.

Hofman, Erik. *The Steam Yachts: An Era of Elegance.* Tuckahoe, New York; John De Graff, Inc., 1970.

Klingel, Gilbert. *Boatbuilding with Steel.* Camden, Maine: International Marine Publishing Co., 1973.

Kunhardt, C.P. *Steam Yachts and Launches: Their Machinery and Management.* New York: Forest and Stream Publishing Co., 1887. (2nd ed. 1891)

In-depth volume written during the "golden age" of the steam launch. Profusely illustrated with drawings. Rare.

Larkin, Larry. *Full Speed Ahead.* 2nd ed. Evanston, Illinois: Larry Larkin, 1972.

Picture history book of the steamers of Lake Geneva, Wisconsin. A well-done private printing.

Mercier, Gilbart B. *Pleasure Yachts of the Thousand Islands circa 1900.* Clayton, New York: The Shipyard Press, 1981.

Morrison, John H. *History of Steam Navigation.* 1903. (Reprint. New York: Argosy-Antiquarian, Ltd., 1967.)

Nicolson, Ian. *Small Steel Craft.* St. Albans, Hertfordshire, England: Adlard Coles Ltd., 1971. New York: Charles Scribner & Sons, 1979.

Pattinson, George H. *The Great Age of Steam on Windermere.* Windermere, England: The Windermere Nautical Trust, 1981.

Phillips-Birt, Douglas. *The Naval Architecture of Small Craft.* New York: Philosophical Library, 1957. London: Hutchinson Publishing Group, 1957.

Pratt, Mike. *Own a Steel Boat.* Camden, Maine: International Marine Publishing Co., 1979. London: Hollis & Carter, 1979.

Scott, Harley E. *Steam Yachts of Muskoka.* Bracebridge, Ontario: Herald-Gazette Press, 1978.

Stanton, John R. *Theory and Practice of Propellers for Aux-*

iliary Sailboats. Centreville, Maryland: Cornell Maritime Press, 1975.

Stapleton, N.B.J. *Steam Picket Boats.* Lavenham, Suffolk, England: Terence Dalton Ltd., 1980.

In-depth account with many illustrations of the British steam picket boats, their duty, armament and traditions.

Steam: Its Generation and Use. 37th ed. New York: The Babcock and Wilcox Co., 1960.

Various editions available since 1879.

Steiner, Kalman. *Fuels and Fuel Burners.* New York: McGraw-Hill Book Co., 1946.

Stephens, William P. *Traditions and Memories of American Yachting, Complete Edition.* Camden, Maine: International Marine Publishing Co., 1981.

Sucher, Harry. *Simplified Boatbuilding: The V-Bottom.* New York: W.W. Norton, 1974.

Thurston, Robert H. *A History of the Growth of the Steam-Engine.* 1878. (Reprint. Ithaca, New York: Cornell University Press, 1939.)

Turbinia: The World's First Turbine Driven Ship. Newcastle-upon-Tyne, England: Museum of Science and Engineering, 1961.

Warren, Nigel. *Metal Corrosion in Boats.* Camden, Maine: International Marine Publishing Co., 1980. London: Stanford Maritime Ltd., 1980.

The following catalog facsimiles have been printed:

A.G. Mumford, Ltd. Catalogue No. 7, 1906.

Reproduced by *Light Steam Power,* Kirk Michael, Isle of Man, United Kingdom. (The current name of the periodical is *Steam Power* and the address is The Midlands Steam Centre, 106A Derby Road, Loughborough, Leics. LE11 0AG, England.)

Charles P. Willard & Co. *Steam Launches and Marine Steam Engines.* Catalogue of 1885.

Reproduced by The Craft Shop, Seattle, from the collection of Ted Middleton, Aberdeen, Washington.

Simpson, Strickland & Co.

Several Simpson, Strickland catalog facsimiles have been printed by British concerns, including *Light Steam Power* and The Steam Boat Association of Great Britain.

The Steam Boat Association of Great Britain offers, in addition to Simpson, Strickland catalog facsimiles, facsimile editions of Liquid Fuel Engineering Co. Ltd., Lune Valley Engineering Co., and other catalogs. Address inquiries to Brian Hillsdon, 72 Marlborough Rd., Ashford, Middlesex, TW15 3PW, England.

Periodicals

Antique Boating published several stories on the steam launch. This magazine, which was based in Cleverdale, New York, is no longer being published.

Funnel is the only magazine the author knows of that deals entirely with the steam launch. It is published by The Steam Boat Association of Great Britain.

Light Steam Power, formerly published by John Walton on the Isle of Man in Great Britain, is a steam hobby magazine going back to the 1940s. Although it deals with every variety of steam application, nearly every issue contains an article on the steam launch. The magazine has now changed its name to *Steam Power,* and the new address is The Midlands Steam Centre, 106A Derby Road, Loughborough, Leics. LE11 0AG, England.

Live Steam, P.O. Box 629, Traverse City, Michigan 49684, is receptive to stories about steam in every form, but each issue contains something of particular interest to the steam launch buff.

Model Engineer, published in England since 1898, publishes occasional stories on the steam launch. Address correspondence to Model & Allied Publishing, Ltd., 13-35 Bridge St., Hemel Hempstead, Herts, England.

Rudder magazine, which started publication before the turn of the century, is perhaps the best historical reference on steam yachts and launches in America, if not in the world. Their articles are too numerous to list. It is no longer being published.

Scientific American issues of 50 to 100 years ago contain many articles on small steamers.

Steamboat Bill, the journal of the Steamship Historical Society of America, with offices at H.C. Hall Building, 345 Blackstone Blvd., Providence, Rhode Island 02906, is primarily oriented toward larger vessels but on occasion publishes articles on the steam launch.

The following articles are of particular interest:

Durham, Bill. "The Steam Launch." *Motor Boating,* July 1958.

Durham, Bill. "More About Steam." *Motor Boating,* November 1958.

Farmer, Weston. "*Diana*: A Steamboat for Today." *Rudder,* December 1975.

A naval architect's plan for the construction of a steam launch.

Farmer, Weston. "Those Wonderful Naphtha Launches." *Yachting,* July 1973.

An in-depth story of how a naphtha launch operates, by an author who remembered them.

Herreshoff, L. Francis. "An Introduction to Yachting." *Rudder,* April through September 1958.

A series of articles on the history of the development of the yacht, both sail and steam, with excellent details and old photographs.

Herreshoff, L. Francis. "N.G. Herreshoff and Some of the Yachts He Designed." *Rudder,* April through July 1949.

A series of articles by the son of the world's most famous yachtsman.

Howland, Wallace. "Steam Returns to the San Juans." *Yachting,* March 1963.

McCready, Capt. L.S. "In the Wake of the *Clermont.*" *Motor Boating,* February 1955.

A story of McCready's trip up the Hudson River in his steam launch, *Little Effie.* Rear Admiral McCready is retired from the United States Merchant Marine Academy.

MacDuffie, Malcolm. "Naphtha Launch — The Missing Link." *National Fisherman,* May 1971.

Mitchell, C. Bradford. "Paddle Wheel Inboard." *American Neptune* 7 (1947).

Some of the history of Oklawaha River (Florida) steamboating and the Hart Line.

Mitchell, R.M. "Build *River Queen.*" *Mechanix Illustrated,* February 1962.

A detailed account of how the author built and operated his own steam launch.

Taylor, William H. "Genuine Antique Yachts and Engines." *Yachting,* December 1952.

White, Richard. "The Naphtha Launch." *Marine Propulsion,* February 1980.

Steam Launch Builders of the Past

The following list of past British and North American builders of engines, boilers, and complete steam launches is not intended to be exhaustive or rigorously selective. It is intended to offer a glimpse of the steam launch industry from about 1870 up to the eclipse of steam launch building in the early 20th century (roughly 1905 in North America and a decade or so later in Britain). British yards frequently built steam launches as a sideline to the major enterprise of building oceangoing vessels. North American manufacturers ranged in size from small shops custom-building a few engines or boats to large concerns with complete lines of stock engines or stock steam launches of various sizes and "grades." The manufacturers listed here represent the whole range. The larger firms, British and North American, developed both domestic and foreign markets with good success. A few firms had the requisite capital and management flexibility to retool for the manufacture of other products as the steam launch market diminished, and some of these companies, transformed beyond recognition, are in business still.

GREAT BRITAIN

Firm	Products
Abbot & Co. Newark-on-Trent England	Boilers
Abercorn Shipbuilding Co. Paisley, Scotland	Launches
G.E. Bellis and Co. Birmingham, England	Supplied engines to John Samuel White before he built his own. Still in business as Bellis & Morcomb Ltd.
Cochran & Co. Birkenhead, England	Launches, engines, boilers
W. Denny & Bros. Govan, Glasgow, Scotland	Launches
G.F.G. Desvignes Strand-on-Green London, England	Designer and builder of high-speed yachts and launches between 1878 and 1900.
Fielding & Platt Atlas Iron Works Gloucester, England	Engines, boilers
E. Gillet & Co., Engineers Hounslow, England	Launches, engines
Edward Hayes Stoney Stratford, England	Launches, engines
R. Hedley and Co. Wharf Road Isle of Dogs, Poplar London, England	Launches. (Hedley established his own works after his partnership with Yarrow was dissolved.)
Jones, Burton & Co. Ltd. Liverpool, England	Launches, engines, and boilers
David J. Lewin Poole, Dorset, England	Launches, engines, boilers
The Liquid Fuel Engineering Co. Works at East Cowes, Isle of Wight, England Office at 20 Abchurch Lane, London, England	Yachts, launches, engines
Lune Valley Engineering Co. Lancaster, England	Boilers, burners
McKie & Baxter Copeland Works Govan, Glasgow, Scotland	Engines, boilers
Matthew Paul & Co. Levenford Works Dumbarton, Scotland	Launches, engines, boilers
Merryweather Greenwich, England	Merryweather boilers
A.G. Mumford, Ltd. Culver Street Iron Works Colchester, England	Engines, WT boilers

Perkins Boilers Derby, England	Perkins boilers
Plenty & Son, Ltd. Newbury, Berkshire, England	Engines, boilers. Still in business.
Redpath and Paris Limehouse, England	Launches, engines, boilers
Thomas Reid & Sons Paisley, Scotland	Engines
Ross & Duncan Ltd. Govan, Glasgow, Scotland	Launches, boilers
T.A. Savery & Co., Ltd. Birmingham, England	Engines, WT boilers
Simpson, Strickland & Co. Dartmouth, England	Launches, engines, boilers. Before 1887, the firm was known as Simpson, Denison & Co. In 1894 they acquired the Kingdon Yacht, Launch and Engineering Co. of Teddington, who manufactured launches and small steam power units including the Kingdon engine and boiler.
W. Sisson & Co., Ltd. Gloucester, England	Engines
R. Smith & Co. Lytham, Lancashire, England	Yachts, launches, engines, boilers
John I. Thornycroft & Co. Church Wharf Chiswick, London, England	Yachts, launches, and torpedo boats. Vosper & Co. later merged with Thornycroft's firm, and Vosper Thornycroft Ltd. is still in business.
Vosper & Co. Portsmouth, England	Launches, engines, boilers
J. Samuel White & Co. Ltd. East Cowes, Isle of Wight England	Launches, engines. Still in business.
Alex Wilson & Co. Vauxhall Ironworks London, S.W., England	Engines
Yarrow & Co. London, England	Yachts, launches, torpedo boats

NORTH AMERICA

Edward S. Clark Boston, Massachusetts	Engines
Clay & Torbensen Camden, New Jersey	Launches, engines, boilers
Clute Brothers & Co. Schenectady, New York	Engines
Colt's Patent Fire Arms Manufacturing Co. Hartford, Connecticut	Colt disc marine engines

Conbrock Brooklyn, New York	Engines and WT boilers
Davis & Smith Dover, New Hampshire	Engines
Davis & Son Kingston, Ontario, Canada	Launches, engines. After 1900, the name was changed to Davis Dry Dock Co., Ltd.
The Davis Boat & Oar Co. Detroit, Michigan	WT boilers
Davis-Farrar Co. Erie, Pennsylvania	Yachts, launches, and engines
Donegan & Swift New York, New York	Engines
Doty Engine Works Goderich, Ontario, Canada	Engines
E.L. Fitzhenry Boston, Massachusetts	Engines
Fore River Engine Co. Weymouth Landing Weymouth, Massachusetts	Yachts, engines. Fore River Engine Co. was founded as a small machine shop by Thomas Watson. (Watson had been associated with Alexander Graham Bell in the development of the telephone until his health failed. He recuperated in Europe, then returned to America determined to be a "gentleman farmer," but failed at this. The machine shop was his next idea.) With another machinist, Wellington, Watson began building small marine engines, and the firm quickly grew to employ 35 to 40 men. Eventually, they began building torpedo boats for the U.S. Navy, and in 1900 they moved to a new yard in Quincy, a few miles away, and changed their name to the Fore River Ship and Engine Co. In 1912 the firm was acquired by Bethlehem Steel Corp., and in 1982 it is the Shipbuilding Division of General Dynamics Corp.
Gas Engine & Power Company and Charles L. Seabury & Co., Consolidated Morris Heights New York, New York	Yachts, launches, engines, boilers, and naphtha launches. At various times this company or component companies also used the names Gas Engine & Power Co.; New York Yacht, Launch, and Engine Co.; and Charles L. Seabury & Co. Seabury and the Gas Engine and Power Co. joined forces and took the name this entry began with; later this was shortened to Consolidated — which, as Weston Farmer said, was "a yachtbuilding name to conjure with."
Herreshoff Manufacturing Co. Bristol, Rhode Island	Yachts, launches, torpedo boats, engines, and boilers
Thos. Kane & Co. Chicago, Illinois, and Racine, Wisconsin	Launches, engines, and boilers
John F. Kemp Quincy, Massachusetts	Engines
George Lawley & Son Corp. South Boston, Massachusetts	Yachts, engines

Lockwood Manufacturing Co. Boston, Massachusetts	Engines
Merwin, Hulbert & Co. New York, New York	Launches
Morris Machine Works Baldwinsville, New York	Engines
Charles D. Mosher Amesbury, Massachusetts	High-speed yachts and launches. Express boilers.
Murray & Tregurtha Co. South Boston, Massachusetts	Engines, boilers. Still exists as the Murray and Tregurtha Division of the Mathewson Corp., Quincy, Massachusetts, and now manufactures powerful outboard marine propulsion units for maneuvering barges, positioning drill rigs, and other such applications.
New York Safety Steam Power Co. New York	Engines
John Paine Boston, Massachusetts	Engines
T. Patchett Stoneham, Massachusetts	Engines
H.W. Petrie Toronto, Ontario, Canada	Engines
Poulson Iron Works Toronto, Ontario, Canada	Engines
Racine Boat Manufacturing Co. Racine, Wisconsin	Yachts, launches
Reeves Machine Co. Trenton, New Jersey	Engines
Rice & Whitacre Manufacturing Co. Chicago, Illinois	Engines, including the Kriebel oscillating engine
Riley & Cowley Brooklyn, New York	Engines
Roberts Safety Water Tube Boiler Co. Works in Red Bank, New Jersey Offices in New York, New York	Boilers. T.E. Roberts served between 1861 and 1865 in the Union navy, gaining much experience with small marine steam plants. Later he invented the Roberts boiler. The first patent was obtained in 1887, and Roberts founded his company in 1890. By 1892, 400 boilers had been sold.
Rochester Machine Tool Works Rochester, New York	The "Buckley" patent safety WT boiler
Salamandrine Boiler Co. Newark, New Jersey	WT boilers
Seattle Talbot Generator Co. Seattle, Washington	Engines, boilers. Known to have been in business in 1912, but not known how much earlier.
Shipman Engine Co. Boston, Massachusetts, and Rochester, New York	Engines, boilers. Early Shipman engines were built by Charles W. Percy at 212 Summer Street, Boston. (Shipman's Boston address was 200 Summer Street.)

Ward B. Snyder New York, New York	Engines
H.R. Stickney Portland, Maine	Engines, boilers
St. Lawrence River Skiff, Canoe and Steam Launch Co. Clayton, New York	Launches
Taylor and Hough St. Paul, Minnesota	Engines
Truscott Boat Manufacturing Co. St. Joseph, Michigan	Launches, engines. Built the Upton high-speed engines.
United States Navy	Built launches and engines in navy yards on both the Atlantic and Pacific coasts.
Vermont Farm Machinery Co. Bellows Falls, Vermont	Engines. This firm was a major manufacturer of cream separators and other farm machinery, serving domestic and foreign markets. Launch engines were built as a sideline.
Ward Engineering Works Charleston, West Virginia	Engines and Ward boilers
Wells Engine Co. New York, New York	Engines
George E. Whitney East Boston, Massachusetts	Engines, boilers, and propellers. Designed hulls.
Charles P. Willard & Co. Chicago, Illinois	Launches, engines, and boilers

Steam Launch Sources

The builders and suppliers listed here are, to the best of the author's knowledge, active as of April 1982. In addition, anyone planning a steam launch should be aware of skilled local artisans, some of whom have extensive steam engine, boiler, or boatbuilding experience.

GREAT BRITAIN

Bossom's Boatyard, Ltd. Medley, via Binsey Lane Oxford, England X02 0NL	19' and 31' steam launches of traditional design, in fiberglass.
Clinker Boats 10 Victoria Street Newark, Notts., England	Various engines and boilers, complete.
Crossmyloof Engineering Gordon Cheape 58 Crossmyloof Gardens Glasgow, G41 4AY, Scotland	Restorations, conversions, and new launch engines to order.
G. Lancaster Jones Owler Mill, Bacup Road Todmorden, Lancs., England	Designer and builder. 16', 20', and 24' fiberglass and wooden fantail hulls, supplied if required with Lune Valley-style WT boiler and Stuart Cygnet or Swan engine.

Hugh A. Jones
Beaumaris Instrument Co. Ltd.
Rosemary Lane
Beaumaris, Anglesey LL58 8EB
Isle of Man

Marine compound engines.

Rupert Latham
Journey's End
Ferry View, Horning
Norwich, Norfolk, England

18' Edwardian-period fiberglass (GRP) launch. Complete with engine and boiler if required, or partly completed suitable for owner to finish as required at home.

McEwen Boilers
Farley Top
Cowling, via Keighley
North Yorkshire, BD22 ONW, England

Solid-fuel-fired boilers in sizes to evaporate 85 to 200 pounds of steam per hour.

Machin, Knight & Sons Ltd.
Medway Bridge Marina
Manor Lane, Borstal
Kent, ME1 3HS, England

Traditional wooden boat construction, of museum quality if required. Specializes in steamboats of all sizes. Used and new machinery usually on hand.

Roger Mallinson
Spruce Howe, Keldwyth Drive
Windermere, England

Builder of small marine steam plants. Windermere kettles, steam whistles, propellers, etc.

Severn Lamb, Ltd.
Western Road
Stratford-on-Avon
Oxon, England

Double-simple launch engines.

Stuart Turner Ltd.
Henley-on-Thames
Oxon, RG9 2AD, England

In business over 75 years supplying castings, drawings, etc., for popular 5A and 6A steam launch engines. Catalog — 60 pence sterling.

Thames Steam Launch Co.
12A Spring Grove
Chiswick, London W4, England

Coal-fired WT boilers to order. Suitable for Stuart Turner 5A and 6A engines or larger engines.

J.E. Whitfield
Beckside North Beverley
North Humberside, HU17 0SU
England

Hulls, engines, and boilers.

NORTH AMERICA

Almy Water Tube Boiler Co.
108 Walnut Street
Warwick, RI 02888

Boiler repairs. Sell the designs and drawings for Almy WT boilers.

Atlantic Boat Works
Belfast Road
Camden, ME 04843

Finished 20' fiberglass fantail launches, engines, and 3-drum WT boilers. Components available separately.

Donald Beckner
291 Mill Street
South Lancaster, MA 01561

23' fiberglass fantail launches, engines, and WT boilers.

Chester River Machine Tool Co.
R.D. #3, Box 326
Chestertown, MD 21620

Materials for copper boilers.

Cole's Power Models, Incorporated P.O. Box 788 Ventura, CA 93001	The complete line of Stuart Turner castings. $3 for Golden Anniversary Catalog, #23.
Elliott Bay Co. 4409 Eastern Avenue North Seattle, WA 98103	20' fiberglass fantail hulls, decks, coamings, and accessories.
Hodge Boiler Works 111 Sumner Street Boston, MA 02128	Boiler building and repair.
David W. Hogan 7040 Alanzo N.W. Seattle, WA 98117	19' fantail fiberglass launch hulls to order.
Hypro Division Lear-Siegler, Inc. 375 Fifth Avenue, N.W. St. Paul, MN 55112	Hypro feed pumps.
Little Engines Box N Lomita, CA 90717	2.5'' x 3.25'' engine castings.
Robert W. Maynard 3825 Virginia Court Cincinnati, OH 45211	Steel boilers to your specifications.
Peter E. Moale Almanor Machine Works 413A Arbutus Drive Lake Almanor Pen., CA 96137	Burleigh 3'' & 5'' x 4'' compound-engine castings, working drawings, and fully machined engines available.
C. William Moore P.O. Box 756 Pleasanton, CA 94566	Steam engine castings.
O'Connor Engineering Laboratories 100 Kalmus Drive Costa Mesa, CA 92626	Engines and WT boilers. Designers and builders.
Pen Models 259 Carolyn Drive Oakville, Ontario, Canada	Stuart Turner castings.
Edwin L. Puryear Route 1, Andrew Corley Rd. Lexington, SC 29072	Marine engines and boilers to order.
Rhode Island Marine Service Classic Yachts Division P.O. Box 209 Wakefield, RI 02880	Bare hulls or complete launches in wood, fiberglass, or aluminum. Engines, boilers. Literature and photos, $2.
Scripps Foundry and Machine Works P.O. Drawer C Fredericksburg, TX 78624	Full-size drawings and castings to build 3'' & 6'' x 5'' Steeple compound, 3'' & 5.5'' x 4'' compound, and 6'' x 5'' marine engines.
Semple Engine Co., Inc. Box 6805 St. Louis, MO 63144	Engine kits, engines, boilers, and fittings. Single and compound steam plants.

Seneca Metal Products, Inc. 1035 Lake Road Webster, NY 14580	Aluminum fantail hulls, compound engines, and cast bronze propellers.
R.J. Spurlock 4409 Eastern Avenue N. Seattle, WA 98103	Materials for 10 sq. ft. VFT steel boilers.
David W. Thompson Box 175 Moultonboro, NH 03254	Roberts-type WT boilers, hull and engine repairs, and installations.
Tiny Power P.O. Box G Tehachapi, CA 93561	Rugged 3'' x 4'' castings designed by Arnold. Other engine castings available. Brochure $1.
David Townsend Route 1, Box 64 Calais, ME 04619	20' fiberglass fantail hulls. Additional information and photos — $2.

Index